Titles in This Series

Volume

1 **Markov random fields and their applications,** Ross Kindermann and J. Laurie Snell

2 **Proceedings of the conference on integration, topology, and geometry in linear spaces,** William H. Graves, Editor

3 **The closed graph and P-closed graph properties in general topology,** T. R. Hamlett and L. L. Herrington

4 **Problems of elastic stability and vibrations,** Vadim Komkov, Editor

5 **Rational constructions of modules for simple Lie algebras,** George B. Seligman

6 **Umbral calculus and Hopf algebras,** Robert Morris, Editor

7 **Complex contour integral representation of cardinal spline functions,** Walter Schempp

8 **Ordered fields and real algebraic geometry,** D. W. Dubois and T. Recio, Editors

9 **Papers in algebra, analysis and statistics,** R. Lidl, Editor

10 **Operator algebras and K-theory,** Ronald G. Douglas and Claude Schochet, Editors

11 **Plane ellipticity and related problems,** Robert P. Gilbert, Editor

12 **Symposium on algebraic topology in honor of José Adem,** Samuel Gitler, Editor

13 **Algebraists' homage: Papers in ring theory and related topics,** S. A. Amitsur, D. J. Saltman, and G. B. Seligman, Editors

14 **Lectures on Nielsen fixed point theory,** Boju Jiang

15 **Advanced analytic number theory. Part I: Ramification theoretic methods,** Carlos J. Moreno

16 **Complex representations of GL(2, K) for finite fields K,** Ilya Piatetski-Shapiro

17 **Nonlinear partial differential equations,** Joel A. Smoller, Editor

18 **Fixed points and nonexpansive mappings,** Robert C. Sine, Editor

19 **Proceedings of the Northwestern homotopy theory conference,** Haynes R. Miller and Stewart B. Priddy, Editors

20 **Low dimensional topology,** Samuel J. Lomonaco, Jr., Editor

21 **Topological methods in nonlinear functional analysis,** S. P. Singh, S. Thomeier, and B. Watson, Editors

22 **Factorizations of $b^n \pm 1$, $b = 2, 3, 5, 6, 7, 10, 11, 12$ up to high powers,** John Brillhart, D. H. Lehmer, J. L. Selfridge, Bryant Tuckerman, and S. S. Wagstaff, Jr.

23 **Chapter 9 of Ramanujan's second notebook—Infinite series identities, transformations, and evaluations,** Bruce C. Berndt and Padmini T. Joshi

24 **Central extensions, Galois groups, and ideal class groups of number fields,** A. Fröhlich

25 **Value distribution theory and its applications,** Chung-Chun Yang, Editor

26 **Conference in modern analysis and probability,** Richard Beals, Anatole Beck, Alexandra Bellow, and Arshag Hajian, Editors

27 **Microlocal analysis,** M. Salah Baouendi, Richard Beals, and Linda Preiss Rothschild, Editors

28 **Fluids and plasmas: geometry and dynamics,** Jerrold E. Marsden, Editor

29 **Automated theorem proving,** W. W. Bledsoe and Donald Loveland, Editors

30 **Mathematical applications of category theory,** J. W. Gray, Editor

31 **Axiomatic set theory,** James E. Baumgartner, Donald A. Martin, and Saharon Shelah, Editors

32 **Proceedings of the conference on Banach algebras and several complex variables,** F. Greenleaf and D. Gulick, Editors

33 **Contributions to group theory,** Kenneth I. Appel, John G. Ratcliffe, and Paul E. Schupp, Editors

34 **Combinatorics and algebra,** Curtis Greene, Editor

Titles in This Series

Volume

35 **Four-manifold theory,** Cameron Gordon and Robion Kirby, Editors

36 **Group actions on manifolds,** Reinhard Schultz, Editor

37 **Conference on algebraic topology in honor of Peter Hilton,** Renzo Piccinini and Denis Sjerve, Editors

38 **Topics in complex analysis,** Dorothy Browne Shaffer, Editor

39 **Errett Bishop: Reflections on him and his research,** Murray Rosenblatt, Editor

40 **Integral bases for affine Lie algebras and their universal enveloping algebras,** David Mitzman

41 **Particle systems, random media and large deviations,** Richard Durrett, Editor

42 **Classical real analysis,** Daniel Waterman, Editor

43 **Group actions on rings,** Susan Montgomery, Editor

44 **Combinatorial methods in topology and algebraic geometry,** John R. Harper and Richard Mandelbaum, Editors

45 **Finite groups–coming of age,** John McKay, Editor

46 **Structure of the standard modules for the affine Lie algebra $A_1^{(1)}$,** James Lepowsky and Mirko Primc

47 **Linear algebra and its role in systems theory,** Richard A. Brualdi, David H. Carlson, Biswa Nath Datta, Charles R. Johnson, and Robert J. Plemmons, Editors

48 **Analytic functions of one complex variable,** Chung-chun Yang and Chi-tai Chuang, Editors

49 **Complex differential geometry and nonlinear differential equations,** Yum-Tong Siu, Editor

50 **Random matrices and their applications,** Joel E. Cohen, Harry Kesten, and Charles M. Newman, Editors

51 **Nonlinear problems in geometry,** Dennis M. DeTurck, Editor

52 **Geometry of normed linear spaces,** R. G. Bartle, N. T. Peck, A. L. Peressini, and J. J. Uhl, Editors

53 **The Selberg trace formula and related topics,** Dennis A. Hejhal, Peter Sarnak, and Audrey Anne Terras, Editors

54 **Differential analysis and infinite dimensional spaces,** Kondagunta Sundaresan and Srinivasa Swaminathan, Editors

55 **Applications of algebraic K-theory to algebraic geometry and number theory,** Spencer J. Bloch, R. Keith Dennis, Eric M. Friedlander, and Michael R. Stein, Editors

56 **Multiparameter bifurcation theory,** Martin Golubitsky and John Guckenheimer, Editors

57 **Combinatorics and ordered sets,** Ivan Rival, Editor

58.I **The Lefschetz centennial conference. Proceedings on algebraic geometry,** D. Sundararaman, Editor

58.II **The Lefschetz centennial conference. Proceedings on algebraic topology,** S. Gitler, Editor

58.III **The Lefschetz centennial conference. Proceedings on differential equations,** A. Verjovsky, Editor

59 **Function estimates,** J. S. Marron, Editor

60 **Nonstrictly hyperbolic conservation laws,** Barbara Lee Keyfitz and Herbert C. Kranzer, Editors

61 **Residues and traces of differential forms via Hochschild homology,** Joseph Lipman

62 **Operator algebras and mathematical physics,** Palle E. T. Jorgensen and Paul S. Muhly, Editors

63 **Integral geometry,** Robert L. Bryant, Victor Guillemin, Sigurdur Helgason, and R. O. Wells, Jr., Editors

64 **The legacy of Sonya Kovalevskaya,** Linda Keen, Editor

65 **Logic and combinatorics,** Stephen G. Simpson, Editor

66 **Free group rings,** Narian Gupta

67 **Current trends in arithmetical algebraic geometry,** Kenneth A. Ribet, Editor

Titles in This Series

Volume

Titles in This Series

Volume

104 Accessible categories: The foundations of categorical model theory, Michael Makkai and Robert Paré

105 Geometric and topological invariants of elliptic operators, Jerome Kaminker, Editor

106 Logic and computation, Wilfried Sieg, Editor

107 Harmonic analysis and partial differential equations, Mario Milman and Tomas Schonbek, Editors

108 Mathematics of nonlinear science, Melvyn S. Berger, Editor

109 Combinatorial group theory, Benjamin Fine, Anthony Gaglione, and Francis C. Y. Tang, Editors

110 Lie algebras and related topics, Georgia Benkart and J. Marshall Osborn, Editors

111 Finite geometries and combinatorial designs, Earl S. Kramer and Spyros S. Magliveras, Editors

112 Statistical analysis of measurement error models and applications, Philip J. Brown and Wayne A. Fuller, Editors

113 Integral geometry and tomography, Eric Grinberg and Eric Todd Quinto, Editors

114 Mathematical developments arising from linear programming, Jeffrey C. Lagarias and Michael J. Todd, Editors

115 Statistical multiple integration, Nancy Flournoy and Robert K. Tsutakawa, Editors

Statistical Multiple Integration

CONTEMPORARY MATHEMATICS

115

Statistical Multiple Integration

Proceedings of a Joint Summer Research Conference
held at Humboldt University, June 17–23, 1989

Nancy Flournoy
Robert K. Tsutakawa
Editors

American Mathematical Society
Providence, Rhode Island

The AMS-IMS-SIAM Joint Summer Research Conference on Statistical Multiple Integration was held at Humboldt University, Arcata, California, on June 17–23, 1989, with support from the National Science Foundation, Grant DMS-8846813 and the U.S. Army Research Office, Grant 26725-MA-CF.

1980 *Mathematics Subject Classification* (1985 *Revision*). Primary 62E30, 62E20, 62H10, 62L20, 65C05, 65C10, 65D30, 62C10.

Library of Congress Cataloging-in-Publication Data

AMS-IMS-SIAM Joint Summer Research Conference on Statistical Multiple Integration (1989: Humboldt University)

Statistical multiple integration: proceedings of the AMS-IMS-SIAM Joint Summer Research Conference held June 17–23, 1989/Nancy Flournoy and Robert K. Tsutakawa, editors.

p. cm.—(Contemporary mathematics/American Mathematical Society; 115)

Includes bibliographical references.

ISBN 0-8218-5122-5 (alk. paper)

1. Sampling (Statistics) 2. Multivariate analysis 3. Numerical integration. 4. Bayesian statistical decision theory. I. Flournoy, Nancy, 1947– . II. Tsutakawa, Robert K. III. Title. IV. Series: Contemporary mathematics (American Mathematical Society); v. 115.

QA276.7.A47 1989 90-27134

519.5—dc20 CIP

Printed in the United States of America.

The paper used in this book is acid-free and falls within the guidelines established to ensure permanence and durability. ∞

This publication was typeset using $\mathcal{A}\mathcal{M}\mathcal{S}$-TEX, the American Mathematical Society's TEX macro system.

10 9 8 7 6 5 4 3 2 1 96 95 94 93 92 91

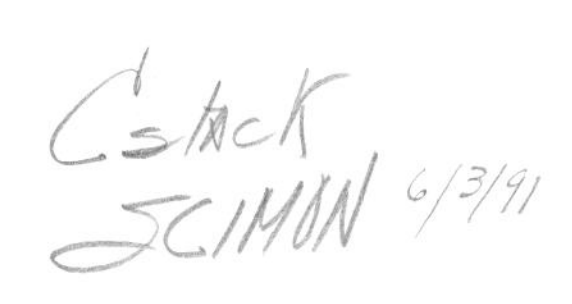

Contents

Preface

The idea for a conference on Statistical Multiple Integration had its genesis during the years 1986 to 1988 when Nancy Flournoy directed the Program in Statistics and Probability at the National Science Foundation (NSF). Directors of NSF programs are frequently called upon to describe "current thrusts", "significant breakthroughs", and "obstacles to progress" on behalf of their field, and thus to keep an ear open for such events while synthesizing grant proposals and reviewers' comments.

Since multiple integration is not considered to be in the mainstream of the field of statistics, associated problems did not receive prominent or highlighted attention in the grant proposals. However, it became apparent that such problems were repeatedly tucked into the midst of a statistical problem, and often, a single statement unobtrusively positioned in the text would assert that multiple integration was *the* major obstacle in the solution to this or that statistical problem.

Some statistical proposals that included a multiple integration component were submitted to the newly created NSF program in Computational Mathematics, and were reviewed by applied mathematicians and numerical analysts as well as statisticians. Reviewers outside the field of statistics invariably lacked an appreciation for the statisticians' problems, commonly suggesting that solutions were available if only the statistician would look in the right place or collaborate with the right people.

Bob Tsutakawa and I had collaborated previously on the design of a Bayesian experiment for clinical trials at the Fred Hutchinson Cancer Research Center in Seattle. This work provided the basis for our joint interest in statistical multiple integration, although through different motivations, and led us to the notion of a conference that would bring together mathematicians, statisticians, and computational scientists to make each constituency more aware of the other's progress.

High dimensional integration arises naturally in two major subfields in statistics: multivariate and Bayesian statistics. Indeed, the most common measures of central tendancy, variation, and loss are defined by integrals over the sample space, the parameter space, or both. Recent advances in computation power have stimulated significant new thrusts in both Bayesian and classical multivariate statistics. Bayesian statisticians are motivated by the need to develop tractable computational methods for the application of Bayesian statistical theory; whereas, multivariate statisticians are motivated by the need to analyze more complex dependencies among variables and objects.

Thus the Joint AMS-IMS-SIAM conference on Statistical Multiple Integration was conceived and this volume speaks to its success. The papers collected together in this volume serve to document the state of the art of multiple integration with respect to problems in statistics, statistical thrusts blocked by problems with multiple integration, and current work directed at expanding our capability to integrate over high dimensional surfaces. Upon reading these works, we trust you will agree that there are many important problems in statistical multiple integration to challenge and stimulate statisticians, numerical analysts, and other computational and algorithmic experts alone and in collaboration. We hope this volume serves to stimulate this effort.

Nancy Flournoy

Contemporary Mathematics
Volume 115, 1991

Introduction

JAMES BERGER

1. Introduction

An introduction to a collection of papers can serve many purposes. First, it can serve as an actual introduction to the subject of the papers, allowing non-experts to more readily access the material. Such a review was not necessary here, because several of the papers herein are already excellent introductions to the subject. In particular, the paper by Genz introduces many of the basic ideas behind the numerical analysis developments, the papers by Geweke and Wolpert and the Discussion by Albert provide considerable background for the statistical Monte Carlo integration ideas, and the paper by Kass, Tierney, and Kadane reviews the Laplace approximation method.

A second possible purpose of an introduction is to compare and contrast the papers in a collection, in order to place them in scientific perspective. The discussions by Albert, Shanmugan, and Olkin are quite successful in this regard, comparing and illustrating the differences among varying approaches to the same problem. There is, however, little discussion of the major differences between the numerical analysis and statistical perspectives on multiple integration; a few comments in this direction are thus given in §2.

A third reason for an introduction is to assist the reader in finding important ideas in the papers. Efforts in this direction are made in §3, but with the serious caveat that it is written mainly for statisticians; in particular, no serious effort is made to identify aspects of the papers that would be of most interest to numerical analysts.

2. Perspective

An interesting facet of the enclosed papers is the different philosophies that are revealed, underlying the numerical analysis and statistical computation literatures. To a large extent, this difference is due to the historical development of the two literatures.

Preparation supported by the National Science Foundation, Grants DMS-8717799 and DMS-8923071.

One fundamental difference in orientation is that, for statistical problems, great accuracy is seldom required, while numerical analysis algorithms are typically developed with the mathematical goal of allowing arbitrary specified accuracy. In statistical problems, obtaining two significant digit accuracy typically suffices, and only rarely are more than three significant digits needed. The reason for this is that the quantities being computed numerically are themselves typically statistical estimates based on data, and the statistical standard deviations of these estimates will typically "swamp" the inaccuracy due to numerical computation of the quantities.

Another fundamental difference between the statistical and numerical analysis approaches to the subject is the considerable influence in numerical analysis of the mathematical desire to successfully deal with "worst cases" and to develop algorithms that will be "guaranteed" to work. This is by no means a bad desire since, in hosts of scientific and engineering problems involving multiple integration, quite nasty functions can be encountered, but it has meant that only lower dimensions are typically considered and that extremely heavy computational burdens are allowed.

In contrast, statistical developments in multiple integration have been driven by statistical problems, most of which have involved integration of "nice" functions. Because of this, the goal has tended to be the development of methods that work quickly on nice functions in possibly very high dimensions, with only passing concern that the algorithms may well fail on more pathological functions (although it is generally hoped that the mode of failure will be a failure to converge, as opposed to convergence to the wrong answer).

Here are some reasons why statisticians concentrate on "nice" integrands. More extensive discussion of these issues can be found in several of the papers (cf., those of Geweke and Wolpert).

(i) In most statistical integrations, the key part of the integrand is the "joint density of the data" or "likelihood function," and the Central Limit Theorem of probability states that, for large amounts of data, this can be well approximated by a multivariate normal distribution. Not only is this a reason to expect the integrand to be nice, but many of the integration methods considered explicitly try to exploit this fact: e.g., the Laplace approximation method discussed in Kass-Tierney-Kadane; use of Monte carlo methods with importance functions related to the normal approximation; and use of product quadrature rules following a linear transformation suggested by the normal approximation.

(ii) Even if the sample size is small, commonly used statistical models (e.g., exponential families) tend to yield integrands that are unimodal or have very few modes.

(iii) In many high dimensional statistical problems, the integrand has a dimensional structure that can be exploited via some iterative scheme of Monte Carlo integration: e.g., Gibbs sampling as discussed in the article by Wolpert. Note that "niceness" here is not necessarily "approximate normality" or unimodality as above, but rather the niceness of the structure of the integrand.

(iv) Statistical practice (especially Bayesian methodology) can require fast integration of a multitude of "similar" integrals on "standard" machines. Also, the integration code must be fairly generic, since little time will be available to devote to the numerical side of the problem. If such cannot be made available, cruder nonintegration methods (such as maximum likelihood) will be employed. Integration software with these properties is possible only if one restricts consideration to "nice" problems.

How do these differences of perspective translate into differences in actual approaches? The fact that only limited accuracy is needed for statistical integration has encouraged concentration on Monte Carlo methods, which can achieve low orders of accuracy relatively cheaply. (Monte Carlo methods employing pseudo-random numbers and variance reduction techniques can provide considerably higher accuracy, if they can be applied.) Monte Carlo methods also are easy to code and can efficiently be used for simultaneous computation of numerous similar integrals. In contrast, the numerical analysis goals are not easily met by Monte Carlo methods, which have thus been less popular in that domain.

Another illustration of the effect of the differing perspectives is the manner in which "adaptive" algorithms are developed. An adaptive algorithm is one that learns about the function being integrated, as it proceeds, and adapts itself to the function in order to improve efficiency. Adaption in the numerical analysis literature virtually always means subdividing the region of integration into smaller regions, in which integration is carried out separately, with the subdivision being based on current estimates of accuracy in the various regions. Such an approach to adaption is sensible and necessary if one is considering functions with possibly strange local behavior.

In the statistical literature, the type of adaption tends to be more global. For instance, there have been efforts to adaptively determine transformations of the variables for which the integrand will be approximately a product integrand to which product quadrature rules can be applied. And in the Monte Carlo approaches, adaptive choice of the importance function is often considered (see, e.g., the papers of Evans and Oh). Gibbs sampling and other iteratively stochastic schemes are also interesting in this regard, in that they can be viewed as automatically self-adapting Monte Carlo schemes.

The above discussion has concentrated on what could be termed statistical inference—the actual process of analyzing data—as opposed to statistical theory—the study of properties of procedures. The papers by Luzar-Olkin and Kaishev are examples of the latter. The distinction here is that integrals encountered in such theoretical problems can be quite arbitrary, and their solution may be more in the spirit of the numerical analysis developments.

3. The papers

In this section, a few words are said about each of the papers. As noted in an earlier apology, the perspective here is mainly that of a statistician, so that the comments about the papers will focus on the main features relating to statistical integration problems. The paper of Hardwick and discussion of Flournoy unfortunately arrived too late for reference herein.

3.1 The numerical analysis papers. The papers by de Doncker-Kapenga, Genz, Kahaner, and Mascagni are written from the numerical analysis perspective.

de Doncker and Kapenga. This paper provides an excellent general review of types of parallel processing machines, and describes the usefulness of the various types for multiple integration. It is observed that the MIMD (Multiple Instruction Multiple Data) machines are best for subregion adaptive integration schemes because constant communication between the processors is not needed. However, these machines would not be optimal for the current globally adaptive statistical integration schemes.

Genz. This paper begins with an excellent review and history of the numerical analysis literature on multiple integration. It then focuses attention on the most extensive subregion adaptive algorithm in use, and indicates that it appears to be successful in dimensions up to 10. This algorithm deserves serious consideration by statisticians.

Kahaner. This paper discusses much of the software that is currently available for multi-dimensional integration. The descriptions are excellent, including discussion of error estimates that are or are not provided, and indication of the dimensions and types of functions for which a given routine is likely to prove effective. Reflecting the numerical analysis literature, most of the routines that are discussed are quadrature routines and many are for just two-dimensional integration. A variety of higher dimensional programs are listed, however. Of most interest are the adaptive quadrature schemes, though an adaptive Monte Carlo program is also available.

Mascagni. This paper focuses on SIMD (Single Instruction Multiple Data) machines and demonstrates their usefulness in product quadrature and Monte Carlo integration (of nonadaptive types). An extensive discussion is given to the problem of using SIMD machines to compute expectations involving stopped Brownian motion. Although this is motivated in the paper as a method of solving P.D.E.s, it interestingly is also a problem encountered in a host of statistical problems, from evaluation of sequential experiments to asymptotics to minimax estimation. Thus this material is of special interest to statisticians and probabilists in its own right.

3.2 The Bayesian–Monte Carlo integration papers. The papers by Evans, Geweke, Müller, Oh, Wolpert, and to some extent Monahan-Liddle are concerned with Monte Carlo integration methods, especially as applied to problems in Bayesian statistical inference.

Geweke and Wolpert. These two papers provide outstanding introductions into the history and ideas behind use of Monte Carlo methods in Bayesian statistical problems. Motivation for the approach and comparisons with alternative integrations methods are given. Extensive discussion of modifications involving variance reduction methods can also be found in these papers. The Wolpert paper includes an interesting applied example of the ideas.

Evans and Oh. These two papers present an apparently successful, fully adaptive algorithm for improving the importance function in Monte Carlo integration. The Evans paper also introduces a different, interesting concept called "chaining" which can be used to determine a good starting point for the adaptive algorithm. The Oh paper also indicates how increasing dimension can affect the accuracy of importance sampling. Both papers discuss examples.

Monahan and Liddle. This paper introduces an interesting modification to stochastic approximation, which is a technique for minimizing a function whose evaluation at any point requires multiple integration. While this is a problem of central interest to Bayesian decision theory, with the integral to be minimized being the posterior expected loss, its domain of applicability is much broader.

Müller. To perform Monte Carlo integration in Bayesian problems, it suffices to have a "sample" of random variables from the posterior distribution (integral estimates then being just appropriate "averages" over this sample). In time series and other problems in which random variables are propagated through stages, it is particularly appealing to approach the integration problem in this way because one can start with a "prior" sample and let it propagate through the stages to obtain the posterior sample. The difficulty in doing this is degeneration of the sample size, through use of necessary accept/reject steps. This paper develops efficient algorithms for implementing the propagation steps and for augmenting the sample at each step so as to avoid degradation. The approach is successfully applied to a complex practical example.

3.3 Analytic approximation papers. The papers of Kass-Tierney-Kadane and Tong are concerned with development and use of analytic approximations to multiple integrals.

Kass, Tierney, and Kadane. This paper considers a version of Laplace's method for approximating integrals, a method based on asymptotic normality of Bayesian integrands for large amounts of data. The approximation is not only of interest in its own right, but also has potential use in determining importance functions in Monte Carlo integration. The paper also contains an interesting example of multiple integration in Bayesian statistics.

Tong. This paper develops exact analytic bounds for the probability of certain rectangles and ellipsoids for a variety of distributions. The bounds are not necessarily quantitatively accurate, but they can be useful when only qualitative judgements are required.

3.4 Case studies.

Kaishev. This paper considers integration, using Gaussian cubature and Monte Carlo methods, of problems involving generalized B-splines. As a number of problems in statistics can be reduced to such integrations, the technical tools derived herein can be quite useful.

Luzar and Olkin. This paper provides an excellent case study comparing the success of various variance reduction techniques in Monte Carlo integration via importance sampling. The problem considered is estimation of means of ordered characteristic roots of covariance matrices, and the paper provides a useful prescription for integration problems of that type.

Tsutakawa. This paper provides an interesting case study of the use of Lindley's version of Laplace's method for analytically approximating a multiple integral. The problem considered is typical of a large group of Bayesian statistical problems called hierarchical Bayes problems.

4. Conclusion

The conference and this collection of papers are extremely valuable for several reasons. Perhaps the primary reason is that research on multiple integration has been performed by heretofor largely separate groups. While there are natural reasons that led to separate research programs, there are clear benefits in maintaining contact. Statisticians can immediately utilize some of the newer adaptive multiple integration routines from the numerical analysis literature for computations of "difficult" integrals encountered in theory and can, at the very least, incorporate new techniques from numerical analysis into their statistical methods. New developments in parallel computation are also highly relevant to statisticians.

Numerical analysts can find value in the statistical literature for several reasons. First, simple recognition of the different nature of many statistical problems may spur numerical-analytic developments in that direction. Second, the extensive developments in Monte Carlo integration that have occurred in the statistical literature may have features that are transportable to the general numerical analysis domain. For instance, the recent work in Gibbs and other iterative sampling schemes (references are given in Wolpert's paper) may be highly successful for general problems because of their strongly "self-adaptive" nature.

Finally, thanks are due the two editors and conference organizers, Nancy Flournoy and Robert Tsutakawa, and the American Mathematical Society for the extensive efforts that led to this proceedings.

DEPARTMENT OF STATISTICS, PURDUE UNIVERSITY, WEST LAFAYETTE, INDIANA 47907

Contemporary Mathematics
Volume **115**, 1991

A Survey of Existing Multidimensional Quadrature Routines

D. K. KAHANER

ABSTRACT. We provide a detailed description of available software to compute multidimensional integrals. More than three dozen routines are surveyed, and their essential characteristics described. This paper was presented at the American Mathematical Society Workshop on Statistical Multiple Integration

I. Introduction

With computers becoming more available, the amount of data about useful software has grown rapidly. The purpose of this paper is to present information of use to the statistical computing community. We focus on software that deals with a specific problem; namely, the numerical evaluation of integrals in several dimensions. In our experience most statistical researches are often too busy with their own activities to keep track of developments in such related fields as numerical analysis. Similarly, experts in numerical integration are often unaware of all the software products that have been written for quadrature. Thus a compendium of useful software can be put to direct use. At the same time most numerical analysts are not familiar with the types of quadrature problems that confronts statisticians. This meeting attempts to bring these specialists together.

In attempting to gather material for this paper we have made extensive use of the NIST GAMS Catalog, **[5]**. GAMS, an acronym for Guide to Available Mathematical Software, is a project originating ten years ago to provide scientists with up to date problem oriented information about software. It is composed of a printed catalog and an on-line interactive consultant, both generated from a database that is kept current by several researchers and a programmer. As new software products are released, individual items are examined and appropriate detailed information records are added to the

1980 *Mathematics Subject Classification* (1985 *Revision*). Primary 65D30.
This paper is in final form and no version of it will appear for publication elsewhere.

database. The catalog is periodically reprinted; the interactive consultant is updated every few months. Locally, consultants are available to answer specific questions about usage. GAMS does not attempt to deal with the difficult problem of ranking software, although the consultants often provide their opinions on specific programs.

The heart of the GAMS system is a hierarchical classification scheme, organized by problem type. Thus users will not find "Gauss Kronrod" quadrature but will find "Automatic quadrature on a rectangle of a function given by a user written procedure." Currently GAMS contains information about more than 6,000 individual pieces of mathematical software. An early design decision was to limit inclusion to general mathematical and statistical software rather than to specific application oriented packages. For example, GAMS maintains information about ordinary differential equation solvers but not about analysis of steam turbine blade wear. Further (and this was a controversial decision) GAMS omits what is usually referred to as "research" software. The database includes only those items that are readily available from sources that are willing to respond to queries, provide support, and whose software meets recognized standards of documentation and testing. Unfortunately, many good algorithmic ideas are never worked through to satisfactory implementation; and hence are lost. We hope that GAMS provides a ready access to information about much of what is good in mathematical software and encourages researchers to "finish" their work.

Figure 1 shows the H2b section of the GAMS classification which deals with multidimensional integrals. The main subcategories are geometric (rectangular or nonrectangular region), within each of these there is a further subdivision into the form of the integrand (available on a grid or as a procedure), and a final level of subdivision for automatic or nonautomatic routine. Classes with a leading bullet, •, are those containing software. The classification scheme was designed with the help of experts in the field. Whenever

	H2b	Multidimensional integrals
	H2b1	Rectangular region or interated integral
	H2b1a	Integrand available via procedure
•	H2b1a1	Automatic (user specifies required accuracy)
•	H2b1a2	Nonautomatic
	H2b1b	Integrand available only on a grid
	H2b1b1	Automatic (user specifies required accuracy)
•	H2b1b2	Nonautomatic
	H2b2	Nonrectangular region
	H2b2a	Integrand available via procedure
•	H2b2a1	Automatic (user specifies required accuracy)
•	H2b2a2	Nonautomatic
	H2b2b	Integrand available only on a grid
	H2b2b1	Automatic (user specifies required accuracy)
	H2b2b2	Nonautomatic
•	H2c	Service routines (computes quadrature weights and nodes)

FIGURE 1. The GAMS scheme for multidimensional integrals

possible software is classified at the finest level of the appropriate tree. We have also tried not to invent too many classes for which there is no software available.

II. Source of routines

Figure 2 lists those categories containing routines, and the routine names and sources. Section III discusses these in detail. Here we only want to comment on the source information. CMLIB **[5]** (Core Mathematics Library) represents the local mathematics library at NIST. It contains most of the well known mathematical software collections (Linpack, Quadpack, etc.) in an organized, coherent form. Readers who want a public domain collection of high quality general purpose software can write for a tape. NAG **[27]** (Numerical Algorithms Group) was originally started in the UK as a government project to develop mathematical software and is now a private company. Now NAG has an active branch in the USA. IMSL **[19]** (International Mathematics and Statistics Libraries) is the oldest of the commercial mathematical software vendors, and many faculty in American universities are familiar with their routines. Recently IMSL reorganized their product, breaking the original IMSL library into an IMSL/MATH library, IMSL/STAT library, etc. Each of these can be purchased separately. The original IMSL library is no longer leased, but some sites still maintain it. We have indicated the specific library associated with each routine. In some cases routines in the IMSL/MATH library are functionally equivalent to something in the original IMSL library although the names may be different. Roughly speaking, the NAG and IMSL libraries are comparable.

TOMS **[2]** (Transactions on Mathematical Software) are the refereed software contributions published in the Journal, "ACM-TOMS." The original name was "Collected Algorithms of the ACM." TOMS contributions are denoted by their algorithm number, such as A657; in a few cases an algorithm has not yet been given a number and we then denote it as *A* ???. The Scientific Desk **[1]** from C. Abaci is a commercial library by the original

ADAPT (CMLIB) [Replace with ADMINT]
D01FCF (NAG) [Equivalent to ADAPT]
ADMINT (A???-TOMS) [Successor to ADAPT]
D01EAF (NAG) [Vector form of ADAPT]
D01DAF (NAG)
D01GBF (NAG)
DBLIN (IMSL)
DMLIN (IMSL)
QAND (IMSL/MATH) [Equivalent toDMLIN]
TWODQ (IMSL/MATH)

FIGURE 2a. Routines and their origin: class H2b1a1

D01FBF (NAG)
D01FDF (NAG)
D01GCF (NAG)

FIGURE 2b. Routines and their origin: class H2b1a2

DBCQDU (IMSL)
BS2IG (IMSL/MATH)
BS3IG (IMSL/MATH)

FIGURE 2c. Routines and their origin: class H2b1b2

CUBTRI (A584-TOMS)
TRIEX (A612-TOMS)
TRISET (A???-TOMS) [Successor toTRIEX]
TWODQ (CMLIB)
H2B2A (Scientific Desk / PC) [Equivalent to TWODQ-CMLIB]
SFERIN (JCAM-A22)
D01JAF (NAG) [Equivalent to SFERIN]
D01DAF (NAG)

FIGURE 2d. Routines and their origin: class H2b2a1

D01PAF (NAG)

FIGURE 2e. Routines and their origin: class H2b2a2

A647-TOMS
IQPACK (A655-TOMS)
GQRCF (IMSL/MATH)
GQRUL (IMSL/MATH)
RECCF (IMSL/MATH)
RECQR (IMSL/MATH)
FQRUL (IMSL/MATH)
D01BBF (NAG)
D01BCF (NAG)
D01GYF (NAG)
D01GZF (NAG)
GAUSQ (PORT)
GQ0IN (PORT)
GQM11 (PORT)

FIGURE 2f. Routines and their origin: class H2c

founders of IMSL, but organized for use on smaller systems such as PCs. JCAM [**20**] (Journal of Computational and Applied Mathematics) is similar to ACM-TOMS. PORT [**3**] is a commercial library marketed by Bell Laboratories.

III

iiiA. Description of the routines in class H2c. Although the miscellany of routines in this class are mostly associated with one-dimensional quadrature they are also useful for multidimensional problems.

A647: Implementation and relative efficiency of quasirandom sequence generators [**13**]. The author discusses sequences of points x_n in R^s that can be used as evaluation points for quadrature sums. The basis of these techniques is the inequality of Niederreiter [**26**], who also shows that the following inequality is the best possible:

$$\left| \int_I f(t)\,dt - \frac{1}{N}\sum_{n=1}^{N} f(x_n) \right| \leq \frac{V(f)D^*(N)}{N},$$

where $V(f)$ is the variation of f and $D^*(N)$ is the discrepancy with respect to the origin. Roughly, random points are thought of as haphazard with large discrepancy. Quasirandom points have low discrepancy; such points do not attempt to emulate randomness. For the sequence generators considered here $D^*(N) = A_s(\log N)^s$ in dimension s. Most of the algorithms that use these generators compute estimates with N, $2N$, etc., points until their relative difference is less than an input tolerance, i.e., $N = 2^k$. The programs here are not so much implementations as they are tests of integral estimates obtained from quasi random sequences generated by the algorithms of Faure, Halton, Sobol, and a uniform random generator. The tests are conducted on a single smooth test integral:

$$\int_0^1 \prod_{i=1}^{s} |4x_i - 2|\,dx_i = 1, \qquad s = 4, \ldots, 40.$$

Fox notes that not much is known about dimensions greater than 40; this author (DKK) has never seen test results for dimensions that high. Fox's conclusion is that if $s < 6$ the Sobol generator produces the most accurate estimates, whereas the Faure generator seems best for $s > 6$. The uniform generator is faster than either of these but is much less accurate.

A655, IQPACK: Weights and nodes for interpolatory quadratures [**12**]. This package contains routines, some very general, for finding weights for interpolatory quadratures of the form:

$$Q(f) = \sum_{j=1}^{s} \sum_{i=0}^{m_j - 1} d_{ji} f^{(i)}(t_j).$$

The quadrature rule $Q(f)$ is designed to be exact

$$Q(f) = I(f) = \int_a^b f(t)w(t)dt$$

whenever f is a polynomial of degree (at least) $m_1 + \cdots + m_s$. The package includes the classical Gauss quadratures as special cases, and for these it can compute the nodes as well as the weights. Routines are also provided to evaluate a quadrature rule once its weights and nodes are determined. Their algorithm generalizes a very successful idea of Golub and Welsch [**17**]. The latter makes use of the eigenvalues of the symmetric tridiagonal matrix associated with the orthogonal polynomials with respect to the weight function $w(t)$. These polynomials satisfy a three term recursion

$$p_n^*(x) = (a_n x + b_n)p_{n-1}^*(x) - c_n p_{n-2}^*(x), \qquad n = 2, 3, \ldots$$

that can be converted to the matrix form

$$xP(x) = JP(x) + [0, 0, \ldots, p_N^*(x)/a_N]^T,$$
$$P(x) = [p_0^*(x), p_1^*(x), \ldots, p_{N-1}^*(x)]^T,$$

where J is the symmetric tridiagonal matrix with

$$J_{i,i} = b_i/a_i, \qquad J_{i,i+1} = 1/a_i.$$

From this one sees that x_i are the zeros of $p_N^*(x)$ if and only if

$$x_i P(x_i) = JP(x_i).$$

In other words x_i are the eigenvalues of J, and $P(x_i)$ are the corresponding eigenvectors [7]. In this formulation the matrix elements depend on the moments of $w(t)$, $\int t^k w(t)\,dt$, and these are required input to this package; (although for some classical weight functions the program explicitly knows the matrix J). The program can be obtained in either single or double precision.

GQRUL: IMSL/MATH. Uses the Golub–Welsch algorithm to compute Gauss Legendre, Gauss Radau, or Gauss Lobotto nodes and weights for various classical weight functions. Double precision version is DGQRUL.

GQRCF: IMSL/MATH. Computes the nodes and weights for a quadrature rule associated with weight function $w(t)$ that has a given input three term recursion. Thus this routine can be used for any weight function as long as the recursion coefficients are known. Double precision version is DGQRCF.

RECQR: IMSL/MATH. Inverse routine to GQRCF. This completes the three term recursion coefficients given the weights and nodes of a quadrature rule. Double precision version is DRECQR.

RECCF: IMSL/MATH. Computes recursion coefficients for the orthogonal polynomials associated with several classical weight functions. Double precision version is DRECCF.

FQRUL: IMSL/MATH. Computes the Fejer quadrature rule for many classical weight functions. Uses an algorithm due to Gautschi [**15**] that works best when N is prime. Given a weight function $\omega(x)$ the Fejer rule associated with ω is

$$Q(f) = \sum_{i=1}^{N} w_i f(x_i) \approx \int_{-1}^{1} f(x)\omega(x)dx,$$

where x_i are the zeros of the Nth degree Tchebycheff polynomial, $T_N(x) = \cos(N \cos^{-1} x)$, and the weights w_i are chosen so that $Q(p) = \int p(x)\omega(x)\,dx$, with p a polynomial of highest possible degree. If we substitute T_l for $f(x)$ above, then using the properties of these polynomials.

$$\int_{-1}^{1} T_l(x)\omega(x)\,dx = \sum_{j=1}^{N} w_j \cos \frac{l(2j-1)\pi}{2N}, \qquad l = 0, \ldots, N-1.$$

This is a system of linear equations for the unknown w_j. It is also the cosine quarter-wave transform for the sequence $w_1, \ldots, w_N$. If the integrals on the left-hand side above are known, then the w_j can be computed very efficiently by FFT techniques. Double precision version is DFQRUL.

D01BBF: NAG. Computes the nodes and weights for Gauss Legendre, Gauss Laguerre, Gauss Hermite, and Gauss Rational quadrature rules. The latter is an N point rule that approximates integrals such as

$$\int_a^\infty f(x)\,dx,$$

and will be exact for any function of the form:

$$f(x) = \sum_{i=2}^{2N+1} \frac{c_i}{(x+b)^i}, \qquad a + b > 0.$$

This requires as input the name of another NAG routine, such as D01xxF that specifies the rule. The number of points N, is restricted to 1, 2, 3, 4, 5, 6, 7, 8, 10, 12, 14, 16, 20, 24, 32, 48, or 64, as the quadrature data is contained in a table. Routines in the NAG library adhere to a naming convention that associates the first three characters (such as D01) with chapters in the NAG manual, and the next two (such as BB) as sequential identifiers indicating when the routine was added to the library. The last character denotes precision. NAG uses F to denote standard precision, and E to denote alternative precision. This is machine dependent; on a 32 bit computer such as a Vax, F means double precision whereas on a Cyber it means single precision. Releases of the library are denoted Mark 8, Mark 9, etc.

D01BCF: NAG. Computes the nodes and weights for one of the classical weight functions by use of the Golub–Welsch algorithm. Because the three term recursion coefficients for the allowed weight functions are known exactly, this routine is capable of generating quadrature rules with an arbitrary number of nodes N. However, the accuracy of the computed eigenvalues (the quadrature nodes) will slowly degrade with increasing N.

GAUSQ: PORT. Computes the quadrature weights and nodes for an arbitrary weight function, given both the three recursion coefficients and the weight function's moments $\int t^k w(t)$. Uses the Sack–Donovan algorithm [**31**]. The number of nodes can be arbitrary. Double precision version is DGAUSQ.

GQ01N: PORT. Computes the quadrature weights and nodes for Gauss Laguerre quadrature,

$$\int_0^\infty \exp(-t)f(t)dt \approx \sum_{i=1}^{N} w_i f(x_i)$$

by generating the three term recursion coefficients and moments and then calling GAUSQ. Double precision version is DGQ01N.

GQM11: PORT. Computes the quadrature weights and nodes for Gauss Legendre quadrature,

$$\int_{-1}^{1} f(t)\,dt \approx \sum_{i=1}^{n} w_i f(x_i)$$

by generating the three term recursion coefficients and moments and then calling GAUSQ. Double precision version is DGQM11.

D01GYF: NAG. Finds Korobov optimal coefficient when the number of points is prime. Korobov quadrature rules are discussed in §IIIc.

D01GZF: NAG. Finds Korobov optimal coefficient when the number of points is a product of two primes.

iiiB. Description of the routines in class H2b1a1

D01DAF: NAG. Approximates a double itergrated integral with general limits,

$$\int_a^b \int_{u(y)}^{v(y)} f(x, y)\,dx\,dy$$

by writing it in the form:

$$\int_b^a F(y)dy \quad \text{where } F(y) = \int_{u(y)}^{v(y)} f(x, y)\,dx$$

and by approximating the integral of F using a one-dimensional algorithm. Each evaluation of F also requires the approximation of a one-dimensional integral which is done in the same way. The one-dimensional algorithm used is due to Patterson **[28]** as follows. Beginning with the three point Gauss rule (of polynomial degree five) four new points are added to give a rule using seven points with polynomial degree 10. To this are added eight new points to give a rule with fifteen points and degree 22. Symbolically,

$$3 + 4 = 7 + 8 = 15 + 16 = 31 + 32 = 63 + 64 = 127 + 128 = 255.$$

The last rule has polynomial degree 382. This sequence of rules has several interesting properties, including the facts that the nodes all interlace and all the rules have positive weights. The program halts when two successive estimates agree to the user's input accuracy. The same tolerance is used on both integrals, which has some theoretical difficulties **[14]**, but is acceptable as long as the region of integration is not too long and thin. Because the algorithm uses rules of very high polynomial degree the program works best for smooth integrands. The general technique of using iterated integrals can also suffer because of the appearance of "fictitious singularities." This is illustrated by trying to integrate a constant on D the positive quadrant of the unit circle,

$$\int_D 1 = \int_0^1 dy \int_0^{\sqrt{1-y^2}} dx = \int_0^1 \sqrt{1-y^2}\, dy.$$

DBLIN: IMSL. Another two-dimensional routine for iterated integrals, such as

$$\int_0^2 \int_{.5-.25x}^{.5+.25x} f(x, y)\, dy\, dx.$$

using a pair of one-dimensional integrators. This uses the IMSL library program DCADRE for each one-dimensional integral. DCADRE is an automatic adaptive quadrature program based on cautious extrapolation [**8**]. No count of integrand evaluations is available as an output, a minor disadvantage. DBLIN is a double precision routine.

DMLIN: IMSL. For integration on a hyper-rectangle using product Gauss quadrature, beginning with the two point Gauss Legendre rule. Sweep through each direction doubling the number of points. If there is no significant change in the integral estimate then back down. Stop if none of the directions improves the result or if 256 points are used in any direction.

QAND: IMSL/MATH. This is equivalent to DMLIN. Double precision routine is DQAND.

TWODQ: IMSL/MATH. Estimates a two-dimensional iterated integral, using iterated calls to the one-dimensional routine QDAG. The latter is essentially equivalent to the globally adaptive Quadpack routine QAG [**29**]. QAG uses the difference between the elements of a Gauss Kronrod rule pair to estimate the quadrature error on each interval. TWODQ gives the user a choice of which rule pair to use in each dimension, (7–15, 10–21, 15–31, 20–41, 25–51, 30–61). Double precision version is DTWODQ.

D01GBF: NAG. Adaptive Monte Carlo for a hyper-rectangle [**24**], that begins by subdividing the original input region into equal volume subregions. In each, the integral and variance are estimated by pseudo-random sampling. All contributions are added together to produce an estimate for the whole integral and total variance. The variance in each coordinate dimension is determined and used to increase the density and widths of subintervals along each coordinate axis so as to reduce the total variance. The total number of subregions is doubled in each iteration, and the program stops when a desired accuracy has been reached or too many integral evaluations are needed for the next cycle.

ADAPT: CMLIB, **[16]** and **D01FCF:** NAG. These are essentially equivalent. This is a globally axis adaptive algorithm, for $N \leq 15$ dimensions on a hyper-rectangle. In each subregion, the integral is estimated using a seventh-degree rule, and an error estimate is obtained by comparison with a fifth-degree rule that uses a subset of the same points. The fourth differences of the integrand along each coordinate axis are evaluated, and the subregion is marked for possible future subdivision in half along that coordinate axis which has the largest absolute fourth difference. If the estimated errors, totalled over the subregions, exceed the requested relative error, further subdivision is performed on the subregion with the largest estimated error, that subregion being halved along the appropriate coordinate axis. Originally based on routine HALF **[33]**.

ADMINT: A-??? TOMS, **[4]**. The successor to ADAPT. This routine has the ability to integrate a vector of integrands on a hyper-rectangle. The same subdivision is used for each integrand, so it may not be efficient unless all integrands have a similar structure. A global axis adaptive algorithm similar in strategy to ADAPT. Experiments suggest that this routine is most appropriate for dimensions 2–10. The user can elect to have the $p/2$ worst subregions subdivided next. This enables the program to be efficiently used in a multiprocessor environment if p is chosen as the number of processors available. The returned estimated quadrature error is the maximum over all integrands. Uses fully symmetric rules for the integral estimates and null rules for error estimates. User can select from various integration rule pairs.

iiiC. Description of the routines in class H2b1a2

D01FDF: NAG. For a product region (in $N \leq 30$ dimensions) that is then transformed into an N-sphere. The technqiue, due to Sag and Szekeres **[32]**, uses the transformation

$$x_i = \frac{y_i}{r} \tanh \frac{ur}{1 - r^2}, \qquad r = \sum y_i^2,$$

that causes the transformed integrand to vanish along with its derivatives at the boundary of the sphere. The sphere is then embedded in a unit cube with the integrand set to zero outside of the sphere. The integral on the cube is estimated using the product trapezoid rule. Because of the Euler Maclaurin formula [7] it is known that the trapezoid rule is very efficient for integrands going smoothly to zero at the endpoints. Moreover, since the integrand approaches zero very rapidly near the boundary of the sphere the program ignores points beyond r_0, a value set by the user; typically $r_0 = 0.8$. The transformation parameter u is also set by the user. It determines how the transformed integrand is distributed between the origin and the surface of the unit sphere; typically $u = 1.5$. The transformation is such that its Jacobian is constant at points with the same value of r and the program takes advantage of this to reduce the work needed to evaluate the integrand at the quadrature nodes. The program has no provision for an input error tolerance and neither does it provide any output error estimate. It does, however, require as input the maximum number of integrand evaluations that are to be permitted and returns as output the actual number of integrand evaluations used.

D01GCF: NAG. Uses the method of Korobov and Conroy as extended by Patterson and Cranley to approximate the integral on a hyper-rectangle [**22**, 6]. Korobov's idea was to translate the input region to the unit cube and approximate it there by an expression of the form

$$\int g(x) \approx \frac{1}{p} \sum_{k=1}^{p} g\left(\left[\frac{a_1}{p}\right], \dots, \left[k\frac{a_n}{p}\right]\right),$$

where p is a prime, a_i are "optimal coefficients," and $[x]$ is the fractional part of x. In this program, as well as in most applications of the method $a_i = a^{i-1} \bmod p$ for an appropriate a. Figure 3 shows a set of Korobov quadrature nodes for a specific selection of a and p. If g is periodic, the right choice of a makes some of the Fourier coefficients of the quadrature error zero. With the optimal a_i it can be shown that the quadrature error E satisfies

$$E \leq Kp^{-\alpha}(\ln p)^{\alpha\beta}$$

with known constants α and β. Here α depends on the convergence rate of the Fourier series, and β depends on the dimension N. For the technique

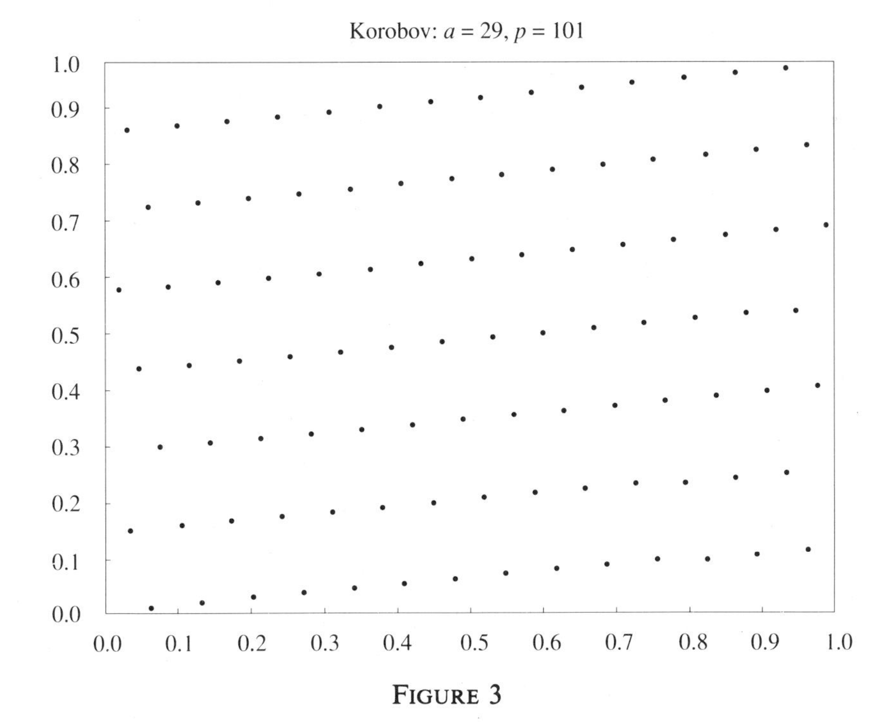

FIGURE 3

to work well the integrand should be periodic in the cube so that its Fourier coefficients go to zero rapidly. The program uses the automatic periodization $x_i = y_i^2(3 - 2y_i)$. In order to generate error estimates Patterson replaced

$$[ka_i/p], \quad \text{with } [\alpha_i + ka_i/p],$$

where α_i is a uniform pseudorandom number. The computation can then be repeated with different α_i and a standard error computed. Six sets of a_i and p are built into this program, or can be input by the user. D01GYF and D01GZF can be used to compute a_i for p a prime or product of two primes.

DBCQDU: IMSL. Bicubic (tensor product) spline quadrature. This program computes the interpolating natural spline (second derivative zero at the endpoints) and integrates it exactly. A useful feature is that the integration rectangle need not coincide with the data.

BS2IG: IMSL/MATH. This program computes the integral of a bivariate tensor product B-spline of arbitrary order. The user must input the spline knots and coefficients, but these are typically computed by another routine, for example after the computation of a B-spline interpolant. Double precision version is DBS21G.

BS3IG: IMSL/MATH. Three dimensional version of DS2IG. Double precision version is DBS3IG.

iiiD. Description of the routines in class H2b1b2

A584: CUBTRI. Automatic global adaptive quadrature on a triangle [**23**]. Each triangle is subdivided into four subtriangles. To compute quadrature and error estimates a 7 node degree 5 Radon rule and 19 node degree 8 Lyness-Jesperson rule are used. This combination is efficient because the 19 node rule reuses all 7 integrand evaluations of the 7 node rule. The difference between the two rules is used to estimate the error on each triangle. Double precision only.

TWODQ: CMLIB. Automatic global adaptive quadrature on a union of triangles [**21**]. The triangle with the largest error estimate is subdivided into four subtriangles. To compute quadrature and error estimates two Lyness-Jesperson rules are used. One feature of this program is that the user can select either of two rule pairs, (a) 19 node degree 9 and 28 node degree 11, or (b) 12 node degree 6 and 16 node degree 8. Pair (a) is closed (evaluates the integrand on the boundary of the triangles) while pair (b) is open (evaluation points are interior to the triangles). Rule pair (a) is of higher polynomial order, but (b) may be more suitable for the integration of functions with singularities, as these often occur at the corners of the input integration region. Double precision version is DTWODQ.

A612-TRIEX A???-TRISET: CMLIB. Automatic global adaptive quadrature on a triangle, TRIEX, [**10**] or on a union of triangles, TRISET, [**11**]. The triangle with largest error estimate is subdivided into four subtriangles. To compute quadrature and error estimates 9th and 11th degree Lyness-Jesperson rules are used. These routines are particularly well suited to integrands with singularities as the programs use nonlinear extrapolation via the ε-algorithm.

iiiE. Description of the routines in class H2b2a1

D01JAF: NAG: **A22-JCAM:** SFERIN, [**30**]. An improvement over D01FDF, but only for dimension $N = 2, 3,$ or 4. The Sag Szekeres method is used but several additional transformations to the sphere are optionally allowed, such as the double exponential transformation of Mori [**25**]. User input tolerance allows node spacing to be halved automatically. Works better than the original Sag Szekeres algorithm for singularities on the surface or at the center of the sphere. Routines such as this one, that perform transformations on the region often pack large numbers of nodes close together. Under certain situations distinct nodes can become computationally equal unless long word length computers are used.

iiiF. Description of the routines in class H2b2a2

D01PAF: NAG. Integration on an N-Simplex given its $N + 1$ vertices. The algorithm combines ideas in [**18**] and [**9**]. The program generates a sequence of quadrature estimates of polynomial degree $2j - 1$. For each it produces an error estimate, but it is not automatic in the sense that the user can request computation to a specified tolerance.

REFERENCES

1. C. Abaci Inc., 208 St. Mary's Street, Raleigh, NC 27605.
2. ACM Algorithms Distribution Service, c/o IMSL Inc., 2500 Park West Tower One, 2500 City West Boulevard, Houston, TX 77042-3020.
3. Bell Laboratories, 600 Mountain Ave., Murray Hill, NJ 07974.
4. J. Bernstein, T. O. Espelid, and A. Genz, *ADMINT: Adaptive multidimensional integration routine for a vector of integrals*, Trans. Math. Software (1989), submitted.
5. R. F. Boisvert, S. E. Howe, and D. K. Kahaner, *GAMS: a framework for the management of scientific software*, Trans. Math. Software **11**, (1985), 313–355.
6. R. Cranley and T. N. L. Patterson, *Randomisation of number theoretic methods for multiple integration*, SIAM J. Numer. Anal. **13** (1976), 904–914.
7. P. J. Davis and P. Rabinowitz, *Methods of numerical integration*, Academic Press, Orlando, 1984.
8. C. de Boor, *CADRE: an algorithm for numerical quadrature*, Mathematical Software (John R. Rice, ed.), Academic Press, NY, 1971.
9. E. de Doncker, *New Euler Maclaurin expansions and their application to quadrature over the s-dimensional simplex*, Math. Comp. **33** (1979), 1003–1018.
10. E. de Doncker and I. Robinson, *An algorithm for automatic integration over a triangle using non-linear extrapolation*, Trans. Math Software **10** (1984).
11. E. de Doncker-Kapenga, D. K. Kahaner, and B. Starkenburg, *TRISET: adpative integration over a triangulated region*, Trans. Math. Software (1989), submitted.

12. S. Elhay and J. Kautsky, *Algorithms* 655 *IQPACK: FORTRAN subroutines for the weights of interpolatory quadratures*, Trans. Math Software **13** (1987), 399–415.
13. B. L. Fox, *Algorithm* 647: *Implementation and relative efficiency of quasirandom sequence generators*, Trans. Math Software **12** (1986), 362–376.
14. F. N. Fritsch, D. K. Kahaner, and J. N. Lyness, *Double integration using one-dimensional adaptive quadrature routines: a software interface problem*, Trans. Math Software **7** (1981), 46–75.
15. W. Gautschi, *Construction of Gauss-Christoffel quadrature formulas*, Math. Comp. **22** (1968), 251–270.
16. A. C. Genz and A. A. Malik, *Algorithm* 019. *Remarks on algorithm* 006: *an adaptive algorithm for numerical integration over an N-dimensional rectangular region*, J. Comput. Appl. Math., **6** (1980), 295–302.
17. G. H. Golub and J. H. Welsch, *Calculation of Gauss quadrature rules*, Math. Comp. **23** (1969), 221–230.
18. A. Grundmann and H. M. Muller, *Invariant integration formulas for the n-simplex by combinatorial methods*, SIAM J. Numer. Anal. **15** (1978), 282–290.
19. IMSL Inc., 2500 Park West Tower One, 2500 City West Boulevard, Houston, TX 77042-3020.
20. J. Computational and Applied Mathematics, c/o R. Piessens, Computer Science Department, Catholic University Leuven, Celestijnenlaan 200A, B-3030 Heverlee, BELGIUM.
21. D. K. Kahaner and O. W. Rechard, *TWODQD*: *an adaptive routine for two-dimensional integration*, J. Comp. Appl. Math. **17** (1987), 215–234.
22. N. M. Korobov, *Number theoretic methods in approximate analysis*, Fizmatgiz, Moscow, 1963.
23. D. P. Laurie, *Algorithm* 584. *CUBTRI*: automatic cubature over a triangle, Trans. Math Software **8** (1982), 210–218.
24. B. Lautrup, *An adaptive multi-dimensional integration procedure*, Proc. 2nd Coll. on Advanced Methods in Theoretical Physics, Marseille, 1971.
25. M. Mori, *An IMT-type double exponential formula for numerical integration*, Publ. Res. Inst. Math. Sci., **14** (1978), 713–729.
26. H. Niederreiter, *Quasi-Monte Carlo methods and pseudo-random numbers*, Bull. Amer. Math. Soc. **84** (1978), 957–1041.
27. Numerical Algorithms Group Ltd., 7 Banbury Rd., Oxford OX2 6NN, England (or NAG Inc., 1311 Warren Ave., Downers Grove, IL 60515).
28. T. N. L. Patterson, *The optimal addition of points to quadrature formulae*, Math. Comp. **22** (1968), 847–856.
29. R. E. Piessens, E. de Doncker-Kapenga, C. W. Uberhuber, and D. K. Kahaner, *QUADPACK*, Springer-Verlag, New York, 1983.
30. D. Roose and E. de Doncker, *Algorithm* 022. *Automatic integration over a sphere*, J. Comput. Appl. Math. **7** (1981), 203–224.
31. R. A. Sack and A. F. Donovan, *An algorithm for Gaussian quadrature given modified moments*, Numer. Math. **18** (1972), 465–478.
32. T. W. Sag and G. Szekeres, *Numerical evaluation of high-dimensional integrals*, Math. Comp. **18** (1964), 245–253.
33. P. Van Dooren and L. De Ridder, *Algorithm* 006: *an adaptive algorithm for numerical integration over an N-dimensional cube*, J. Comput. Appl. Math. **2** (1976), 207–217.

National Institute of Standards and Technology (formerly National Bureau of Standards), Gaithersburg, Maryland 20899

Contemporary Mathematics
Volume **115**, 1991

Subregion Adaptive Algorithms for Multiple Integrals

ALAN GENZ

ABSTRACT. After a brief survey of adaptive and nonadaptive methods for numerical multiple integration, a certain class of adaptive algorithms is described. In this class the adaptivity is accomplished by using a sequence of subdivisions of the integration region. A globally adaptive algorithm is described in detail. Extensions, refinements, and tests of this algorithm are reviewed.

Introduction

The problem. The purpose of this paper is to give an overview of certain classes of algorithms for the approximate computation of integrals in the form

$$\mathbf{I}(f) = \int_{a_1}^{b_1}\int_{a_2}^{b_2}\cdots\int_{a_n}^{b_n} f(\mathbf{x})d\mathbf{x}.$$

Here $\mathbf{x} = (x_1, x_2, \dots, x_n)$ and the integrand f($\mathbf{x}$) is assumed to be computable at any point $\mathbf{x}$ within the n-dimensional integration region $R = [a_1, b_1] \times [a_2, b_2] \times \cdots \times [a_n, b_n]$. If a problem is given with any of the endpoints a_i or b_i infinite, then there are a variety of standard transformations that allow the problem to be put into finite limit form. Problems also occur where the limits are functions of lower indexed variables. A standard transformation also reduces this type of problem to a constant limit form. Good general references for this problem are the books by Stroud (1971), and Davis and Rabinowitz (1984).

Practical multiple integration problems in statistics and other areas require approximate values for integrals with n that can range from two or three up into the hundreds, and with integrands that are often complicated and therefore expensive to compute. Because the methods that are available for numerical computation often require the computation of the integrand

1980 *Mathematics Subject Classification* (1985 *Revision*). Primary 65D30, 65D32.
Key words and phrases. Numerical integration, multiple integrals.
This paper is in final form and no version of it will be submitted for publication elsewhere.

$f(\mathbf{x})$ at a large number of points $\mathbf{x}$, the computational cost is often very high. It is often hard to tell at the beginning of a computation whether a particular problem will be easy or difficult. Research in this area has focussed on finding methods that can provide an acceptably accurate result using a limited amount of computation, or if this is not possible, can provide a result that includes an error estimate and some information about how much work is required to obtain a more accurate result.

Early methods. The first methods to be considered for multiple integration were product rules. With these methods, an appropriate one-dimensional formula (or *rule*)

$$\int_a^b f(x)dx \simeq \sum_{i=1}^{N} w_i f(x_i)$$

is used for each dimension. These one-dimensional rules are combined in product form

(1)

$$I(f) \simeq B_N(f) = \sum_{i_1=1}^{N} w_{1,i_1} \sum_{i_2=1}^{N} w_{2,i_2} \cdots \sum_{i_n=1}^{N} w_{n,i_n} f(x_{1,i_1}, x_{2,i_2}, \ldots, x_{n,i_n})$$

to give an "integration rule" for a multiple integral. The one-dimensional rules are usually polynomial integrating rules, in the sense that the points x_i and weights w_i are chosen so that for a particular N, the rule will integrate exactly all polynomials with degree $\leq d$, for some d which increases with N (the "Gauss" rules have the optimal $d = 2N-1$). This polynomial integrating property is inherited by the product rules, which therefore converge very rapidly to $I(f)$ when $f(\mathbf{x})$ is sufficiently smooth.

The problem with the simple use of product rules is often referred to as the "curse of dimensionality"; the product rule (1) requires N^n values of the integrand. An associated problem with these product rules is error estimation. A common strategy is to use $B_N(f)$ with an increasing sequence of integers $N_1, N_2, \ldots$ until the difference between the results $D_k(f) = |B_{N_k}(f) - B_{N_{k-1}}(f)|$ is sufficiently small. In order to be robust this strategy must also be combined with checks on the rate of convergence, and so at least three successive estimates of $I(f)$ are needed. $D_k(f)$ is really an error estimate for $B_{N_{k-1}}(f)$, so when n is larger than two or three a significant amount of extra computation is needed for $B_{N_k}(f)$ (and sometimes $B_{N_{k+1}}(f)$ in order to decide that $B_{N_{k-1}}(f)$ is good enough.

Monte Carlo methods (see Stroud (1971)) were developed as a partial answer to "the curse". These methods use a rule of the form

$$B_N(f) = \frac{V}{N} \sum_{i=1}^{N} f(\mathbf{x_i}),$$

where V is the volume of the integration region and the integrand evaluation points $\mathbf{x_i}$ are chosen from R with random uniform distribution. For large classes of integrands these rules converge to $I(f)$ with an error that is $O(\frac{1}{\sqrt{N}})$. The simple Monte Carlo rules allow an easily controlled increase in the computational effort, and the sample variance provides a robust error estimate, but their slow convergence severely limits their usefulness.

Lattice rules (originally called "number theoretic" or "quasi-random" methods; see Stroud (1971) or, a more recent Lyness (1989) have the same form as the Monte Carlo rules, but the points $\mathbf{x_i}$ are chosen from R to be as equidistributed as possible for a particular N. The determination of the parameters that are used to generate the optimal point sets involves a computational intensive search procedure, but this search needs only be done once for each N for a standard region R (e.g. the n-dimensional unit hyper-cube). The optimal rule for a particular N can then be linearly transformed to any other finite hyper-rectangular region. For significantly large and useful classes of integrands these rules converge with an error that decreases almost as rapidly as $O(\frac{1}{N})$. However, the statistical error estimates used for Monte Carlo rules cannot be used for the lattice rules, so even though the lattice rules often produce better results for a particular N, it is not easy to determine how much better these results are. A randomization technique developed by Patterson and Cranley (1976) uses random samples of these rules to provide an error estimate based on the sample variance, but this increases the total number of points. Another problem with these rules is that the optimal points set generators are available only for a limited range of N and n values.

Better methods. Progress with developing better methods for numerical multiple integration has occurred in two main directions: better integration rules and adaptive methods. The subregion adaptive methods that are discussed in later sections are based on the use of good polynomial integrating "basic" rules that take the form

$$B(f) = \sum_{i=1}^{N} w_i f(\mathbf{x_i}) \simeq I(f).$$

Research work (see Berntsen and Espelid (1982), Genz and Malik (1983), Genz (1986), Keast (1979), and Keast and Lyness (1979) for a selection of references) in this area has focused on finding, for each polynomial degree d and dimension n, the points and weights for rules with small N for a fixed standard region like the unit hyper-cube. Many of the rules that have been found and are now in use have N values for a particular d that are much smaller than the N values for product rules. A minor difficulty with the low N rules is that the weights w_i are not always positive. Continuing research work also has produced better lattice rules (see Lyness (1989) and Sloan and Lyness (1989)), but these rules are still not widely used in practical calculations.

Adaptive methods do not use integrand evaluation points that are fixed at the beginning of a calculation. These methods are designed to dynamically concentrate the evaluation points in areas of the integration region where the integrand is most irregular. Most practical problems involve integrands that have one or more somewhat isolated irregular features like peaks or oscillations, so methods that can adjust or adapt the computation to focus on these features can be much more efficient than methods that do not. A number of different adaptive Monte Carlo methods have been developed (see Stroud (1971) and Davis and Rabinowitz (1984) for some examples); with these methods the distribution of the random points is dynamically constructed to model the integrand behavior. This can significantly reduce the sample variance and therefore the error. With *subregion* adaptive methods, a finer and finer subdivision of the original integration region is dynamically constructed, with smaller subregions concentrated where the integrand is most irregular. Within each subregion a basic rule is used to provide a local estimate for the integral and these results are combined to produce the global estimate. Although various basic rule types ccould be used, polynomial integrating basic rules are usually used with the subregion adaptive methods because they can provide rapid convergence to an accurate result once the subdivision has been refined enough to make the integrand locally smooth.

Subregion adaptive algorithms

Overview. All of the subregion adaptive algorithms for numerical integration have a general structure that takes following form:

> **Repeat**
> choose some subregion(s) from a set of subregions;
> subdivide the chosen subregion(s);
> apply a basic integration rule to new subregions;
> update subregion set, and global integral and error estimates;
> **Until** Convergence.

The components that distinguish these adaptive algorithms are a data structure for the subregion set, a subdivision strategy and a basic integration rule $B(f)$ that includes an error estimator.

There are many different algorithms that have been proposed for subdivision of R in adaptive integration algorithms. We will restrict ourselves to subdivision algorithms that are defined in the following way. A subdivision of R is defined as a set of disjoint subregions $S_i = \{R_1, R_2, \dots, R_{M_i}\}$ with $\cup R_j = R$ and $S_0 = \{R\}$. We assume S_{i+1} is obtained from S_i by taking one or more subregions R_j from S_i and dividing each of these subregions into two or more pieces. S_{i+1} consists of the new pieces, together with the original undivided subregions from S_i. Given a subdivision S_i, an integral estimate $I_{S_i}(f)$, for $I(f)$, is obtained by applying an appropriately transformed basic rule $B(f)$ to each of the subregions in S_i and summing the results. An estimate $E_{S_i}(f)$, for the absolute error $E(f) = |I_{S_i}(f) - I(f)|$, is obtained in a similar way by summing the error estimates for each of the subregions.

One method for modeling adaptive algorithms using subdivisions of this type associates the subdivisions with a weighted digraph. The nodes of the digraph are the allowed subdivisions of R. The directed edges indicate a step in the adaptive algorithm from one subdivision to a finer subdivision. The weight for such an edge is the sum of the costs for determining the new subdivision, computing basic rule and error estimates for the new subregions in the finer subdivision and updating the global estimates for the integral and error. The goal of a subregion adaptive integration algorithm is to dynamically follow a low cost path through this graph to a node with an global error estimate that is less than some user specified tolerance ε.

A second model for adaptive algorithms of this type uses a rooted tree of subregions. The root node of the tree is the initial integration region R. At the beginning of each stage in the adaptive algorithm, there is a leaf node in the tree for each subregion in the current subdivision of R. New leaves in the tree are produced when the adaptive strategy causes a subdivision of one or more of the current subregions. Global estimates for $I(f)$ and $E(f)$ are obtained by summing over the respective local estimates for the integral and error for each of the leaves in the tree.

Two main types of subregion adaptive strategies have been used algorithms development: locally adaptive and globally adaptive. Locally adaptive algorithms are characterized by decisions to subdivide particular subregions which do not use results from other subregions. A common locally adaptive strategy, described in terms of the subregion tree, divides the leftmost leaf subregion R_j that has an error greater than $\varepsilon V_j/V$, where V_j is the volume of R_j and V is the volume of R. Termination occurs when no leaf in the tree satisfies this condition. This strategy corresponds to a type of depth first search in the subregion tree, where the actual depth of the search along each branch is determined by the error test above. Locally adaptive algorithms have been very popular for one dimensional calculations, where good results are often obtained in a short time. However, for multidimensional integrals, it is often unknown at the beginning of a calculation whether a result accurate to within a given ε can be found in a reasonable amount of time. In this case locally adaptive algorithms can often spend most of some allotted amount of time working in a small part of the integration region R, trying to achieve an unreasonably small error. When the time runs out, all that is available globally is an inaccurate result. Although this can also be a problem with one-dimensional integrals, much smaller times are usually involved.

Globally adaptive algorithms subdivide using information about all of the current subregions. A popular globally adaptive strategy (originally proposed by van Dooren and de Ridder (1976)) always subdivides the subregion with the largest error, until the global error sum is less than ε, or a time limit has been reached. This strategy corresponds to a priority first search in the subregion tree. Although globally adaptive algorithms usually require more memory space to maintain the current subregion list than do locally adaptive algorithms, and globally adaptive algorithms can take more time to select subregions for subdivisions, at each stage in the calculation the global estimate for $I(f)$ is in some sense the best one available using the computation that has been done so far. For difficult problems, early termination of this type of algorithm usually gives a much more accurate result than that obtained with a locally adaptive algorithm. For certain types of problems globally adaptive algorithms have optimal convergence rates (see Rice and deBoor (1982)). For these reasons, globally adaptive algorithms have been gaining popularity for the approximate computation of multiple integrals. The rest of this paper will describe the details and features of the van Dooren and de Ridder algorithm and its successors.

A globally adaptive algorithm. The original van Dooren and de Ridder algorithm (Fortran implementation HALF) used a degree seven polynomial integrating basic rule $H_7(f)$. The difference between the result for this rule in a particular subregion, and the result from a degree five rule $H_5(f)$ (that used a subset of the integrand values from the degree seven rule) provided an error estimate for that subregion. Some of the $H_7(f)$ integrand values were also used to determine fourth difference approximations to the fourth derivatives for the integrand along directions parallel to each of the coordinate axes. If the subregion is selected at a later stage for further subdivision that subregion is divided in half in the direction of largest absolute fourth difference. This clever strategy for halving in only one direction, using fourth difference information to measure integrand irregularity is probably one of the main reasons for the practical effectiveness of the algorithm. Some earlier algorithms subdivided a particular subregion into 2^n congruent pieces, by halving along each axis, but this strategy suffered from the "curse". Other measures instead of the fourth difference could be used to measure irregularity, but this measure allows the efficient re-use of already computed integrand values, and is of high enough order to be sensitive to the magnitude of the derivative terms in the actual error. The algorithm usually proceeds to produce more and more accurate results in a controlled manner that adapts to the difficulties of the integrand.

HALF was initialized by applying the integration rule to the integrand on the whole integration region. In the original algorithm, the subregion set was an ordered list with the subregions ordered according to the corresponding error estimates, and the algorithm proceeded from one stage to the next by always choosing one subregion with largest estimated error for subdivision. This subregion was divided in half along one coordinate axis to produce two new subregions, and then the algorithm continued.

An improved version of HALF was developed by Genz and Malik (1983), and called ADAPT. ADAPT had a better degree 7/5 basic rule pair that improved the performance of the algorithm on an extensive series of tests.

Following the recommendations made by Malcolm and Simpson (1975) for one dimensional globally adaptive algorithms, Genz and Malik also changed the subregion set data structure to an error keyed heap, in order to reduce the overheads associated with maintenance of the set. If the current subregion set has M subregions the heap always has the largest error subregion as its first element, and new subregions can be inserted into the heap in $O(\log(M))$ time. A slightly modified version of ADAPT was placed in the NAG (1981) library (D01FCF).

In an even more extensive performance study (Genz, (1984)), ADAPT was compared with a simple Monte Carlo method, a lattice rule method based on randomized Korobov rules (see Korobov (1957) and Patterson and Cranley (1976)) and two adaptive Monte Carlo methods (Freedman and Wright (1981) and Lautrup (1971)). In these tests ADAPT had generally superior performance for several families of test integrands for n in the range 2–6. The test integrand families had a variety of features like peaks, oscillations and discontinuities that were randomly placed in integration region to generate particular integrands for the individual tests. Some tests were carried out for n as large as 15, but these were somewhat inconclusive, although the lattice rule method had apparent better performance on many of the higher dimension tests, with ADAPT in second place. An obvious problem with ADAPT that emerged from the tests was that the simple error estimation procedure was very often too conservative,and also was sometimes unreliable.

The vectorization and parallelization of ADAPT (see Genz (1989) for a review) has also been studied. Many practical problems involve a set of similar integrands that differ only through a few parameters, but have a common integration region. With these problems, it often happens that a significant part of the computation required for each integrand is the same for all of the integrands. These common calculations need be done only once for each integrand evaluation point. An additional saving with this type of computation may occur because all of the integral calculations use the same subdivision of the integration region, and so the work for finding a good subdivision can be shared among all of the integral calculations. The use of this common subdivision may also have the effect of smoothing the error vector calculations. An early vector version of ADAPT was placed in the NAG library. More recently, a considerably improved algorithm has been developed by Berntsen, Espelid, and Genz (1990). Both vector algorithms use the same basic subdivision strategy as ADAPT, except that the errors that key the heap of subregions are determined for each subregion by taking the maximum of the estimated errors for all on the basic rule results for that subregion, and the fourth differences used for subdivision axis choice are computed using the scalar function defined by taking the sum of the absolute values of the integrands. The Berntsen, Espelid, and Genz algorithm also contains a more complicated and more robust error estimator for the basic rule, allows a choice from a selection of different basic rules, and is structured in a way that allows easy modification of the implementation for efficient use on parallel computers.

Concluding remarks

Subregion adaptive algorithms integrals have been described as a partial answer to the "curse of dimensionality" for the numerical computation of multiple integrals. The algorithms are currently most useful for n in the range 2–10, and have been implemented in computer software that efficiently handles groups of integrals and can be used on the most modern computers. Adaptive methods of this type can never be completely reliable and must be used with caution. There continues to be a need for research that will produce more reliable and discriminating error estimation methods. There also continues to be a need for research that will produce better integration rules, particularly for regions other than hyper-rectangles, because the transformations from other regions to hyper-rectangles referred to at the beginning of this paper often increase the difficulty of the integrals involved.

References

J. Berntsen and T. O. Espelid (1982), *On the construction of higher degree three dimensional embedded integration rules*, SIAM J. Numer. Anal. **25**, 222–234.

J. Berntsen, T. O. Espelid, and A. Genz (1990), *An adaptive algorithm for the approximate calculation of multiple integrals*, ACM Trans. Math. Software (to appear).

P. J. Davis and P. Rabinowitz (1984), *Methods of numerical integration*, Academic Press, New York.

P. van Dooren and L. de Ridder (1976), *An adaptive algorithm for numerical integration over an N-dimensional rectangular region*, J. Comp. Appl. Math. **2**, 207–217.

J. H. Freedman and M. H. Wright (1981), *A nested partitioning procedure for numerical multiple integration*, ACM Trans. Math. Software **7**, 76–92.

A. Genz (1984), *Testing multiple integration software*, Tools, Methods and Languages for Scientific and Engineering Computation (B. Ford, J-C. Rault, and F. Thomasset, eds.), North Holland, New York, pp. 208–217.

A. Genz (1986), *Fully symmetric interpolatory rules for multiple integrals*, SIAM J. Numer. Anal. **23**, 1273–1283.

A. Genz (1989), *Parallel adaptive algorithms for multiple integrals*, Mathematics for Large Scale Computing (J. C. Diaz, ed.), Marcel Dekker, Inc., New York. pp. 35–48.

A. Genz and A. A. Malik (1980), *An adaptive algorithm for numerical integration over an N-dimensional rectangular region*, J. Comp. Appl. Math. **6**, 295–302.

A. Genz and A. A. Malik (1983), *An imbedded family of fully symmetric numerical integration rules* SIAM J. Numer. Anal. **20**, 580–587.

P. Keast (1979), *Some fully symmetric rules for product spaces*, J. Institute of Math. Appl. **23**, 251–264.

P. Keast and J. N. Lyness (1979), *On the structures of fully symmetric multidimensional quadrature rules*, SIAM J. Numer. Anal. **16**, 11–29.

N. M. Korobov (1957), *Approximate calculation of multiple integrals using number theoretic methods*, Dokl. Acad. Nauk. SSSR. **115**, 1062–1065.

B. Lautrup (1971), *An adaptive multidimensional integration procedure*, in *Proceedings of the 2nd colloquium on advanced computing methods in theoretical physics*, C.N.R.S., Marseille.

J. N. Lyness (1989), *An introduction to lattice rules and their generator matrices*, IMA J. Numer. Anal. **9**, 405–419.

M. A. Malcolm and R. B. Simpson (1975), *Local versus global strategies for adaptive quadrature*, ACM Trans. Math. Software **1**, 129–146.

NAG Library Mark 11 (1981), Numerical Algorithms Group Limited, Mayfield House, 256 Banbury Road, Oxford OX2 7DE, United Kingdom.

T. N. L. Patterson and R. Cranley (1976), *Randomization of number theoretic methods for multiple integration*, SIAM J. Numer. Anal. **13**, 904–914.

J. R. Rice and C. deBoor (1982), *An adaptive algorithm for multivariate approximation giving optimal convergence rates*, MRC Technical Report no. 1773, Univ. of Wisconsin, Madison, Wisconsin.

I. N. Sloan and J. N. Lyness (1989), *The representation of lattice quadrature rules as multiple sums*, Math. Comp. **52**, 81–94.

A. H. Stroud (1971), *The approximate calculation of multiple integrals*, Prentice Hall, Englewood Cliffs, New Jersey.

DEPARTMENT OF ELECTRICAL ENGINEERING AND COMPUTER SCIENCE, WASHINGTON STATE UNIVERSITY, PULLMAN, WASHINGTON 99164-2752

Contemporary Mathematics
Volume **115**, 1991

Parallel Systems and Adaptive Integration

E. DE DONCKER AND J. A. KAPENGA

ABSTRACT. Many difficult integration problems can be solved reasonably by resorting to adaptive methods, if the integration region is standard or can be reduced to one or more standard regions and the dimensionality is low to average. These methods are appropriate for parallel implementation, despite the irregular, problem dependent subdivision patterns and the need to update and check global quantities. First we will investigate what parallel architectures can be put to use and give a brief survey of current machines.

The adaptive strategies fit the model of task partitioning algorithms, which were implemented successfully on shared memory multiprocessors. This paper describes the design and an implementation of global adaptive integration strategies on loosely coupled systems, based on the distributed task pool concept and asynchronous processing of the quadrature domain subregions.

Introduction

The capabilities of parallel architectures have lead to an increasing effort in the design and implementation of parallel algorithms. For procedures of high computational expense, it is natural to aim at the design of algorithms that can be implemented on parallel computers. Numerical integration, at the basis of many large scale computations, is often computationally demanding, especially in cases of high dimensionality, unspecified irregular integrand behavior, or irregularity of the integration region (Davis and Rabinowitz (1984), Piessens et al.(1983), Kapenga et al. (1987), Mullen and Ennis (1987), and Mullen et al. (1988)). This paper deals with the design of accurate and efficient parallel algorithms for adaptive numerical integration on Multiple Instruction Multiple Data (MIMD) architectures, in particular on local memory ("loosely coupled") machines.

1980 *Mathematics Subject Classification* (1985 *Revision*). Primary 65D30, 65V05, 65W05, 68M05.

Key words and phrases. Adaptive numerical quadrature, parallel architectures, task partitioning algorithms.

This paper is in final form and no version of it will be submitted for publication elsewhere.

With respect to adaptive strategies, multivariate integration shares the fundamental techniques with its univariate special case where to date, the *Quadpack* algorithms (Piessens et al.(1983)), adopted by major numerical libraries (of the Numerical Algorithms Group (NAG), IMSL and of the Sandia, Los Alamos, Air Force Weapons Laboratory Technical Exchange Committee (SLATEC)), have the leading edge. Adaptive integration methods classify as region partitioning and more generally as task partitioning algorithms, with the multivariate problems naturally requiring considerably more tasks and more work within each task. Task partitioning algorithms were studied in a more general context in Kapenga and de Doncker (1988b), with respect to implementations and performance analysis on shared memory multiprocessors. Applications to numerical integration were given in de Doncker and Kapenga (1987) and de Doncker and Kapenga (1988).

The basic problem is the calculation of a numerical approximation Q to an integral

$$I = \int_{\mathcal{D}} f(\vec{x}) d\vec{x} \tag{1}$$

and of an error estimate or upper bound E_a satisfying

$$|I - Q| \leq E_a \leq \varepsilon = \text{Max}\{\varepsilon_a, \varepsilon_r |I|\} \tag{2}$$

where ε_a and ε_r are the tolerated absolute and relative error respectively and ε_r should not be less than the relative machine accuracy. It should be flagged by an integration algorithm if (2) cannot be satisfied. $\mathcal{D}$ either is a "standard" region (N-cube or simplex), can be transformed into a standard region, or can be split into regions that can be transformed into a standard region.

In adaptive integration algorithms, a sequence of approximations to (1) is generated, until the estimated error $E_a \leq$ the tolerated error ε. Although complicated nonlinear sequence transformations may be formed, the basic approximations are linear combinations of integrand values. The evaluation points depend on the integrand function as well as the region and on the requested accuracy; they tend to be clustered in the neighborhood of irregular integrand behavior. This is obtained by a progressive partitioning of the integration region.

An adaptive integration method is characterized by its selection, at each step of the algorithm, of one or more regions from the current partition, subdivision of which results in their replacement with the subregions and possible insertion of the latter back into the pool of regions, until a termination criterion is met. The adaptivity component results from the fact that regions are selected according to their assumed priority. As a consequence, the course of the computations differs for individual problems, since the priorities issued during the integration process are problem dependent. The definition and assignment of priorities, the method of selecting regions and the termination criterion used determine the adaptive strategy at hand. The local and global adaptive strategies are well known (Malcolm and Simpson (1975)), but modifications and hybrids are possible (Shapiro (1984)). In Piessens et al.(1983) and in de Doncker (1978), we considered a staged global adaptive strategy, where each new stage delivered the next item of a sequence used for extrapolation.

For an efficient parallelization of these algorithms, it is essential that the subdivision steps are performed asynchronously by the processors. This is further suggested and made possible by the fact that the task performed on each individual region is self contained, i.e. does not depend on the tasks related to the other regions in the pool. Note that each region together with the information pertaining to it can be identified with the corresponding task.

On Multiple Instruction Multiple Data machines with shared memory, the task pool can be kept and updated in the memory shared by the processors (de Doncker and Kapenga (1987), Kapenga and de Doncker (1988b), de Doncker and Kapenga (1988), and Kapenga and de Doncker (1988a)). The use of monitor macros as basic tools for handling the synchronization in task pool problems on shared memory multiprocessors was advocated in Lusk and Overbeek (1983), Lusk and Overbeek (1984), Boyle et al. (1987), and Kapenga and de Doncker (1988b). In de Doncker and Kapenga (1987), de Doncker and Kapenga (1988), and Kapenga and de Doncker (1988b), we dealt with parallel stage adaptive strategies, requiring loose synchronization at the end of each stage. On local memory MIMD machines, the pool is distributed over the nodes and information is communicated between the processors via message passing. On these systems, rough parity of the local pools has to be retained through load balancing of the processors. We believe that a similar scheme can be derived for use on a distributed system such as a network of workstations, for problems where the computational efforts dominate data communications significantly.

In the next section we shall define some terms in parallel computing and give a survey of current machines. Section 3 outlines a methodology for adaptive integration through task partitioning on loosely coupled systems. The model for load balancing used is covered in §4. The adaptive strategy discussed in §3 and §4 was implemented on an Intel hypercube machine (iPSC/1). A modification of the methodology to support global staged adaptive strategies is proposed in §5.

Current parallel machines and taxonomy

There has been a great increase in the variety of high performance commercial computer architectures available in the last six years. Prior to this time the only common successful architectures were the Single Instruction Single Data (SISD) processors and vector processors (VP). Now over a dozen vendors provide programmable machines which can be considered in some sense parallel. Most of these are classified with features from five types of architecture: Multiple Instruction Multiple Data (MIMD), either tightly coupled with shared memory (MIMD-SM) or loosely coupled with distributed memory (MIMD-DM), Single Instruction Multiple Data (SIMD), Very Long Instruction Word (VLIW) and the venerable vector processor (VP). In addition, Distributed Systems, DS, now are being built which can connect heterogeneous machines to cooperate on problem solving.

Actual machines often have features from more than one of six given architectures, which are by no means as distinct as they might at first appear. This presentation contains some gross generalizations; it is intended to provide some context for someone new to architectures. However, this is a starting point and these architectural categories do fairly well classify the common approaches used by programmers in application development seeking high performance. For a more precise presentation see Almasi and Gottlieb (1989) and Quinn (1987).

On many of these machines UNIX is the main or only operating system available. C is often the most robust and supported language, except on the vector processors, where FORTRAN still dominates. Special parallel languages are only beginning to become widespread, mainly for logic programming (Foster and Taylor (1990), Gregory (1987), and Bal et al. (1989)). As parallel machines hopefully mature, hardware solidifies first, followed by the operating system and finally languages and support environments.

First we mention some of the problems with comparing performance between machines. Then a very light overview of each architecture's characteristics will be given, as well as reference to some current commercial examples and some commercial speculations. The book (Babb (1988)) contains brief introductions to eight commercial parallel computers. Examples of several important noncommercial high performance research machines may be found in Almasi and Gottlieb (1989).

Measuring computer performance. It is now within the lore of parallel processing that a manufacturer's rating of his machine is "the performance the manufacturer absolutely guarantees you will not achieve." One attempt to rate machine performance is the MIPS, Millions of Instructions Per Second, the machine can execute. For many reasons this can be a very poor number with which to compare machines. The Millions of Floating point Operations Instructions Per Second (MFLOPS) rating, is only a little more useful then the MIPS rating. It is often obtained by adding the performance of all the floating point hardware units in the machine together, even if they cannot be fed simultaneously on the machine. We call these mystical ratings the machine's raw or peak performance.

The simplest reason the MIPS can be misleading is that different machines may require far different numbers of instructions to do the same thing. In an attempt to compensate for these problems with MIPS many people quote VAX-MIPS, which assumes a VAX-780 executes 1 MIPS and rates a machine

X by

$$\text{VAX-MIPS} = \frac{\text{Machine X time}}{\text{VAX time}}, \tag{3}$$

for some program run on both systems. The selection of an appropriate benchmark program is critical to the applicability of VAX-MIPS and any other comparison. Different benchmarks can easily provide orders of magnitude differences in a machine's rating, especially when architectures vary.

Speedup is another dangerous term used with abandon in parallel processing (Flatt and Kennedy (1989) and Quinn (1987)).

Single Instruction Single Data (SISD) architectures. The SISD architecture uses a single stream of instructions. Each instruction operates on a "small" single data item, usually a byte or word in memory. Examples are most minicomputers, such as the the single processor VAXs from Digital Equipment Corporation, and Workstations, such as the SUN from SUN Microsystems. This class also includes most microcomputers, such as the families of IBM/PCs and Macintoshs.

The very rapid advance of microprocessors has increased performance of the low end workstations to that provided by mainframes only a few years ago. Current microprocessors such as the Motorola 68040, Intel 80886, Sun SPARC and Mips MIPS now provide from 20 to 60 MIPS and 3 to 20 MFLOPS (Various Authors (1989)). In the next seven years it is a fairly safe to predict that the performance of microprocessors can be increased by a factor of three: some predictions are much higher. This will be near the power provided by the first generation supercomputers (Wilson (1989)). Utilizing this increase in raw performance, that is building a moderate cost complete system, is likely to be a bigger engineering challenge than producing the chips themselves (Andrews (1990)).

Vector Processors (VP). Classic VP machines use a single instruction stream as well. They have special vector hardware which allows a single instruction to operate on vector objects, at far greater speed than can a sequence of single instructions to accomplish the same task. Typical operations include: multiplying a vector by a constant, adding two vectors and adding a scalar times a vector to a vector. These types of operations were done on the earliest supercomputers, such as the Cray 1, and provided orders of magnitude speedups over conventional SIMD machines on scientific problems whose main loops require heavy numerical linear algebra on large matrices.

The hardware for the top VP machines has always pushed the state of the art in technology and been very expensive. These machines tend to be faster than conventional SIMD machines on scalar operations, however the speedups they provide is often not as dramatic as it is when strictly vector operations are involved.

The compiler technology for automatic vectorization is very mature, stable, and successful, compared to automatic parallelization on MIMD machines.

Although some artistic talent is required to get the last cycles form a vectorizing compiler, FORTRAN is usually the best supported language here; most serial code programming, using vectorizable operations, is treated fairly well by compilers. These machines also tend to have large libraries of highly optimized subroutines available.

One trend in the supercomputer market has been to connect a few, typically 2 to 16, fast vector processors together, often with shared memory. This provides both the fast vector performance and speedup on some problems suitable for MIMD parallelization.

Taking full advantage of the vector facilities on these machines is hard in adaptive integration. They are well suited to application of rules over regular meshes, using high numbers of points, if the processor supports efficient evaluation of vectors of any special functions needed in integrand evaluation. Experience with vectorization of one-dimensional quadrature codes is reported on in Gladwell (1987).

Multiple Instruction Multiple Data, Shared Memory (MIMD-SM) architectures. The MIMD-SM architecture allows a moderate number of processors, typically 2 to 32, to quickly access the same memory. There are several ways this feature can be used. On machines like the Sequent Symmetry and Encore Multimax, the machine looks much like a SISD Unix machine, with processes being run independently by multiple users. The difference is that, unlike the SISD machine which provided multiprocessing by context switching between processes to give the illusion that the processes are running at the same time, on these machines there actually can be a process running on each processor at the same time. If you have more processors than processes this can result in no process switching at all! In any case the total process switching is usually greatly reduced and the speedup in the amount of useful work the machine will potentially do is more than the number of processors.

The raw performance of these mini-supercomputers about equals that of a collection of a like number of workstations combined. In most cases the processors are the same microprocessor chips used in workstations. Of course it is far more likely to approach the raw performance rating on one of these machines than it is on a distributed system of workstations. The software environment is usually very stable and reliable. The cost exceeds that of a like number of workstations, caused by R and D together with the special hardware needed to interconnect the processors to the shared memory.

The Alliant FX series is an interesting combination of up to eight fairly fast workstation class processors, each with modest vector processing capabilities, in a tight shared memory system.

These MIMD-SM machines can be expected to keep pace with the new microprocessor improvements, with a little longer time lag than the workstations, as they are incorporated into products.

To use more than about 32 processors, it appears they must be connected in clusters, with intercluster memory access time slower than intra cluster access time. Although some such systems exist, it is too early to tell if they will become widespread. The problems with parallel software development on multiple clusters of processors is much harder than for a single cluster.

To use an MIMD-SM machine as a parallel machine in a single problem, the programmer must usually make his parallelizations explicit in the program. This is unlike the vector processing case. It is typically harder to parallelize for an MIMD-SM machine then for a VP machine, however many problems, such as branch and bound searching, which cannot be vectorized, can get good speedups here.

Ten years ago, many predicted that compiler technology would advance to the point where automatic parallelization for MIMD-SM machines would be as easy as for the vector processors. This has not proved to be the case. There are several special environments and programming paradigms emerging that will improve the process of parallel application development on these machines (Wouk (1989) and Boyle et al. (1987)).

These are by far the easiest machines to parallelize adaptive integration on, excluding the VLIW machines. The processors can all access the next subdomain to process from a common task pool and access or update global information, using synchronization primitives provided (de Doncker and Kapenga (1987), Kapenga and de Doncker (1988b), and de Doncker and Kapenga (1988)). No special load balancing or data movement is needed. The only limiting factor in performance is the moderate number of processors on such a machine.

A version of adaptive multivariate integration, using compiler directives, was described in Genz (1987).

Multiple Instruction Multiple Data, Distributed Memory (MIMD-DM) architectures. The MIMD-DM machines are also usually based on standard microprocessor technology. There are more custom microprocessor chips used here than in the MIMD-SM case. These machines remove the limit on the number of processors, which constrains MIMD-SM machines. There are often between 16 and 1042 processors in such a machine. Thus the largest machines have more raw power than the largest VP supercomputers (Fox (1989)).

Each processor has its own memory, making the memory interconnect much cheaper and simpler than in the MIMD-SM case. The drawback is that to get processes to cooperate, the programmer must usually explicitly pack and pass both information and data in a message from one processor and explicitly receive it and unpack it in another. This is a much heavier programming burden than faced on most MIMD-SM or VP machines.

In general the development environment on MIMD-DM machines is much weaker than on MIMD-SM machines, debugging being a case in point. There are several efforts in improving MIMD-DM environments underway; however, one should expect the relationship between MIMD-DM and MIMD-SM to remain: more expandable, lower cost per processor, more total memory, and harder development.

It can be noted that if the interprocessor communication time was small enough, then a compiler could make a MIMD-DM machine look like a shared memory machine. This can be done by sending a message to get or store the contents of any memory located on another processor any time that memory is to be accessed. Such environments exist, one for the BBN Butterfly, though communication can sometimes cripple attempts to use them on some applications.

These machines offer the potential for outstanding performance in parallelizing adaptive integration. The problem is that load balancing and data transfer between processors, as well as distribution and collection of global information updates must be explicitly managed, making program development much harder than on MIMD-SM machines. This is a quite general observation. There are many attempts underway to develop machine independent environments to ease this burden (EXPRESS (1987) and Seitz et al. (1989)).

For an excellent introduction to using MIMD-DM machines see Fox et al. (1988), and for a recent review of some large problems solved on MIMD-DM machines see Fox (1989). For a description of simple distributed integration without load balancing see Genz (1987).

Single Instruction Multiple Data (SIMD) architectures. A SIMD machine has a single instruction stream which operates on multiple copies of the data at the same time, unlike the previous machines where each processor could have a separate program. Each processor concurrently receives identical instructions but has its own local memory to which to apply them. Processors can use their own local data in conditional tests, hence the results of the tests may vary from processor to processor. One outcome of a conditional test being false can be to make the processor inactive, ignoring all instructions it receives until a reset is received. This allows executions of conditionals such as the if-then, if-then-else and while loops. In the case of the while all instructions in the body of the loop must be broadcast and received by all processors, until all the processors become inactive. Then a reset can be issued and the program can proceed.

SIMD computers usually have instructions for moving data between neighboring processors. Processors usually have limited memory, compared to the MIMD machines, but the potential for more processors at a lower per processor cost exists.

SIMD machines like the DAP are best suited to applications like linear algebra and image processing, where the majority of the computation can be done in a regular manner on a static data structure, using the operations provided. The Connection Machine, from Thinking Machines Inc., is another

SIMD machine; it is more flexible than the DAP and can often be used almost like an MIMD-DM machine. It can have up to 64K processors. For an introduction to these SIMD machine we suggest Almasi and Gottlieb (1989).

The parallelization of integration on these machines carries almost the same feeling that is found on VPs, with the exception that any required special function evaluation is more certain to be supported. However, some potential for adaptive integration does exist. It requires the processors to select regions, subdivide regions, process subregions, return processed subregions to the local pools and load balance local pools in lockstep. Here global updates would be done synchronously, after all processors have processed regions for some time. These concepts will be discussed for MIMD-DM machines in the following sections.

Very Long Instruction Word (VLIW) architectures. The idea behind a VLIW machine is to put multiple functional units in and allow them to all process and exchange information at the same time. Thus there may be several adders, multipliers, dividers, and boolean operators in the machine, with a programmable interconnection network between them. Each machine instruction may contain separate subinstructions for each functional unit and the interconnect. So when an instruction is fetched it is split and subinstructions are concurrently passed to each functional unit.

VLIW functional units are typically faster than workstations functional units, though much slower and cheaper than supercomputer functional units. There may be between four and about thirty functional units in a machine. Machines of this type are made by Multiflow, Floating Point Systems, and Prevec.

Compiler technology for VLIW machines is very good. Only local information is needed to effectively use multiple function units, as long as only a moderate number exist. These machines can speed up most programs and almost no special action on the part of an application programmer is needed, unlike the VP machines where getting good performance is a little more of an art and harder on nonvector oriented code. To the user these almost look exactly like SISD machines, which is their strong point. They may provide performance in the 50 MFLOP range.

The question of whether the rapid improvements in microprocessor performance will still leave a market niche for these machines is an open question. They are much more expensive than workstations, cheaper than supercomputers and have performance in between as well.

Adaptive integration on these machines consists of compiling the serial code for the machine, and occasionally making a modification indicated by a profiler or compiler to enhance performance. Next to the serial machine, this is as simple as possible. The drawback of these, as in all applications, is the modest maximum speed expected. However, for some applications this will be adequate.

Distributed Systems (DS). At the present time there are several interesting

(1) languages such as SR, "Synchronous Resources" (Andrews and Olsson (1987) and Andrews et al. (1988));
(2) complete systems such as ISIS (Birman et al. (1988));
(3) environments such as CE, the "Cosmic Environment" (Seitz et al. (1989)); and
(4) direct process communication facilities such as RPC, "Remote Procedure Call" (Coulouris and Dollimore (1989) and Bal et al. (1989)).

These are all rather portable and designed to allow a computation to be distributed on a collection of processors. These processors may include workstations, mainframes, and parallel machines. In some sense the DS looks like an MIMD-DM machine.

The current limitation of this type of system is that it usually uses a standard ethernet to connect the processors. Network bandwidth has not kept pace with the increased performance of workstations. The limited bandwidth of the ethernet and high delay in the system communications software presents a bottleneck for parallelizing many applications. As networks switch to higher bandwidth carriers, these system will become more popular. They can provide near supercomputer performance on some problems, using existing workstations.

Distributed adaptive integration

In this section we shall outline an adaptive task pool partitioning scheme for numerical integration, using asynchronous processes on MIMD-DM machines. On these systems, an application program consists typically of a part that is run on the host processor, responsible for communicating with the outside world, and a part for the nodes which do the actual computations.

Figure 1 represents a "generic" adaptive integration algorithm. In our application, it is run by all the nodes involved in the adaptive partitioning of regions, in the sequel referred to as "task processors". One of the nodes, designated as "control", is responsible for global initializations, monitoring updates of global quantities from the task processors, testing termination criteria and flagging termination to the task processors. Since these operations can be performed asynchronously with region partitioning, the designation of a single control presents no serial bottle neck.

Our application relies on a "global adaptive strategy" (Malcolm and Simpson (1975)), where the acceptance criterion of Figure 1 is based on the global estimated absolute error, E_a, calculated as the sum of the local estimated absolute errors over all subregions in the distributed pools. All regions remain active or pending until the end of the computations, but modifications are possible where regions with very small error estimates can be discarded.

As implemented at present, partition() performs a set number of subdivision steps on the local pool of regions followed by a load balancing step, as given in Figures 2 and 3.

```
adaptive_integration_algorithm() {
        initialize;
        while(acceptance criterion is not satisfied) {
          partition();              /* perform subdivisions */
          update_global_results();
        }
}
```

FIGURE 1. Adaptive intergration algorithm.

```
partition() {              /* subdivisions and load balancing */
        for(i=0;i < n;i++)
          subdivision_step();
        load_balance();
}
```

FIGURE 2. Partitioning loop.

```
subdivision_step() {
        get_region(region_list,k ) ;  /* select next region */
        work_region();       /* subdivide and integrate over subregions */
        put_region(region_list ) ;    /* add new regions to the pool */
        update_results();      /* update; flag if termination conditions */
}
```

FIGURE 3. Subdivision step.

The *get_region*() and *put_region*() primitives represent manipulations on the local pool of regions. A priority queue is used as the underlying data structure for managing the pool. The *region_list* argument of *get_region*() and *put_region*() is a pointer to a linked list of regions to be removed from or added to the pool respectively and k is the number of regions to be removed.

If R_0 and E_0 are the current approximation of the integral and error estimate over the local pool, then the improvements in these quantities incurred through work_region() are applied via update_results(). Assuming that the sequence of subdivision steps in partition() yields the corresponding values R and E at the end of partition(), then the differences $R - R_0$ and $E - E_0$ are updates to the global results and will be either buffered, or sent to the control node (if significant) using update_global_results() in Figure 1. Since local partitioning may cause imbalance of the distributed pools, partition() incorporates a load balancing step. The model for load balancing used will be discussed in the next section together with some details of the implementation.

Dynamic load balancing strategy

A local load balancing operation, termed "local" here in the sense that it is restricted to each processor and its immediate neighbors in the architecture, is

```
load_balance() {
        receive and buffer work and load values from neighbors;
        add work buffer to local pool;
        remove work from local pool to accommodate neighbors with
lowload;
        send work and current load value to neighbors;
}
```

FIGURE 4. Load balancing step.

given in Figure 4. Assume that the "load" of a processor is defined as L. The model used is based on the processor's load L minus the average A of its load and that of its immediate task partitioning neighbors. A neighbor with load smaller than A is determined to have "low" load. If the difference $D = L - A$ is significant, a load of this size is subdivided among and transferred to the neighbors with low load. To avoid that too much work would be sent, D is replaced by $\min\{D, L/2\}$. Let ℓ be the number of low neighbors, L_i, $i = 1, \ldots, \ell$, their loads and

$$S = \sum_{i=1}^{\ell} L_i. \tag{4}$$

Then the neighbor with load L_i is sent an amount $S_i = f_i D$, where the fraction f_i is given by

$$f_i = \frac{(S - L_i)}{(L - 1)S}. \tag{5}$$

Note that $\sum_{i=1}^{\ell} f_i = 1$. Note furthermore that the load values L_i communicated are in general predictions of what the actual loads will be by the time they are processed by the neighbor.

This model gives the initiative for load balancing to the processors with heavy load. Alternatively, it may occur at the request of processors with low load, and other viable schemes are possible. Various models for load balancing are available in the literature (Chou and Abraham (1982), Ni et al. (1985), Lin and Keller (1987), Barhen et al. (1988), Fox and Furmanski (1988), Cybenko (1989), and Bertsekas and Tsitsiklis (1989)).

For the sake of example, the function $1/\sqrt{x}$ is integrated over the unit square, by subdividing the square into the eight triangles pictured in Figure 5. Nodes 0, 1, 2, and 3 of a hypercube are used for the integration. Node 0 is used as control and will not participate in the region partitioning; nodes 1, 2, and 3 are the task processors in this example. During the initialization, each processor is provided with the triangles which are numbered correspondingly in Figure 5, indicating the initial composition of the local pools, and integrates over each of its triangles. This provides an initial quadrature and error approximation over the original pools, which is communicated to the control and contributes to the initial global quadrature

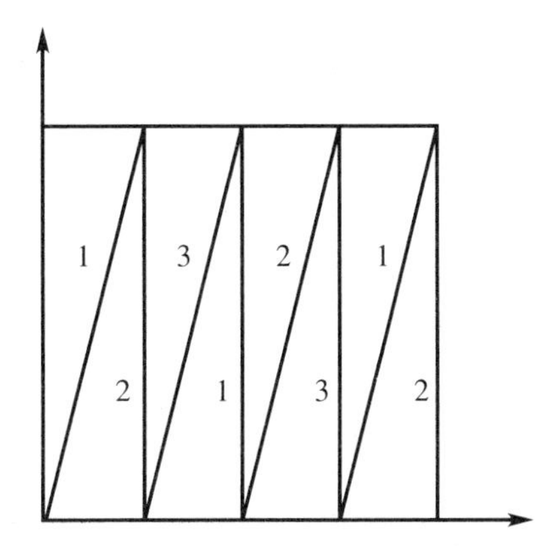

FIGURE 5. Domain of example problem.

result Q and the absolute error estimate. Furthermore it allows initializing the global tolerated error $\varepsilon = \max\{\varepsilon_a, \varepsilon_r|Q|\}$. The user-tolerated error for the example was $\varepsilon = \varepsilon_a = 10^{-6}$ ($\varepsilon_r = 0$).

For the adaptive integration, we used a simplified version of the integrator in de Doncker et al. (1990), without extrapolation. In each subdivision step, as represented in Figure 3 for $k = 1$, one triangle is selected subdivided into 4 similar triangles. The latter are integrated over using a Lyness and Jespersen rule (Lyness and Jespersen (1975)) of degree 11 for the integral approximation and one of degree 9, used together with the degree 11 rule to calculate an error estimate. The maximum number of subdivisions n in partition() was set to 5, so that 15 new triangles were generated in the successive subdivisions, unless the accuracy over the local pool was estimated $\leq \varepsilon_p = \varepsilon/P$ or an abnormal error condition such as roundoff was flagged before n subdivisions were performed; $P =$ the number of participating task processors. In general, the number n should not necessarily be a constant. It is an important parameter since it determines how frequently load balancing is allowed; as such it is a determining factor in the amount of communication between the processors. As outlined in the previous section, the updates $R - R_0$ and $E - E_0$ obtained locally as a result of partition() are either communicated with the control node or buffered.

We let triangles contribute to the load of the processor if their error estimate exceeded ε_p and set the load value L of the processor to its number of triangles with this property in the local pool. In general, however, the load value should be a prediction of the load at the time of its use by a neighbor.

Supported by the simple definition of load given above, the load balancing scheme of Figure 4, applied in partition() after each subdivision loop, is demonstratively effective. Figure 6, on p. 46, indicates the result of the load balancing operations performed in this example. In view of the singularity along $x = 0$, processor 1 generates an overload of subtriangles with large errors in the subdivision process. As a result of the load balancing procedure,

partition	loads			load balancing op.		
	P_1	P_2	P_3	from	to	#
1	11	1	0			
2	21	0	0	P_1	P_3	10
	11		10			
3	19	0	10	P_1	P_3	9
	10		19			
4	20	0	20	P_3	P_2	14
		14	6			

FIGURE 6. Load balancing operations for example problem.

load is transferred from node 1 to its neighbor 3 and from there to 2.

Asynchronous communications are an important aspect in the efficiency of this scheme. This implies that a processor does not wait after issuing a request to receive or send data. To make this work, care has to be taken that data does not get scrambled, as can happen with reuse of a buffer before the previous operation on it is complete.

To supply this facility an extra layer is built, isolating the system communication primitives from the application. This layer provides queuing of both send and receive requests, on systems which require it, until such time as they can be executed without blocking. This provides a fully asynchronous interface appropriate to this application. On the Intel iPSC/1 this is done by careful probing before receives and status checks before transferring received data to the application level. Status checks are also required before issuing send calls on data in the send queue.

Global staged adaptive strategy

In this section we propose a fairly straightforward modification of the global adaptive strategy for loosely coupled systems as presented in the previous sections. Figures 7 and 8 give a "generic" representation of a global adaptive algorithm where the region selection process is organized in stages. In de Doncker and Kapenga (1987), Kapenga and de Doncker (1988b), and de Doncker and Kapenga (1988), we dealt with this type of problem on MIMD-SM machines and described a set of macros supporting its implementation.

For its application to the staged adaptive strategy in numerical integration, the stages correspond to levels of subdivision. At all times, the pool of regions consists of an active and a passive set, which are disjoint and correspond respectively to the "large" regions at or below and the "small" ones above the current level of subdivision. All regions are in the active set at the beginning of a stage. In the course of the stage, regions are selected from the active set only, until an acceptance condition is met for the large regions.

```
staged_adaptive_algorithm() {
        initialize;
        while(acceptance criterion is not satisfied) {
          stage();
          merge_pools();
        }
}
```

FIGURE 7. Stage adaptive task partitioning algorithm.

```
stage() {
        while(active set acceptance criterion is not satisfied)
          partition();      /* subdivisions on the active set */
        stage_end();/* inter stage operations and termination checking */
}
```

FIGURE 8. Task partitioning stage.

Subdivision of a region may deliver new large or small subregions. The integral approximation obtained at the end of a stage is used to calculate an extrapolated result and the remaining active set is merged with the passive set in between stages.

The implementation on MIMD-SM machines described in de Doncker and Kapenga (1987), Kapenga and de Doncker (1988b), and de Doncker and Kapenga (1988) synchronizes the processes at the stage ends for the control process to do the inter stage operations. For a modest number of processors, the bottleneck caused does not affect the performance of the algorithm. However, a modified approach allowing overlapping stages is more attractive with distributed pools and potentially more processors.

In the new strategy, the algorithm of Figures 7 and 8 is executed by the task partitioning processors. Stages are considered mainly global features. One of the task processes or yet a different one may be designated as control. The latter is used to keep track of the stage succession globally and of the global data associated with each stage, in order to check global end of stage criteria and termination. Global end of stage and termination conditions are flagged to the task processors.

An individual region has a subdivision level associated with it. A region at a particular subdivision level may contribute to the global results of the present and future stages. To fix the idea, assume that stage $s-1$ was the last stage finished by the control, i.e. the current global stage is s. Globally, the regions at subdivision levels $\leq s$ are active. Stage s will be finished if a global criterion calculated over this set is satisfied. For this criterion we use that the sum of the absolute error estimates over all large regions does not exceed the tolerated absolute error or a fraction thereof.

However, the set is distributed in the local pools and it should be avoided that its regions are subdivided unnecessarily by processors forced to work at stage s until it is finished globally. For that reason, the participating task processors are assigned a local end of stage criterion, such that achieving the local end of stage s condition by each task processor also implies that the global condition is met. As a trivial example of such a criterion, request the sum of the error estimates over the subregions at subdivision levels $\leq s$ in the the local pool not to exceed $\varepsilon_p = \varepsilon/P$. Consequently, a task processor is allowed to carry out work for later stages after attaining the condition for the local end of stage s. However, it has to keep testing the criterion, since load balancing may bring in new work that changes the condition.

If it is no longer met, the attempt to finish stage s receives priority. One can actually envision the local pool having a priority queue for each stage that is not yet finished globally. These queues interleave, since a region at some subdivision level can be used to improve upon the global results of all later stages as long as they are in effect. The *merge_pools*() primitive in Figure 7 performs a merge when the end of the global stage s is flagged in the processor.

A global result update communicated to the control is a difference with respect to a previous result and will serve as an update to the global results for all the stages that are currently in effect. In view of the fact that the number of stages is small as well as the amount of data associated with each global stage, keeping track of the information pertaining to the global stages currently in effect is well feasible.

Concluding remarks

In this paper we describe a global adaptive strategy for numerical integration on a loosely coupled system, as implemented on an Intel hypercube machine. We also discuss a modification of the strategy which deals with global staged adaptive algorithms, which appear in several of the Quadpack integrators (Piessens et al. (1983)), the two-dimensional integration algorithms over a triangle (de Doncker and Robinson (1984b) and de Doncker and Robinson (1984a)) and over a triangulated region (de Doncker et al. (1990)), and a multivariate routine over an N-dimensional hyperrectangle (de Doncker and Kapenga (1988)). We note that many applications of these strategies exist in areas other than numerical integration.

At present we are attempting a systematic implementation and performance analysis of the algorithms on loosely coupled systems. The intent is to build the implementations using the Reactive Kernel/Cosmic Environment (RK/CE) as the underlying communication protocol (Seitz et al. (1989)). This is a message-passing programming environment for multicomputer nodes (RK) and for networked hosts running Unix (CE). As a consequence, applications are transportable between systems that support RK/CE, including Intel hypercubes (iPSC/1/2), the Symult 2010 Series and "ghost cubes". The latter may consist of a network of workstations, so that the same effort will provide us with a version for a distributed system. We expect the use of distributed sequential machines to be advantageous for problems where the subdivision steps done locally are very expensive and many steps are expected. This will be true in moderate dimension multivariate problems where a large number of integrand evaluations is typically needed in the local integrations.

Acknowledgment

The authors acknowledge the support of the Advanced Computer Research Facility at Argonne National Laboratory and of the Caltech Concurrent Supercomputing Facility.

References

G. Almasi and A. Gottlieb (1989), *Highly Parallel Computing*, Benjamin/Cummings.

G. Andrews and R. Olsson (1987), Revised report on the SR programming language, Technical report, Dept. of Computer Science, The Univ. of Arizona, TR 87-27.

G. Andrews, R. Olsson, M. Coffin, I. Elshoff, K. Nilsen, T. Purdin, and G. Townsend (1988), *An overview of the SR language and implementation*, ACM Trans. on Prog. Lang. and Systems, **10**(1).

W. Andrews (1990), *Bridging today's busses to futurebus*, Computer Design **29**(3), 72–84.

R. Babb II (1988), *Programming parallel processors*. Addison-Wesley, Reading, MA.

H. Bal, J. Steiner, and A. Tanenbaum (1989), *Programming languages for distributed computing systems*, ACM Computing Surveys **21**(3), 261–322.

J. Barhen, S. Gulati, and S. Iyengar (1988), *The pebble crunching method for load balancing in concurrent hypercubes ensembles*, Proc. of The Third Conference on Hypercube Concurrent Computers and Applications (G. C. Fox, ed.), vol. I, ACM Press, NY, pp. 189–199.

D. Bertsekas and J. Tsitsiklis (1989), *Parallel and Distributed Computation-Numerical Methods*, Prentice-Hall, Englewood Cliffs, NJ.

K. Birman, T. Joseph, K. Kane, and F. Schmuck (1988), *The ISIS system manual*, Technical report, The ISIS Project, Department of Computer Science, Cornell University.

J. Boyle, R. Butler, T. Disz, B. Glickfield, E. Lusk, R. Overbeek, J. Patterson, and R. Stevens (1987), *Portable programs for parallel processors*. Holt, Rinehart, and Winston.

T. Chou and J. Abraham (1982), *Load balancing in distributed systems*, IEEE Trans. on Software Engineering **8**, 401–412.

G. Coulouris and J. Dollimore, *Distributed systems, concepts and designs*, Addison-Wesley.

G. Cybenko (1989), *Dynamic load balancing for distributed memory multiprocessors*, J. Parallel and Distributed Computing **7**, 279–301.

P. Davis and P. Rabinowitz (1984), *Methods of numerical integration*, Academic Press, New York.

E. de Doncker (1978), *An adaptive extrapolation algorithm for automatic integration*, SIGNUM Newsletter **13**, 12–18.

E. de Doncker, D. Kahaner, and B. Starkenburg (1990), *An algorithm for numerical integration over a set of triangles* (submitted).

E. de Doncker and J. Kapenga (1987), *Parallelization of adaptive integration methods*, Proc. of NATO Workshop on Numerical Integration; Recent Developments, Software and Applications (P. Keast and G. Fairweather, eds.), NATO ASI Series C: Mathematical and Physical Sciences 203, Reidel, pp. 207-218.

——(1988), *A portable parallel algorithm for multivariate numerical integration and its performance analysis*, Proc. of the Third SIAM Conference on Parallel Processing for Scientific Computing, Dec. 1987, Los Angeles (G. Rodrigue, ed.), SIAM, pp. 109–113.

E. de Doncker and I. Robinson (1984a), *Algorithm 612, TRIEX: Integration over a triangle using nonlinear extrapolation*, ACM Trans., Math. Software **10**(1), 17–22.

——(1984b), *An algorithm for automatic integration over a triangle using nonlinear extrapolation*, ACM Trans. Mathematical Software **10**(1), 1–16.

EXPRESS (1987), *An operating system for parallel computers*, Parasoft Corporation, Pasadena, California.

H. Flatt and K. Kennedy (1989), *Performance of parallel processors*, Parallel Computing **12**(1), 1-20.

I. Foster and S. Taylor (1990), *Strand, new concepts in parallel programming*, Prentice Hall, Engelwood Cliffs, NJ 07632.

G. Fox (1989), *Parallel computing comes of age: Supercomputer level parallel computations at Caltech*, Concurrency Practice and Experience **1**(1), 63–104.

G. Fox and W. Furmanski (1988), *Load balancing loosely synchronous problems with a neural network*, Proc. of The Third Conference on Hypercube Concurrent Computers and Applications (G. C. Fox, ed.), vol. I, ACM Press, NY, pp. 241–278.

G. Fox, M. Johnson, G. Lyzengs, S. Otto, J. Salmon, and D. Walker (1988), *Solving problems on concurrent processors*, vol. I. Prentice Hall, Engelwood Cliffs, NJ 07632.

A. Genz (1987), *The numerical evaluation of multiple integrals on parallel computers*, Proc. of NATO Workshop on Numerical Integration; Recent Developments, Software and Applications (P. Keast and G. Fairweather, eds.), NATO ASI Series C: Mathematical and Physical Sciences 203, Reidel, pp. 219–229.

I. Gladwell (1987), *Vectorisation of one dimensional quadrature codes*, Proc. of NATO Workshop on Numerical Integration; Recent Developments, Software and Applications (P. Keast and G. Fairweather, eds.), NATO ASI Series C: Mathematical and Physical Sciences 203, Reidel, pp. 231–238.

S. Gregory (1987), *Parallel logic programming in paralog*, Addison-Wesley, Reading, MA.

J. Kapenga and E. de Doncker (1988a), *Concurrent management of priority queues*, Proc. of the Third SIAM Conference on Parallel Processing for Scientific Computing, Dec. 1987, Los Angeles (G. Rodrigue, Ed.), SIAM, pp. 347–351.

——(1988b), *A parallelization of adaptive task partitioning algorithms*, Parallel Computing **7**, 211–225.

J. Kapenga, E. de Doncker, K. Mullen, and D. Ennis (1987), *The integration of the multivariate normal density function for the triangular method*, Proc. of NATO Workshop on Numerical Integration; Recent Developments, Software and Applications (P. Keast and G. Fairweather, eds.), NATO ASI Series C: Mathematical and Physical Sciences 203, Reidel, pp. 321–328.

F. Lin and R. Keller (1987), *The gradient model load balancing method*, IEEE Trans. Software Engineering **13**, 32–37.

E. Lusk and R. Overbeek (1983), *Implementation of monitors with macros: A programming aid for the Help and other parallel processors*, Technical report, Argonne National Laboratories, MCS ANL-83-97.

——(1984), *User of monitors in FORTRAN: A tutorial on the barrier, self-scheduling do-loop and askfor monitors*, Technical report, Argonne National Laboratory, Report MCS ANL-84-51.

J. Lyness and P. Jespersen (1975), *Moderate degree symmetric quadrature rules for triangles*, J. Inst. Maths. Applics. **15**, 19–32.

M. Malcolm and R. Simpson (1975), *Local versus global strategies for adaptive quadrature*, ACM Trans. Mathematical Software **1**, 129–146.

K. Mullen and D. Ennis (1987), *Mathematical formulation of multivariate Euclidean models for discrimination methods*, Psychometrica **52**(2), 235–249.

K. Mullen, D. Ennis, E. de Doncker, and J. Kapenga (1988), *Multivariate models for the triangular and duo-trio methods*, Biometrics **44**, 1169–1175.

L. Ni, C.-W. Xu, and T. Gendreau (1985), *A distributed drafting algorithm for load balancing*, IEEE Trans. Software Engineering **11**, 1153–1161.

R. Piessens, E. de Doncker-Kapenga, C. Überhuber, and D. Kahaner (1983), QUADPACK, *A subroutine package for automatic integration*, Springer Series in Computational Mathematics, Springer-Verlag.

M. Quinn (1987), *Designing efficient algorithms for parallel computers*, McGraw-Hill, New York, NY.

C. Seitz, J. Seizovic, and W.-K. Su (Jan. 1988, Rev. 1–April 1989), *The C programmer's abbreviated guide to multicomputer programming*, Technical report, California Institute of Technology, Department of Computer Science, Caltech-CS-TR-88-1.

H. Shapiro (1984), *Increasing robustness in global adaptive quadrature through interval selection heuristics*, ACM Trans. Mathematical Software **10**(2), 117–139.

Various Authors (1989), *RISC, enough for the next generation?* Computer Design Supplement **28**(22).

R. Wilson (1989), 80860 *CPU positions Intel to take on minisupers*, Computer Design **28**(7), 20–24.

A. Wouk, ed. (1989), *Parallel processing and medium-scale multiprocessors*, SIAM.

COMPUTER SCIENCE DEPARTMENT, WESTERN MICHIGAN UNIVERSITY, KALAMAZOO, MICHIGAN 49008

Contemporary Mathematics
Volume **115**, 1991

High-Dimensional Numerical Integration and Massively Parallel Computing

MICHAEL MASCAGNI

ABSTRACT. We consider the efficient implementation of numerical algorithms for high-dimensional quadrature on massively parallel computers. We will consider both Monte Carlo and deterministic methods of numerical quadrature. It will be shown that the accumulation of a quadrature estimate which is a sum of N terms can be accomplished in $O(\log_2 N)$ time on a massively parallel computing machine. We then consider a problem requiring the numerical evaluation of an infinite-dimensional (Wiener) integral, and its massively parallel implementation.

1. Introduction

The advent of massively parallel computing machines has begun to change the way many computations are undertaken. Similarly, massive parallelism has led to the consideration of algorithms which scale in computational complexity at fairly modest rates with problem size when the number of processing elements used in the computation is assumed arbitrarily large. In particular, the evaluation of sums involving N terms, the common computational task in numerical quadrature, can be accomplished in $O(\log_2 N)$ asymptotic time with a massively parallel computing machine which has $O(N)$ processing elements. Of the two major classes of parallel computing machines, multiple instruction stream multiple data (MIMD) and single instruction stream multiple data (SIMD), we will focus on the implementation details and algorithm design specific for SIMD machines. (For a discussion of these same issues for MIMD machines, see the article by Dedonker in this volume.)

We will begin with an introduction to SIMD parallel computing machines and aspects of their programming. The concrete example of a SIMD

1980 *Mathematics Subject Classification* (1985 *Revision*). Primary 65D30; Secondary 65C05, 68A99.

Key words and phrases. Numerical quadrature, elliptic PDE, parallel computing.

This work was supported by an NIH-NRC Associateship.

This paper is in final form and no version of it will be submitted for publication elsewhere.

machine which we shall use throughout this discussion is Thinking Machines Corporation's Connection Machine, Hillis (1985) . We will then discuss the evaluation of formulas for nonadaptive numerical quadrature on SIMD machines. Included in this discussion is consideration of Monte Carlo methods of quadrature. Obvious extensions of these implementations for adaptive quadrature then conclude the discussion of numerical quadrature in high-dimensional spaces.

We then conclude with a brief introduction to infinite-dimensional (Wiener) integration for the numerical solution of discrete elliptic Dirichlet boundary value problems (BVPs). This infinite-dimensional quadrature is explored with a Monte Carlo method which has been explained in great detail elsewhere, Mascagni (in press). We will consider the SIMD implementation and performance details of this algorithm as well as the relation of this infinite-dimensional quadrature method with more classical methods of solving discrete elliptic BVPs.

2. SIMD parallel computers

Massively parallel SIMD computers are, perhaps, the easiest class of massively parallel computers to program. The reason for this is, as their name implies, that there is but a single instruction stream to prescribe. A typical SIMD parallel computer is a cluster of computer processors each with its own local processor memory connected to each other with some interconnection topology, and all connected to a single "master" processor. This single "master" processor broadcasts instructions to the entire cluster of processors. The clustered processors then all execute the broadcast instruction in synchrony and await the next broadcast instruction. It is important to point out that the common broadcast instruction stream will often request that processors interchange data between their local memories or compute a boolean expression which then determines which processors execute the subsequent group of broadcast instructions. A diagram illustrating the general structure of SIMD parallel computers is given in Figure 1.

It is obvious from the nature of SIMD computers that only a single instruction stream must be supplied by the programmer and that all issues of synchronization, which are so vexing when programming MIMD machines, are therefore implicitly handled. The programming paradigm for SIMD computers is often called the "data parallel" programming style, where parallelism is implicitly achieved after a "mapping" of problem data into the local memories of the cluster of processors is accomplished. For any given problem, there are many possible alternative mappings of problem data onto a SIMD computer. For many reasons, an obvious mapping to use for numerical integration is to map individual points in the domain of integration onto processors. For deterministic quadrature formulas, it is often convenient, if possible, to configure the cluster of processors into a high-dimensional grid topology of appropriate dimension and extent in each dimension to facilitate

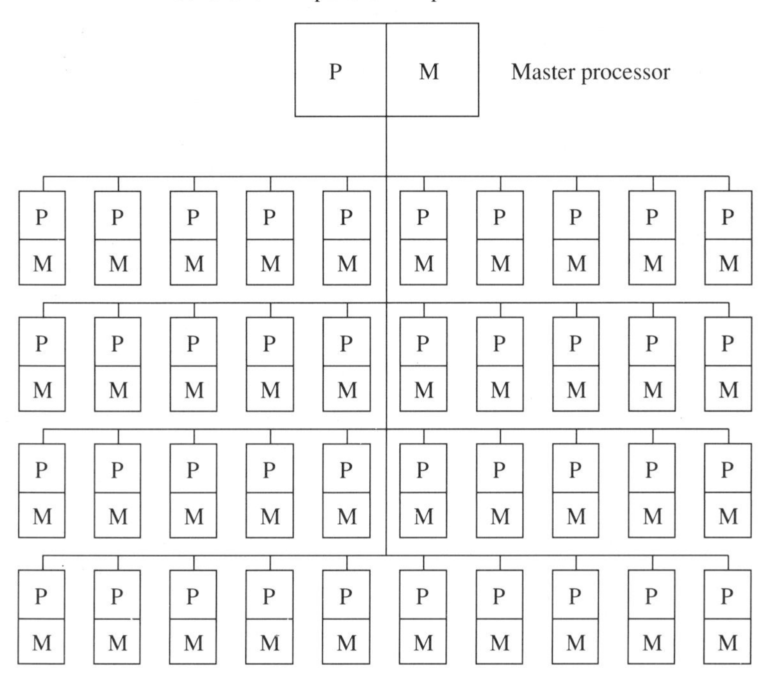

FIGURE 1. The general configuration of SIMD parallel computing machines.

the evaluation of the elements in the formulas. For Monte Carlo methods of quadrature no special processor geometry is necessary, only the mapping of, now random, points in the integration domain onto processors.

Although SIMD machines are fairly new to the computing marketplace, it is relatively easy for researchers to gain access to such a machine from one of several national parallel computing resources. At at least six of these sites one can find a Connection Machine, by Thinking Machines Corporation, which is a massively parallel SIMD computer. The Connection Machine is a parallel computer composed of up to 65532 bit serial processors with up to 131072 bytes of local memory each. This cluster of processing elements is configured into a 16 dimensional hypercube ($2^{16} = 65532$) and supports "virtual" processors via time-slicing replicated processor cycles multiplexed over different segments of the local memory which is partitioned into independent segments for each virtual processor. The Connection Machine supports three high level computing languages which have the data parallel paradigm as a naturally expressible part of them. These are *Lisp (pronounced star-lisp), C*, and Connection Machine Fortran (CMF). These Connection Machine specific languages are extensions of the more familiar Lisp, C, and Fortran 8X languages.

The Connection Machine supports two modes of interprocessor communication, one is called the "router" and allows arbitrary processor to processor communication via specialized hardware and software. The second mode "gets" and "sends" messages to nearest neighbors on the hypercube via fast hypercube wires. It is generally a good idea to use the less general but more direct hypercube connections for interprocessor communications as the "router" is 20 to 40 times slower than a single hypercube communication. Since the Connection Machine is physically a high-dimensional hypercube, it is easy to imbed high-dimensional grids and tori very naturally into the machine. Since it is often the case that no sharing of information between different points in a numerical integration formula is necessary, the parallel mapping of physical points in the domain onto processors leads to no interprocessor communication until the results are to be centrally accumulated. In some cases nearest neighbors on a regular grid need exchange information, which is readily implemented without recourse to the slow "router".

3. Complexity of massively parallel numerical integration

With the parallel mapping introduced in the previous section let us discuss some examples of numerical integration on massively parallel machines. Assume that the numerical value of the integral

$$\int_{\Omega} f(\mathbf{x})\, d\mathbf{x} \tag{3.1}$$

is desired. Here $\Omega \subset \mathcal{R}^d$ and $\mathbf{x} \in \Omega$. The numerical methods of Newton–Cotes and the quadrature formulas of Gauss and Legendre all give formulas that may be commonly denoted as

$$\int_{\Omega} f(\mathbf{x})\, d\mathbf{x} \approx \sum_{i=1}^{N} w(\mathbf{x}_i) f(\mathbf{x}_i), \tag{3.2}$$

where $\mathbf{x}_i \in \Omega$ is a finite number of points in Ω appropriate to the particular method of quadrature and $w(\cdot)$ is the appropriate weight, see Stoer and Bulirsch (1980).

When Ω is the cross product of closed intervals, i.e. $\Omega = \otimes_{j=1}^{d} [a_j, b_j]$, where a_j and b_j are extended real numbers, we have a particularly easy enumeration of the $\mathbf{x}_i$. We have that

$$\mathbf{x}_i = [p_{1\, i_1}, p_{2\, i_2}, \dots, p_{d\, i_d}], \tag{3.3}$$

where $p_{j\, i_j}$ is the i_j th point required in the one-dimensional quadrature formula chosen to be used in the interval $[a_j, b_j]$. Since we have assumed Ω to be a cross product of intervals, we have that $[i_1, i_2, \dots, i_d] \neq [k_1, k_2, \dots, k_d]$ if $i \neq k$. As an example, consider a numerical integration where in the finite one-dimensional intervals, uniformly spaced points are used, and in the infinite or semi-infinite intervals Hermite or Laguerre points are used. Then the finite subset of points is well defined and results in a quadrature formula as in equation (3.2).

Given that the quadrature formula of equation (3.2) is to be evaluated with points enumerable via equation (3.3) we begin our Connection Machine (SIMD) implementation by configuring the Connection Machine into a d-dimensional grid, where each dimension, j, has as many processors as the interval $[a_j, b_j]$ has points for quadrature. To evaluate the general quadrature formula in equation (3.2) there are three separate computations that must be undertaken. The first is the evaluation of $w(\mathbf{x}_i)$ for each $\mathbf{x}_i \in \Omega$. Second we must evaluate the integrand at each point, i.e. $f(\mathbf{x}_i)$. Lastly we must accumulate the sum in equation (3.2), which is really an inner product of the vectors $w(\mathbf{x}_i)$ and $f(\mathbf{x}_i)$. We will consider the numerical evaluation of these three subcomputation separately below.

The evaluation of $w(\mathbf{x}_i)$, the weights, often requires the knowledge of the distance between $\mathbf{x}_i$ and all of its neighbors on the topologically d-dimensional lattice. This is so terms involving the spatial spacing in the lattice can be computed. Recall that in 1-dimension the familiar Newton–Cotes formulas have weights that depend on h, the spatial spacing. This is also true, in a modified form, for the formulas for Gaussian, Hermite, Legendre, and Laguerre quadrature. The exchange of information required to compute the local spacing takes at most $O(d)$ computational time units as each of the $2d$ information exchanges may be done in parallel on the SIMD computer. On the other hand, it may be argued that in laying out the grid, the local grid information is known, so that this information exchange is unnecessary. However, even if this information is known in the "setup" stage of the computation, $w(\cdot)$ must still be computed, which depends at most on all of the d local grid spacings.

The complexity situation is similar when we evaluate $f(\mathbf{x}_i)$. At best, this evaluation is $O(1)$ in the case that either $f(\cdot)$ is constant or can be found in a lookup table. At worst, the complexity is $O(d)$, as each component of $\mathbf{x}_i$ many require the evaluation of a function of a single variable. If $f(\cdot)$ is a rather complex function, not expressible as quotients of elementary functions, then the situation may be much worse. However, let us assume that $f(\cdot)$ is expressible as a quotient of elementary functions, so that it can be thought of as computable through a series of calls to some function subroutine library.

Lastly we turn to computing efficiently the sum in equation (3.2) when each of the individual terms is located in the local memory of different processors in the SIMD computer. We can form the product of $w(\mathbf{x}_i)$ and $f(\mathbf{x}_i)$ in each processor in $O(1)$ time. We must then accumulate these N numbers in a single memory location. For the moment let us forget that the processors in our SIMD machine are configured in a d-dimensional grid and consider them enumerated from 1 to N in single linear array of processors. For simplicity assume that $N = 2^p$ for some p. At this point each processor contains its contribution to the sum, which is a kind of partial sum. If we first send our partial sum to processor number $2^0 = 1$ greater than our own, and add our sent partial sum to partial sum stored in the other processor, we will have a new partial sum. We now repeat this procedure with processor numbers 2^j greater for $j = 1, 2, \ldots, p-1$. This recursively defines a new partial sum in each processor and places the total sum in processor number N in $O(p) = O(\log_2 N)$ time. More generally, this procedure places the partial sum $\sum_{i=1}^{k} w(\mathbf{x}_i)f(\mathbf{x}_i)$ in processor k. An example of the implementation of this algorithm on a SIMD machine with $N = 2^3 = 8$ processors is illustrated in Figure 2.

It is important to emphasize the generality of this algorithm and its particular suitability to efficient SIMD implementation on machines with hypercube interprocessor connectivity topology, such as the Connection Machine. This algorithm is a general way to evaluate an expression involving the N-fold application of a commutitive operation in $O(\log_2 N)$ time. And, as with the summation example from numerical integration, the algorithm also furnishes the partial operation applications, which may be desirable. As for the efficient SIMD implementation on topologically hypercube machines, it turns out that when one has imbedded any d-dimensional grid, with each dimension a power of two in length, into a high-dimensional hypercube, that communication with processors a power of two away in any dimension can be accomplished along a single hypercube wire. Thus in our example, where we have chosen to think of the processors as part of a 1-dimensional grid, implementation would only require the use of the fast hypercube communication channels. Similarly, if we relax our conceptual requirement that the processors be configured in a 1-dimensional grid, but remain in a d-dimensional grid, the above algorithm is still implementible on machines such as the Connection Machine using only fast hypercube wires for interprocessor communication. Thus we conclude that the asymptotic time complexity

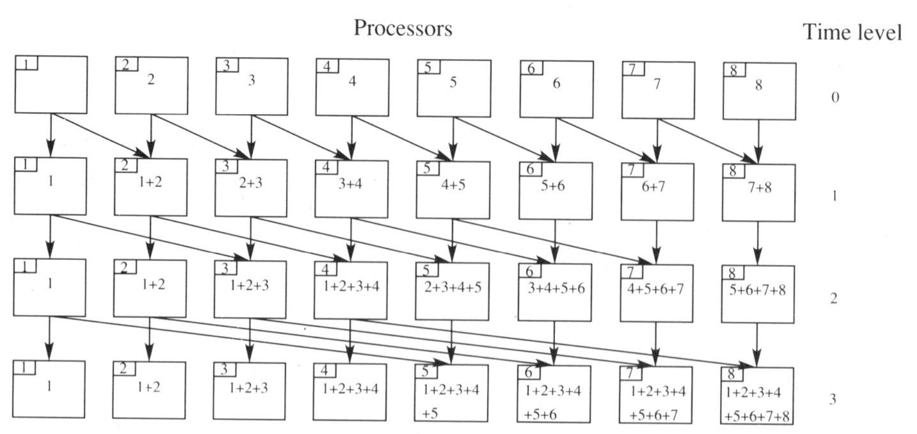

FIGURE 2. A small example of the SIMD implementation of an $(\log_2 N)$ algorithm for the partial summation of a series.

for the numerical evaluation of numerical quadrature formulas of the form of those in equation (3.2) is $O(\log_2 N)$ on a SIMD computer with $O(N)$ processors, as the dominant portion of the computation is the evaluation of the sum in equation (3.2).[1]

Now let us consider the SIMD implementation of a Monte Carlo method for the numerical evaluation of the integral in equation (3.1). Monte Carlo numerical integration has a very long history going back to the experimental determination of π via a geometrical probabilistic realization of the Buffon needle problem, Hall (1873) . Methods for Monte Carlo integration also lead to the numerical evaluation of formulas like those in equation (3.2). In the Monte Carlo case, however, the points $\mathbf{x}_i$ are chosen randomly, with some a priori probability distribution, and $w(\mathbf{x}_i)$ is a weight based on either an importance function, requirements of conditional Monte Carlo, or another variance reduction technique, see Hammersley and Handscomb (1964). Since the points, $\mathbf{x}_i$, are randomly chosen in Monte Carlo integration, no specific interconnection topology of processors need be assumed, and we may consider the SIMD machine layed out as a 1-dimensional grid of processors. The computation then can be broken down into: first, the determination of a suitably random point $\mathbf{x}_i \in \Omega$; second, the evaluation of $w(\mathbf{x}_i)$; third, the evaluation of $f(\mathbf{x}_i)$; and fourth evaluation of the sum in equation (3.2).

The third and fourth parts of the Monte Carlo computation are exactly the same as in our discussion of deterministic quadrature. The first part of the Monte Carlo integration, determination of the random $\mathbf{x}_i \in \Omega$, is a very different type of computation than anything discussed for deterministic quadrature. With $\Omega \subset \mathcal{R}^d$, we will require a minimum of d random numbers per processor to determine $\mathbf{x}_i$, if Ω is a cross product of generalized intervals. If Ω is a more complicated region or if points are to be chosen with a rather complicated probability density, then the accept/reject Metropolis algorithm is often used to pick points, Metropolis et al. (1953). Let us call the fraction of accepted $\mathbf{x}_i$'s p, and so $1 - p = q$ is the fraction of rejections. When one uses the Metropolis algorithm it is always desirable to use it in a situation where the fraction of rejections is small. This is especially true when the Metropolis algorithm is used on a SIMD computer. Since there is but a single instruction stream, rejections do one of two things, depending on the implementation. If we choose to use those values of $\mathbf{x}_i$ accepted after r attempts, then only $1 - q^r$ of the processors will participate in the subsequent computation. Similarly if we continue to use the Metropolis algorithm until all the processors have accepted suitable values of $\mathbf{x}_i$, then we will require, on the average, r iterations where r is the smallest

[1]On a serial computer this computation has $O(N)$ optimal asymptotic time complexity, while on a parallel computer with P processors and using the fast partial summation algorithm has $O(\lfloor N/P \rfloor \log_2 N)$ asymptotic complexity. When $P = O(N)$ we again regain the above $O(\log_2 N)$ asymptotic time complexity.

integer satisfying $q^r < 1/(2N)$, and N is the number of processors. If q is not small, these estimates are rather pessimistic, and so it behooves us to try to make r as small as possible, even at the expense of other simplifying considerations. Thus, given $1+q$, it will require $O(-d(1+\log_2 N)/(\log_2 q))$ time, on the average to choose random $\mathbf{x}_i$ in a d-dimensional space with a rejection frequency of q.[2]

Given that we have a value of $\mathbf{x}_i$ in each processor, the calculation of $w(\mathbf{x}_i)$ is rather straight forward. In all of the standard techniques of variance reduction, $w(\cdot)$ is a known, easy to calculate function, Hammersley and Handscomb (1964). Thus we compute $f(\mathbf{x}_i)$, and use the above mentioned algorithm to accumulate the sums in $O(\log_2 N)$. Thus the overall time complexity of the SIMD implementation of Monte Carlo integration is asymptotically $O(-d(1+\log_2 N)/(\log_2 q))$, for N samples. One added virtue of using the $O(\log_2 N)$ partial summation algorithm, is that the partial sums give a qualitative feeling for the statistical convergence. Also, one can use the partial summation algorithm to accumulate not only the mean but the variance of the computation, and monitor the statistical sampling error as a function of the number of samples.

Before we end our discussion of finite-dimensional numerical quadrature, we should mention SIMD implementations of adaptive numerical quadrature. In point of fact, MIMD machines are much better suited to adaptive numerical quadrature than SIMD machines, since local estimates of truncation error dictate which regions of the computational grid require refinement. This kind of truncation error driven local grid refinement is an intrinsically asynchronous computation. Thus, certain compromises must be made in order to carry out adaptive quadrature efficiently on SIMD computers. One must first carry out an integration with a given set of points $\mathbf{x}_i \in \Omega$ as was described above. The evaluation of a local truncation error function at each $\mathbf{x}_i$ is then accomplished in $O(d)$ time, as nearest neighbor information is required to estimate local errors. This local truncation error function is then used to generate a new set of $\mathbf{x}_i$ and a new set of weights, $w(\cdot)$, for the evaluation of a new sum. Since we want to be able to use the values of $f(\cdot)$ at the previous $\mathbf{x}_i$'s, this subsequent iteration of SIMD adaptive quadrature involves the evaluation of a sum as in equation (3.2) involving both the old and the new $\mathbf{x}_i$'s with a suitably modified weight function. Thus it is desirable to store the previous values of $f(\cdot)$. The difficult part in SIMD adaptive quadrature is using the local truncation error information from the previous approximate integration to select a new set of points to be included in the quadrature. This is further complicated if one is constrained to maintain the same d-dimensional processor configuration throughout the computation.

[2]The problem of providing high quality psuedorandom numbers to every processor in a massively parallel computing machine is a nontrivial problem. See the articles by Durst and Marsaglia in this volume for a more detailed discussion of this problem.

4. Infinite-dimensional (Wiener) numerical integration

There is a classical probabilistic representation theory of the solutions to boundary value problems (BVPs) for linear elliptic and initial-boundary value problems (IBVPs) associated with linear parabolic PDEs, see Freidlin (1985) and Itô and McKean (1965). This representation theory is based on integration with respect to a measure (Wiener measure) placed upon the space of continuous functions defined over a domain, Itô and McKean (1965) and Wiener (1923). The relationship of integration with respect to this measure to PDEs arises from the use of the fundamental solution of linear parabolic PDEs to define the measure of cylinder sets of continuous functions. In particular, these results allow the representation of solutions to elliptic BVPs as mathematical expectations over Banach spaces of Brownian motion sample paths.

As a simple example of this probabilistic representation theory, let us consider the BVP for Laplace's equation:

$$\begin{aligned} \tfrac{1}{2}\Delta u(x) &= 0, \qquad & x \in \Omega, \\ u(x) &= g(x), \qquad & x \in \partial\Omega, \end{aligned} \tag{4.1}$$

then the Wiener integral representation is denoted by:

$$u(x) = E_x[g(\beta(\tau_{\partial\Omega}))]. \tag{4.2}$$

This equation states that the value of the solution at an interior point, x, is the Brownian motion expectation (integral with respect to Wiener measure) of the boundary value at the first hitting location of the sample (Brownian motion) path $\beta(\cdot)$ starting from the point x. This is equivalent to the more familiar Green's function representation, John (1982), as (4.2) can be viewed as a boundary integral of a boundary mass against the desired boundary values. In the Green's function representation, the boundary mass is the value of the Green's function along the boundary, and it can be shown that the probability of a Brownian motion started at x first striking the boundary at a given boundary point, y, is the Green's function, $G(x, y)$, see Courant, Friedrichs, and Lewy (1928). Thus (4.2) is a probabilistic representation of the Green's function solution to these elliptic Dirichlet problems.

The use of these and related probabilistic representations as the basis of algorithms for the numerical solution of problems arising from PDEs has a rich history. Algorithms based on random walks have been used to solve problems in neutron diffusion, Ulam, Richtmeyer, and von Neumann (1947); determine spectral quantities of Schrödinger equations, Donsker and Kac (1951); and solve linear algebraic equations, Curtiss (1953, 1956), Forsythe and Leibler (1950). In fact, the paper that many believe to be seminal in the field of the numerical solution of PDEs by the method of finite-differences has an extensive section devoted to Monte Carlo solutions of both linear elliptic and parabolic equations, Courant, Friedrichs, and Lewy (1928). This interest in using Monte Carlo algorithms in a wide variety of problems was an outgrowth of the success of the Monte Carlo method in the Manhattan project. The particular interest in the Monte Carlo method for the solution of linear algebraic equations, especially those related to difference approximations of elliptic PDEs, waned when the computational cost of these methods was found to be greater than for the stationary iterative or direct methods of solution, Curtiss (1956).

These pessimistic comparisons of the complexity of Monte Carlo methods for the solution of elliptic BVPs need no longer be considered valid in the light of the development of massively parallel SIMD computers. These massively parallel computing machines make the concurrent accumulation of statistics from a Monte Carlo method very easy. Another development in favor of Monte Carlo methods for elliptic problems is a modification of the random walk algorithm to avoid the inherent inefficiency in sampling statistics based on the first hitting time, $\tau_{\partial\Omega}$. Since $\tau_{\partial\Omega}$ is known only after the entire random trajectory is executed, a single sample is obtained in $O(\tau_{\partial\Omega})$ steps if we use the näive algorithm suggested by the probability theory. Below we present a class of algorithms, motivated by their efficient implementation on the Connection Machine, which utilize "adjoint" random walks allowing the rapid accumulation of statistics based on random walk first hitting times.[3]

If we replace the continuous BVP in equation (4.1) with a finite-difference analog, the resulting discrete equations have an analogous representation theory based not on Brownian motion but on random walks. In fact, if we replace the region Ω with a discretized region Ω_h, replace the derivatives in equation (4.1) with their well-known $O(h^2)$ finite-difference approximations, and replace the continuous Brownian motion paths with isotropic random walks on Ω_h, the description of the representation theory in the previous paragraph carries over to this discrete class of approximations. For example, if Ω is the unit cube in $\mathcal{R}^d$, then Ω_h can be chosen to be the uniform grid of points in Ω with grid spacing $h = 1/M$ for some M. If we then wish to solve the discrete Laplace equation we can use the analog of equation (4.2) to do so via a random walk method. The discrete interpretation of (4.2) is that the solution to the problem at a grid point, x, is the average of the boundary values of the boundary point first encountered by random walkers starting out from x. Reinterpret equations (2.2) and (2.3) as expectations of quantities over the space of random walks in order to solve discretizations of the original equations.

This reduction of discretizations of the Dirichlet problem to expectations of statistics over the space of random walks has interesting implications for the construction of algorithms which can exploit the inherently parallel nature of independent statistical sampling. This notion, coupled with the recent

[3]A version of the "adjoint" algorithm is briefly discussed by Hammersley and Handscomb (1964) in connection with the solution of systems of linear equations. They also mention that it can be used to solve the Dirichlet problem, however they did not analyze it convergence behavior.

availability of massively parallel SIMD machines, was the motivation for the reexamination of these classical links between probability and elliptic PDE. As such, the "adjoint" random walk algorithm will be described in the context of an efficient implementation of the above random walk representations on a massively parallel SIMD machine. The first step in constructing an algorithm to implement these representations on a parallel computing machine is to determine a mapping of information in the computational task onto the parallel machine. On a massively parallel SIMD machine, which has a relatively small amount of memory per processing element, such as the Connection Machine, it is prudent to map each grid point in Ω_h onto a single processing element. The alternative mapping which exploits the parallelism of independent sampling is to associate individual random "walkers" with processors. This approach is impractical when either Ω_h has many grid points or when individual processing elements in the parallel computing machine have small memories.

With the mapping of grid points to processors, the näive algorithm is to start random walkers from each grid point simultaneously. The walkers then walk until first striking the boundary, at which point they will have a sample of the Wiener integral (4.2) to communicate to their starting point. While this algorithm is appealing in its simplicity, it has two serious drawbacks. The first is that each sample requires the execution of a random walk of length $\tau_{\partial\Omega_h}$. The second is that when the walkers reach the boundary, the act of communicating their sample to the starting point of the walk can be very costly. For example, in the Connection Machine, if Ω_h is a subset of a regular grid in $\mathcal{R}^d$, then communication between adjoining grid points required for the walkers to execute their random excursion can be done via the high speed hypercube connections between processors. However, when the walkers must communicate their accumulated statistic to the starting point of their walk, communication will instead take place over the general "router" hardware. Unfortunately, the "router" is relatively slow compared to the hypercube connections. Thus an implementation which abolishes the need for more than nearest neighbor grid point communication and reduces the amount of work required per sample to less than $O(\tau_{\partial\Omega_h})$ is desirable.

A solution to the communication problem is to use the random trajectory of the walker, which involves only nearest neighbor communication, to retrace the path from the boundary point to the starting point. Let us illustrate this solution to the communication problem by considering the Wiener integral integral representation for the ordinary Laplace equation (4.2). In this case the statistic that is averaged over the space of random walks is simply the boundary value of the first hitting location. The problem is that since the hitting location is random, we either must realize the entire trajectory or choose boundary points appropriately in order to sample each random walk with the proper probability. When we have realized an entire trajectory, the näive algorithm uses only the starting and ending points of the walk. However, the walker has visited a number of points in between, and it is the case that the walks from these intermediate points to the boundary are being generated with the proper probability, as these random walks are generated by a stationary Markov process and so are completely determined by the matrix of transition probabilities from grid point to grid point.[4] Thus when the boundary is reached, the boundary value can be scored not only at the starting point but at all of the intermediate points in the walk. This means that the solution to the communication problem of retracing the walk, when augmented with scoring the encountered boundary value at each point along the retraced trajectory reduces the work required per sample to $O(1)$ from $O(\tau_{\partial\Omega_h})$.

This retracing algorithm has many advantages over the original näive algorithm; moreover, it is easy to see that the initial "forward" computation of the trajectory of the random walker can be totally eliminated. All of the scoring of the encountered boundary value occurs during the retracing of the "forward" random walk trajectory, so that if we start a walker from the boundary and advance it with the correct probability, scoring its boundary value at each encountered point will be equivalent. In this case the correct probability will be that determined by the adjoint of the original "forward" random walk transition probability matrix, $P^T = [p_{ji}]$. In order to simulate walks which start at interior points and walk to an absorbing boundary, it is important to the "adjoint" random algorithm that "adjoint" walkers are removed when they reencounter the boundary. This description is what will be called the "adjoint" random walk algorithm.

Before concluding our discussion of the relation of the "forward" and "adjoint" random walk methods we must consider a very important practical detail. In defining the "adjoint" random walk transition probability as the transpose of the "forward" random walk matrix of transition of probabilities we must interpret how we computationally impose a given p_{ij}^T when $x_i \in \partial\Omega_h$. This particular p_{ij}^T is the probability of a walker stepping from a boundary point, an otherwise absorbing point, to one of its neighboring grid points. One way to impose this initial probability in the "adjoint" random walk algorithm is to start out walkers with an initial density along the boundary proportional to p_{ij}^T. This notion of starting off walkers in the "adjoint" random walk algorithm with a nonuniform density is necessary when either discrete elliptic operator is nonsymmetric or when Ω_h lacks the necessary regularity for all of the p_{ij}^T, when $x_i \in \partial\Omega_h$, to have a common value.

When we use the "adjoint" random walk method to solve equation (4.1) numerically, there are three types of errors which arise. The first is the discretization error involved in replacing Ω with Ω_h and subsequently

---

[4]A rigorous demonstration of this can be found in Mascagni (in the Press).

defining a discrete analog of the Laplacian on Ω_h. These discretization errors are the same as we encounter when using finite-difference approximation to solve PDEs and so are well understood, Richtmeyer and Morton (1967). Statistical sampling error involved with the estimation of the expected value representations of our solution. Since the underlying stochastic process is a finite state Markov chain, and our statistics have a finite second moment, the central limit theorem holds, Spitzer (1964). Thus we know that the statistical sampling error decreases at worst as $K^{-1/2}$, where K is the number of independent samples of the statistic taken over the Markov chain. The third error is of a somewhat different nature; it is due to the fact that certain computational considerations add a bias to the way we sample the space of random walks. We will consider this error in detail below.

In the numerical computation of the discrete Wiener integrals, depending on the mapping of the problem onto the computational machine, there are two basic criteria which can be used to terminate the computation. When we follow an individual walker along its trajectory, a natural termination occurs when the walker reencounters the boundary at which point it is removed from the computation. In this case, we execute a number of complete walks until the statistical sampling error is tolerable. There is no bias in the sampling from the space of random walks in this case. However, if we use the mapping described above for placing this algorithm onto a massively parallel SIMD machine, it may be impractical to allow all of the random walkers to reencounter the boundary before terminating the computation. A reasonable halting criterion in this case is that all of the walkers in the ensemble be allowed to take up to a given number of steps in their respective random walks. Since the Connection Machine is SIMD in design, this corresponds to computing a given number of iterations, where each iteration moves each walker one step, with walkers still removed upon reencountering the boundary. This stopping criterion does introduce bias because it allows us to sample only from the space of random walks up to a given length. We will refer to the error introduced in sampling only from random walks up to a given length as the "random walk approximation" error.

To analyze this "random walk approximation" error we must find an expression for the expected value of our Wiener integral statistic if we sample only from this truncated space of random walks. Our statistic is the quotient of the running total of the discrete Wiener integral values sampled divided by the number of samples obtained. If we call the vector of expected values of the statistics gathered from random walks up to length n, S^n, and the expected value of the numerator and denominator W^n and N^n respectively, then the strong law of large numbers assures us that $S^n = W^n/N^n$, where here we mean componentwise division of the vectors. It is easy to compute what W^n and N^n are as follows. W^n is the expected value of the cumulative sum of Wiener integral values, thus we have that for the "adjoint" random walk algorithm:

(4.3a)
$$W^n = \sum_{i=0}^{n} [P^T]^i E_W .$$

In this geometric sum the vector E_W is the vector of weighted boundary values: each boundary value weighted by the initial number or density of walkers. This sum just states that the expected value of the statistic for random walks up to length n is the sum over the contributions from random walks of length $i \leq n$, where walks of exactly length i contribute $[P^T]^i E_W$. If we use (4.3a) to compute a recursion for W^{n+1} we get:

(4.3b)
$$W^{n+1} = P^T W^n + E_W .$$

This has the same form as a Jacobi iteration in equation, except for the boundary term contribution of E_W. In a similar fashion it can be shown that the mean number of visits obeys the following recursion:

(4.4)
$$N^{n+1} = P^T N^n + E_N ,$$

where E_N is the initial mass distribution of the walkers.

An alternative interpretation of the above results can be found by recognizing that the Jacobi solution to finite-difference discretizations of elliptic BVPs is equivalent to an explicit finite-difference solution to the corresponding parabolic initial-boundary value problems (IBVPs), Garabedian (1956). Thus the expected value of our Wiener integral statistic over random walks up to length n is the the *quotient* of two linear parabolic IBVPs, as we have determined that the numerator and denominator of this statistic each satisfy linear parabolic IBVPs themselves. In our case the statistic is the quotient of explicit finite-difference (Jacobi) solutions as u_W/u_N, where:

(4.5)
$$\begin{aligned} u_{W_t} &= \tfrac{1}{2}\Delta^T u_W , & x \in \Omega,\ t > 0, \\ u_W(x, t) &= g(x)E(x), & x \in \partial\Omega,\ u_W(x, 0) = 0, \\ u_{N_t} &= \tfrac{1}{2}\Delta^T u_N , & x \in \Omega,\ t > 0, \\ u_N(x, t) &= E(x), & x \in \partial\Omega,\ u_N(x, 0) = 0. \end{aligned}$$

Here $g(x)$ is the desired boundary value, and $E(x)$ is the boundary mass required to use "adjoint" random walks with $\frac{1}{2}\Delta^T$ in order to solve the desired original problem with $\frac{1}{2}\Delta$. Since in our case $\frac{1}{2}\Delta = \frac{1}{2}\Delta^T$, $E(x) = 1$ in equation (4.5).

Next we consider the computational complexity of the "adjoint" random walk algorithm. In the previous section we established a strong relationship between this algorithm and one of the stationary iterative methods for the solution of difference equation analogs of elliptic BVPs, i.e., the Jacobi method. Thus we compare this Monte Carlo algorithm to the well-known iterative Jacobi, Gauss–Seidel, and successive over-relaxation (SOR) methods. We will consider two approaches at comparison. The first is the a priori complexity of the methods, and the second is by comparison of the time dependent problem which can be associated with each of the methods.

A simple a priori measure of the computational complexity of the "adjoint" random walk method is $\bar{\tau}_{\partial\Omega_h}$, the mean first hitting time for random walks in Ω_h. Since all of the statistics for the Wiener integrals depend on walkers reaching the interior from the boundary, information from the boundary will reach the interior after $\bar{\tau}_{\partial\Omega_h}$ steps of an "adjoint" walk. Because the "adjoint" walk is the realization of a Markov process, if we take a given region Ω_h, and find two regions, W_h and w_h, where $\bar{\tau}_{\partial W_h}$ and $\bar{\tau}_{\partial w_h}$ can be computed, and such that $w_h \subset \Omega_h \subset W_h$ we will have that:

$$\sup_{x_i \in w_h} \bar{\tau}_{\partial w_h}(x_i) < \sup_{x_i \in \Omega_h} \bar{\tau}_{\partial\Omega_h}(x_i) < \sup_{x_i \in W_h} \bar{\tau}_{\partial W_h}(x_i), \tag{4.6}$$

due to the Markov property. An obvious Monte Carlo computation of $\tau_{\partial\Omega_h}$ for the points in a region is to start walkers from the boundary and increment the statistic one for each step taken by each "adjoint" walker. From this description it is clear that $\tau_{\partial\Omega_h}$ is the "adjoint" random walk solution to the following discrete elliptic BVP, given that L_h is our notation for the discrete elliptic operator:

$$\begin{aligned} L_h \mathcal{T} &= -H^{-1}, \qquad x_i \in \Omega_h, \\ \mathcal{T} &= 0, \qquad x_i \in \partial\Omega_h, \end{aligned} \tag{4.7}$$

where H is a diagonal matrix corresponding to the discretization of Ω_h, and in the case of the Laplacian, is just h^{-2} times the identity, and $\mathcal{T}$ is the vector of values of $\bar{\tau}_{\partial\Omega_h}(x_i)$.

It is particularly useful to have the exact solution to a representative family of problems for comparison. For the discrete Laplacian in $\mathcal{R}^d$, and Ω_h the uniformly discretized unit d-cube, we can show that:

$$\sup_{x_i \in \Omega_h} \bar{\tau}_{\partial\Omega_h}(x_i) = O(N^{2/d}). \tag{4.8}$$

Here N is the number of unknowns (grid points) in Ω_h. This is the same order of complexity as both the Jacobi and Gauss–Seidel methods for the unit d-cube and the discrete Laplacian; however, the SOR method with the optimal relaxation parameter values is but $O(N^{1/d})$ for this family of problems.

Another useful comparison of the number of iterations required for these methods to reach an acceptable level of error is to determine an expression which the iterations satisfy and analyze the convergence properties of these transient problems. A variation of this technique was proposed by Garabedian (1956) in order to choose an optimal relaxation parameter for SOR. One can associate a parabolic PDE to each of the stationary iterative methods we have considered. Equation (4.5) defines the nonlinear relationship between the "adjoint" random walk algorithm and a time dependent problem. For the ordinary Laplace equation, the stationary iterative methods are explicit finite-difference solutions to:

(4.9)

$$\begin{aligned}
u_t &= \tfrac{1}{2}\Delta u, && x \in \Omega,\, t > 0,\\
u(x, t) &= g(x), && x \in \partial\Omega,\, u(x, 0) = 0, \text{ for Jacobi},\\
u_t + \sum_{i=1}^{d} u_{x_i t} &= \Delta u, && x \in \Omega,\, t > 0,\\
u(x, t) &= g(x), && x \in \partial\Omega,\, u(x, 0) = 0,\\
& && \qquad \text{for Gauss–Seidel},\\
(1 - \omega/2)u_t + \omega/2 \sum_{i=1}^{d} u_{x_i t} &= \omega/2\Delta u, && x \in \Omega,\, t > 0,\\
u(x, t) &= g(x), && x \in \partial\Omega,\, u(x, 0) = 0, \text{ for SOR}.
\end{aligned}$$

Here ω is the relaxation parameter required for the method of SOR. It is believed that this will provide the clearest comparison of the efficiency of the "adjoint" random walk method; however, since equation (4.5) gives a nonlinear time dependent problem, it remains an open problem to quantify the rate of convergence of the solution to (4.5) to the solution of the desired elliptic BVP given in (4.1).

The final comparison of the methods will be an empirical one. The a priori analysis of the "adjoint" random walk algorithm yielded rather unpromising results, however, empirically measuring the numerical efficiency of the four algorithms compared in the previous section gives unexpectedly good convergence behavior for the "adjoint" random walk algorithm. We have implemented a massively parallel version of the "adjoint" random walk algorithm on the Connection Machine for the case of Ω_h the uniformly discretized unit cube in $d = 2$ dimensions with $N = 1024 = 32^2$ grid points, and the discrete Laplacian. We have chosen the boundary conditions to be $g(x, y) = (x + y)/2$ so that the solution to this BVP is $u(x, y) = (x + y)/2$, which is used to check the accuracy of the solution.[5] Below we will plot the L^2 norm of the error as a function of the number of iterations. In this case the L^2 norm of the error after n iterations, E^n, is given by:

$$E^n = \sqrt{\sum_{i=1}^{N} (u(x_i) - u_h^n(x_i))^2 h^2}, \tag{4.10}$$

where $u(\cdot)$ is the value of the exact solution, N is the number of grid points in Ω_h, and $u_h^n(\cdot)$ is the value of the iterative solution after n iterations. In the case of the "adjoint" random walk algorithm, $u_h^n(\cdot)$ is the value of statistic sampled with random walks up to length n.

[5] Note that $\frac{x+y}{2}$ is the exact solution to both the differential and difference equations.

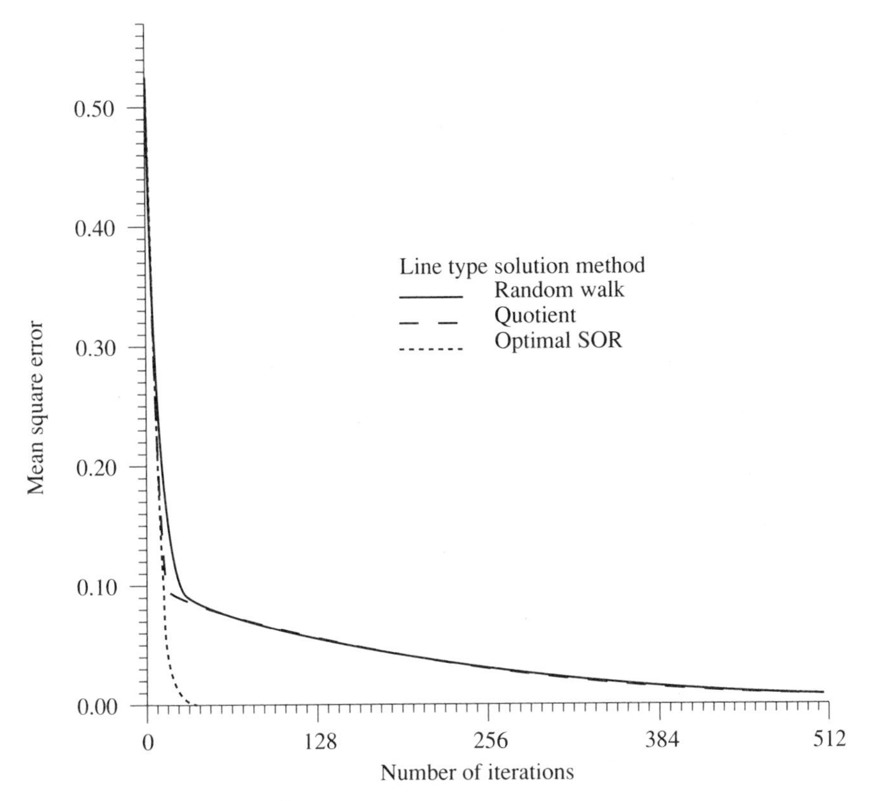

FIGURE 3. L^2 error as a function of iteration number for the Jacobi, Gauss–Seidel, and optimal SOR stationary iterative methods, and the "adjoint" random walk algorithm.

In Figure 3 we plot E^n versus n for the Jacobi, the Gauss–Seidel, SOR, and the "adjoint" random walk method. These computations were carried out on a Connection Machine, and so to maximize parallelism the Gauss–Seidel and SOR computations were executed with a red–black ordering of the unknowns. In addition, the SOR computation was carried out with the optimal choice of the relaxation parameter, ω, for a uniform discretization of the Laplacian on the unit square, see Stoer and Bulirsch (1980).

We notice in Figure 3, that for small iteration numbers the random walk algorithm performs almost as well as the optimal red–black SOR method, and performs better than both the Jacobi and Gauss–Seidel methods. In Figure 4 we further compare the behavior of SOR to the random walk algorithm by plotting these two curves along with the curve for the nonlinear deterministic algorithm which was shown to describe the evolution of the expected value of our "adjoint" random walk statistic, i.e., the quotient of explicit

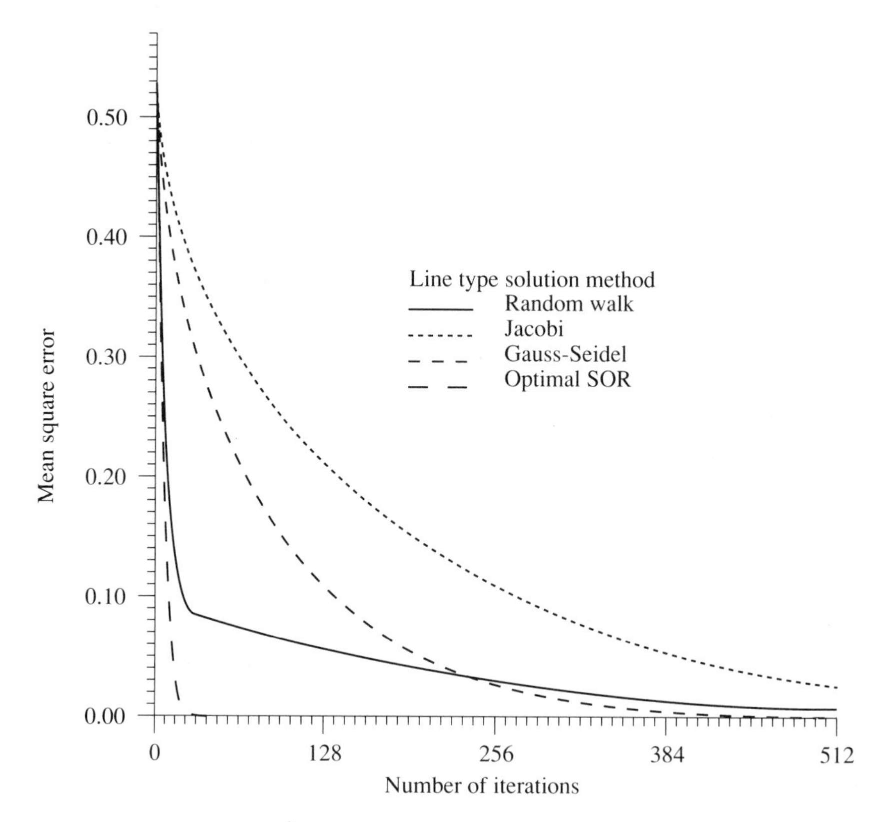

FIGURE 4. L^2 error as a function of iteration number for the optimal SOR method, the "adjoint" random walk algorithm, and the quotient of two Jacobi method solutions, which is the expected value of the Monte Carlo method.

finite-difference solutions to the problems in (4.5). In Figure 4, we refer to this nonlinear iteration as the "quotient" curve.

We see that the similarity of optimal SOR to the "quotient" curve for small iteration numbers is even more pronounced than the similarity between SOR and the "adjoint" random walk algorithm curves. Since the "quotient" curve is the expected value of the curve for the "adjoint" random walk algorithm, any discrepancy must be due to statistical sampling error. It is interesting to note that the "adjoint" random walk and "quotient" curves have an abrupt change at approximately $n = 16$ iterations. In this particular computation, 16 is the half diameter, in grid points, of Ω_h, and so is also the longest distance which information must travel from the boundary before all interior points have been visited. It appears that both the"adjoint" random walk and the "quotient" algorithms achieve an accelerated convergence due to the fact that the interior grid points which are visited less frequently have their statistic value boosted compared to more frequently visited interior grid points, since the mean, viewed as the quotient given in equation (4.5), has a relatively small denominator value. It is also important to point out that both the "adjoint" random walk and "quotient" algorithms achieve optimal SOR-like convergence behavior without the need for either a relaxation parameter and that parallelism can be exploited without the need for grid point coloration associated with the Gauss–Seidel algorithm.

Closely related to the computational complexity of an implementation of a given algorithm on a parallel computing machine is load balancing. The parallel implementation of either the"adjoint" random walk algorithm, the deterministic "quotient" algorithm, or any of the stationary iterative methods, where grid points in Ω_h are mapped onto processing elements will suffer from the initial load imbalance associated with the information flow from the boundary into the interior of Ω_h. On parallel computing machines which have relatively powerful processing elements which allow the association of several grid points per processor, we can circumvent this initial load imbalance by assigning both boundary and interior grid points to each processing element. With the "adjoint" random walk algorithm, the mapping of random walkers onto processing elements offers an implementational alternative with almost perfect a priori load balancing properties. Another interesting consequence of this parallel implementation of the "adjoint" random walk algorithm, is that regardless of the number of processors, any or all of them can be used efficiently in the computation to accumulate the statistic samples. Similarly, when we map grid points onto processors we can employ as many processors as there are grid points; however, when there are many times fewer grid points than processors, we can replicate Ω_h many times and compute statistics concurrently in the replicates, accumulating the values as is necessary.

We have presented a Monte Carlo algorithm for the evaluation of Wiener integral representations of solutions to elliptic BVPs. The algorithm is based on "adjoint" random walks, which allows the evaluation of one Wiener integral sample in $O(1)$ as opposed to $O(\tau_{\partial\Omega})$ random walk steps. The algorithm has been shown empirically to be as effective as optimal red–black SOR for small iteration numbers. It has been also shown that the "adjoint" random walk algorithm is equivalent to a deterministic algorithm which is a nonlinear iterative method. This nonlinear iterative method can be computed as the quotient of two Jacobi (explicit finite-difference) method iterations of related elliptic BVPs.

There are many possible variations to be considered of both the "adjoint" random walk and "quotient" algorithms. A variation that is applicable to both the random and deterministic algorithms is to use the algorithms in a residual correction scheme. Since both the algorithms have very desirable behavior for small iteration numbers, we can compute the solution for a small number of iterations, on the order of where the corner in Figure 4 occurs, compute the residual for problem, and restart the computation with this new residual. Similarly, we can combine these algorithms with a third algorithm in a hybrid to take advantage of the rapid initial convergence of these two algorithms by starting the computation with the random walk based algorithm, and concluding the computation with the third method.

Another area where these algorithms hold great promise is nonsymmetric elliptic BVPs. As we previously observed, when L, the continuous elliptic operator, is nonsymmetric, L_h will also be nonsymmetric. The numerical solution of nonsymmetric systems of linear equations is considerably more difficult than symmetric systems. Intuition suggests, however, that the "adjoint" random walk method should work better with nonsymmetric problems. The "random walk approximation" error associated with sampling the entire space of random walks with walks only up to length n will be smaller when L_h is nonsymmetric. This is because the nonsymmetric terms all are interpreted as sampling from random walks with drift. When we add a drift to a random walk, the mean first hitting time drops, and so long random walks become greatly less probable. Preliminary computations with nonsymmetric elliptic BVPs is in qualitative agreement with this intuition.

Another application of both these methods is the solution of systems of linear equations. Given a linear equation, $Ax = b$, where we can factor A as $A = M(I - P)$, where M is easy to invert and P is a stochastic matrix, then we can set up a random walk algorithm for the solution of $Ax = b$ based on the transition probability matrix P, Forsythe and Leibler (1950). Since P will define a random walk with absorbing states, an "adjoint" version of these random walks can be constructed, where the walkers emanate from the absorbing states, Hammersley and Handscomb (1964). These "adjoint" walks then define both the random walk and "quotient" algorithm analogs for $(I - P)x = M^{-1}b$.

A final area for future work is the construction of multigrid algorithms based on random walks. The basic limitation of the stationary iterative and the "adjoint" random walk algorithm is that information which initially resides at $\partial\Omega_h$ must be transmitted to the interior. This observation limits the optimal complexity for this class of methods to $O(n^{1/d})$. Multigrid algorithms overcome this rate of information spread by employing a variety of exponentially coarser grids to achieve a $O(\log n)$ asymptotic complexity, Brandt (1977). These same ideas may be incorporated into the random walk algorithm by allowing the random walkers to take exponentially increasing length steps and modifying the transition probabilities and statistics suitably. Also, the deterministic "quotient" algorithm may provide an interesting nonlinear smoothing procedure for the standard multigrid algorithm, as the deterministic "quotient" algorithm has SOR-like convergence behavior without the necessity of either choosing a value of ω, the relaxation parameter, or of determining a coloration of the grid for parallelism, as it is based on Jacobi iterations.

REFERENCES

A. Brandt (1977), *Multi-level adaptive solutions to boundary value problems*, Comput. Math. **31**, 333–390.

R. Courant, K. O. Friedrichs, and H. Lewy (1928), *Über die partiellen Differenzengleichungen der mathematischen Physik*, Math. Ann. **100**, 32–74.

J. H. Curtiss (1953), *Monte Carlo methods for the iteration of linear operators*, J. Mathematics and Physics **32**, 209–232.

____(1956), *A theoretical comparison of the efficiencies of two classical methods and a Monte Carlo method for computing one component of the solution of a set of linear algebraic equations*, Symposium on Monte Carlo methods, (H. A. Meyer, ed.), Wiley, New York, pp. 191-233.

M. D. Donsker and M. Kac (1951), *A sampling method for determining the lowest eigenvalue and the principle eigenfunction of Schrödinger's equation*, Journal of Research of the National Bureau of Standards **44**, 551–557.

G. E. Forsythe and R. A. Leibler (1950), *Matrix inversion by a Monte Carlo method*, Mathematical Tables and Other Aids to Computation **4**, 127–129.

M. Freidlin (1985), *Functional integration and partial differential equations*, Princeton Univ. Press, Princeton, New Jersey.

P. R. Garabedian (1956), *Estimation of the relaxation factor for small mesh size*, Mathematical Tables and Other Aids to Computation **10**, 183–185.

A. Hall (1873), *On an experimental determination of* π , Messeng. Math. **2**, 113–114.

J. M. Hammersley and D. C. Handscomb (1964), *Monte Carlo methods*, Chapman and Hall, London.

D. Hillis (1985), *The connection machine*, M.I.T. University Press, Cambridge, MA.

K. Itô and H. P. McKean (1965), *Diffusion processes and their sample paths*, Springer-Verlag, New York, Berlin, Heidelberg.

F. John (1982), *Partial differential equations*, Fourth Edition, Springer-Verlag, New York, Berlin, Heidelberg.

M. Mascagni, *A Monte Carlo method based on Wiener integration for the numerical solution of elliptic boundary value problems*, SIAM J. Sci. Statist. Comput. (submitted).

N. Metropolis, A. W. Rosenbluth, M. N. Rosenbluth, A. H. Teller, and E. Teller (1953), *Equation of state calculations by fast computing machines*, J. Chemical Physics **21**, 1087–1092.

R. D. Richtmeyer and K. W. Morton (1967), *Difference methods for initial value problems*, Second Edition, Interscience Publishers, division of John Wiley and Sons , New York, London, Sydney.

F. Spitzer (1964), *Principles of Random Walks*, D. Van Nostrand, Princeton, New Jersey.

J. Stoer and R. Bulirsch (1980), *Introduction to numerical analysis*, Springer-Verlag, New York, Berlin, Heidelberg.

S. Ulam, R. D. Richtmeyer, and J. von Neumann (1947), *Statistical methods in neutron diffusion*, Los Alamos Scientific Laboratory report LAMS-551.

N. Wiener (1923), *Differential space*, Journal of Mathematics and Physics **2**, 131–174.

MATHEMATICAL RESEARCH BRANCH, NIDDK, THE NATIONAL INSTITUTES OF HEALTH, BUILDING 31, ROOM 4B-54, BETHESDA, MARYLAND 20892 AND APPLICATIONS RESEARCH GROUP, SUPERCOMPUTING RESEARCH CENTER, 17100 SCIENCE DRIVE, BOWIE, MARYLAND 20715-4300

E-mail address: mascagni@super.org

Contemporary Mathematics
Volume **115**, 1991

Multiple Integration in Bayesian Psychometrics

ROBERT K. TSUTAKAWA

ABSTRACT. The current movement towards introducing Bayesian hierarchical models into psychometric research has created the need for efficient numerical methods to evaluate multiple integrals. The dimensions of these integrals are generally in the hundreds and the conventional quadrature methods are generally inadequate. This paper uses data from a standardized test to illustrate the type of numerical problems encountered and presents some crude approximations which are available for routine use.

Introduction

During the last decade there has been considerable interest in Bayesian and empirical Bayes techniques in psychometrics and, in particular, item response analysis. Much of the development in this area began with the introductionof the EM algorithm, which was promoted by Dempster, Laird, and Rubin (1976).

I chose to talk on this topic, not because the statistical and numerical problems to be considered are unique to psychometrics, but because they are quite typical in a large number of other areas in which statisticians have been recently working. I expect many of you will immediately recognize and recall related problems, while others of you will become more aware of the importance of multiple integration in the newly developing areas of statistics. I believe that the next big breakthrough in statistics will come from efficient tools for numerical multiple integration and not from the theory of statistics.

In recent years educational researchers and testing agencies have made extensive use of psychometric models to analyze data from standardized tests such as Scholastic Aptitude Test (SAT), Graduate Record Examination (GRE), those by the American College Testing Program (ACT), etc. One example of such a model is the three-parameter logistic curve (3PL) proposed by Birnbaum (1968). It represents the probability that an examinee with real

1980 *Mathematics Subject Classification* (1985 *Revision*). Primary 62F15, 62P15, 65D30.
Key words and phrases. Item response theory, multi-dimensional integration.
The final version of this paper has been submitted for publication elsewhere.

valued ability θ in some given area will respond to a given test item correctly and is defined by

$$p_\xi(\theta) = c + \frac{1-c}{1+\exp\{-a(\theta - b)\}}, \tag{1}$$

where $\xi = (a, b, c)$ is called the item parameter, $0 < a < \infty$ (discrimination parameter), $-\infty < b < \infty$ (difficulty parameter), $0 < c < 1$ (guessing parameter), and $-\infty < \theta < \infty$. The parameter space for ξ will be denoted by Ω. As defined here there is a nonuniqueness problem in the parameterization. A constant may be added to θ and to b without changing the value $p_\xi(\theta)$. Similarly a constant may be multiplied to a and then θ and b may be divided by the same constant without changing the value of $p_\xi(\theta)$. This problem may be easily avoided by the use of a suitable prior distribution for θ and need not concern us at this time.

Estimation of item parameter

Now suppose we have a K-item test with parameters $\underset{\sim}{\xi} = (\xi_1, \dots, \xi_K)$. Let $\underset{\sim}{x} = (x_1, \dots, x_K)$ denote binary responses to the K items where

$$x_i = \begin{cases} 1 & \text{if the response to item } i \text{ is correct,} \\ 0 & \text{if the response to item } i \text{ is incorrect.} \end{cases}$$

The assumption of local independence, commonly made among psychometricians, states that the joint probability function of $\underset{\sim}{x}$ satisfies

$$P(\underset{\sim}{x} \mid \underset{\sim}{\xi}, \theta) = \prod_{j=1}^{K} p_{\xi_j}(\theta)^{x_j}[1 - p_{\xi_j}(\theta)],^{1-x_j} \tag{2}$$

where $\underset{\sim}{\xi} = (\xi_1, \dots, \xi_K)$ are parameters for the K items. When the K-item test is given to a sample of n individuals, with abilities $\underset{\sim}{\theta} = (\theta_1, \dots, \theta_n)$, let $y_{ij} = 0$ or 1 if the ith individual's response to item j is incorrect or correct, respectively, and let

$$\underset{\sim}{y} = \begin{pmatrix} y_{11} & \cdots & y_{1K} \\ \vdots & & \vdots \\ y_{nl} & \cdots & y_{nK} \end{pmatrix}. \tag{3}$$

Then the joint probability function of $\underset{\sim}{y}$ is

$$P(\underset{\sim}{y} \mid \underset{\sim}{\xi}, \underset{\sim}{\theta}) = \prod_{i=1}^{n} \prod_{j=1}^{K} p_{ij}^{y_{ij}} (1 - p_{ij})^{1-y_{ij}}, \tag{4}$$

where $p_{ij} = p_{\xi_j}(\theta_i)$. The following problem is considered in this talk.

Given $\underset{\sim}{y}$, what can we say about a new individual's θ whose response is $\underset{\sim}{x} = (x_1, \dots, x_K)$?

Given a population distribution with pdf ϕ for the θ's, the marginal probability function of $\underset{\sim}{y}$ is

$$(5) \qquad P(\underset{\sim}{y} \mid \underset{\sim}{\xi}) = \prod_{i=1}^{n} \int \left(\prod_{j=1}^{K} p_{ij}^{y_{ij}} (1 - p_{ij})^{1-y_{ij}} \right) \phi(\theta_i)\, d\theta_i.$$

When ϕ is assumed to be a normal distribution, then without loss of generality we may restrict it to be $N(0, 1)$. With this restriction, the nonuniqueness of the parameterization in (1) which was mentioned earlier vanishes.

In a fully Bayesian model we assume some prior distribution $p(\underset{\sim}{\xi})$ on $\underset{\sim}{\xi}$. If the item parameters have independent priors, $p(\underset{\sim}{\xi}) = \prod_{j=1}^{K} p(\xi_j)$, and the joint posterior of $(\underset{\sim}{\xi}, \underset{\sim}{\theta})$ is

$$p(\underset{\sim}{\xi}, \underset{\sim}{\theta} \mid \underset{\sim}{y}) \propto \left\{ \prod_{i=1}^{n} \phi(\theta_i) \right\} \left\{ \prod_{j=1}^{K} p(\xi_j) \right\} \left\{ \prod_{i=1}^{n} \prod_{j=1}^{K} P(y_{ij} | \theta_i, \xi_j) \right\},$$

where $P(y_{ij} | \theta_i, \xi_j) = p_{ij}^{y_{ij}} (1 - p_{ij})^{1-y_{ij}}$ and the marginals of $\underset{\sim}{\xi}$ and $\underset{\sim}{\theta}$ are, respectively,

$$(7) \qquad p(\underset{\sim}{\xi} \mid \underset{\sim}{y}) \propto \left\{ \prod_{j=1}^{K} p(\xi_j) \right\} \prod_{i=1}^{n} \int \left[\prod_{j=1}^{K} P(y_{ij} | \theta_i, \xi_j) \right] \phi(\theta_i)\, d\theta_i$$

and

$$(8) \qquad p(\underset{\sim}{\theta} \mid \underset{\sim}{y}) \propto \left\{ \prod_{i=1}^{n} \phi(\theta_i) \right\} \prod_{j=1}^{K} \int \left[\prod_{i=1}^{n} P(y_{ij} | \theta_i, \xi_j) \right] p(\xi_j)\, d\xi_j.$$

Note that the integrals in (7) are one-dimensional, whereas those in (8) would be three-dimensional for the 3PL model. Although the posterior means of these parameters can be expressed in terms of these pdf's, the numerical evaluations require computing very high-dimensional integrals, for which techniques are lacking at this time. For example, if there are $K = 40$ items, the dimension would be 121 for evaluating the posterior means of the item parameters for the 3PL model. The current Bayesian practice is to avoid the integration and compute the posterior mode. This is commonly done in variance components analysis as seen in Lindley and Smith (1972). When the EM algorithm is used to compute the posterior mode of $\underset{\sim}{\xi}$, the maximization with respect to the $3K$ parameters can be reduced to a series of maximizations with respect to 3 parameters at a time. (See Bock and Aitkin (1981) and Tsutakawa and Lin (1986) for further details.) Although single integrals must still be evaluated by some approximation, this is not a major computational problem.

Bayesian estimation of θ given x

First suppose that $\underset{\sim}{\xi}$ is known. The posterior mean and variance of θ for a given individual whose response is $\underset{\sim}{x}$ is

$$\mu_{\underset{\sim}{\xi}} = E(\theta | \underset{\sim}{x}, \underset{\sim}{\xi}) = \int_{-\infty}^{\infty} \theta p(\theta | \underset{\sim}{x}, \underset{\sim}{\xi}) \, d\theta \tag{9}$$

and

$$\sigma^2_{\underset{\sim}{\xi}} = \text{Var}(\theta | \underset{\sim}{x}, \underset{\sim}{\xi}) = \int_{-\infty}^{\infty} (\theta - \mu_{\underset{\sim}{\xi}})^2 p(\theta | \underset{\sim}{x}, \underset{\sim}{\xi}) \, d\theta, \tag{10}$$

where

$$p(\theta | \underset{\sim}{x}, \underset{\sim}{\xi}) \propto P(\underset{\sim}{x} | \underset{\sim}{\xi}, \theta) \phi(\theta), \tag{11}$$

the posterior pdf of θ given $\underset{\sim}{x}$.

Suppose next that $\underset{\sim}{\xi}$ is unknown. Then given $\underset{\sim}{x}$ and the previous data $\underset{\sim}{y}$, the posterior mean of θ is given by

$$\mu = E(\theta | \underset{\sim}{x}, \underset{\sim}{y}) = \int_{-\infty}^{\infty} \theta p(\theta | \underset{\sim}{x}, \underset{\sim}{y}) \, d\theta, \tag{12}$$

where $p(\theta | \underset{\sim}{x}, \underset{\sim}{y})$ is the posterior pdf of θ, which may be expressed as

$$p(\theta | \underset{\sim}{x}, \underset{\sim}{y}) = \int p(\theta | \underset{\sim}{x}, \underset{\sim}{y}, \xi) p(\underset{\sim}{\xi} | \underset{\sim}{x}, \underset{\sim}{y}) \, d\underset{\sim}{\xi}, \tag{13}$$

$$p(\underset{\sim}{\xi} | \underset{\sim}{x}, \underset{\sim}{y}) = \frac{P(\underset{\sim}{x} | \underset{\sim}{\xi}) P(\underset{\sim}{y} | \underset{\sim}{\xi}) p(\underset{\sim}{\xi})}{P(\underset{\sim}{x} | \underset{\sim}{y}) P(\underset{\sim}{y})} = \frac{P(\underset{\sim}{x} | \underset{\sim}{\xi}) p(\underset{\sim}{\xi} | \underset{\sim}{y})}{P(\underset{\sim}{x} | \underset{\sim}{y})}, \tag{14}$$

and

$$p(\theta | \underset{\sim}{x}, \underset{\sim}{y}, \underset{\sim}{\xi}) = P(\underset{\sim}{x} | \theta, \underset{\sim}{\xi}) \phi(\theta) / P(\underset{\sim}{x} | \underset{\sim}{\xi}). \tag{15}$$

The posterior variance of θ,

$$\sigma^2 = \int (\theta - \mu)^2 P(\theta | \underset{\sim}{x}, \underset{\sim}{y}) \, d\theta, \tag{16}$$

may be similarly expressed.

In order to appreciate the Bayesian approach to measuring the uncertainty in the unknown θ, it is interesting to look at a decompostion of the variance in light of the other methods of approximating the uncertainty. Using a well-known variance decompostion (DeGroot, 1986) and the conditional independence of $(\theta, \underset{\sim}{x})$ and $\underset{\sim}{y}$ given $\underset{\sim}{\xi}$ it is easily shown that

$$V(\theta|\underset{\sim}{x}, \underset{\sim}{y}) = E\{V(\theta|\underset{\sim}{\xi}, \underset{\sim}{x}, \underset{\sim}{y})|\underset{\sim}{x}, \underset{\sim}{y}\} + V\{E(\theta|\underset{\sim}{\xi}, \underset{\sim}{x}, \underset{\sim}{y})|\underset{\sim}{x}, \underset{\sim}{y}\} \tag{17}$$

$$= \int V(\theta|\underset{\sim}{\xi}, \underset{\sim}{x}) p(\underset{\sim}{\xi}|\underset{\sim}{x}, \underset{\sim}{y}) d\underset{\sim}{\xi} \tag{18}$$

$$+ \int \{E(\theta|\underset{\sim}{x}, \underset{\sim}{\xi}) - E(\theta|\underset{\sim}{x}, \underset{\sim}{y})\}^2 p(\underset{\sim}{\xi}|\underset{\sim}{x}, \underset{\sim}{y})\, d\underset{\sim}{\xi}.$$

Under current practices (Lord, 1980), θ is estimated by first computing the joint maximum likelihood estimate $(\hat{\underset{\sim}{\xi}}_{ML}, \hat{\underset{\sim}{\theta}}_{ML})$ of $(\underset{\sim}{\xi}, \underset{\sim}{\theta})$ from $\underset{\sim}{y}$ and then estimating θ of a new individual by the maximum likelihood estimate, $\hat{\theta}$, of θ given $\underset{\sim}{x}$, while assuming that $\underset{\sim}{\xi} = \hat{\underset{\sim}{\xi}}_{ML}$. The variance of θ is then approximated by the reciprocal information evaluated at $\theta = \hat{\theta}$, while treating $\underset{\sim}{\xi}$ as again known and equal to $\hat{\underset{\sim}{\xi}}_{ML}$. The method thus fails to account for the uncertainty in the estimated $\underset{\sim}{\xi}$.

On the other hand the standard empirical Bayes method is to first compute the marginal maximum likelihood estimate $\hat{\underset{\sim}{\xi}}_{MML}$, which maximizes (7) with respect to $\underset{\sim}{\xi}$ and then computes the posterior mean and variance of θ given $\underset{\sim}{x}$, conditionally on $\underset{\sim}{\xi} = \hat{\underset{\sim}{\xi}}_{MML}$. This approach is a first order approximation to $V(\theta|\underset{\sim}{x}, \underset{\sim}{y})$ since it estimates the first term in (18) by its integrand evaluated at $\underset{\sim}{\xi} = \hat{\underset{\sim}{\xi}}_{MML}$. It makes no attempt to approximate the second term in (18).

For the Bayesian approach, consider the representation of the posterior mean of θ as the posterior expectation of a function of $\underset{\sim}{\xi}$; namely,

$$E(\theta|\underset{\sim}{x}, \underset{\sim}{y}) = \int w(\underset{\sim}{\xi}) p(\underset{\sim}{\xi}|, \underset{\sim}{x}, \underset{\sim}{y})\, d\underset{\sim}{\xi}, \tag{19}$$

where

$$w(\underset{\sim}{\xi}) = E(\theta|\underset{\sim}{x}, \underset{\sim}{y}, \underset{\sim}{\xi}). \tag{20}$$

Although $w(\underset{\sim}{\xi})$ and its derivatives are single integrals, they are relatively easy to work with numerically.

One approximation to the posterior expectation of $w(\underset{\sim}{\xi})$ is that of Lindley (1980), which is given by

$$E(\theta|\underset{\sim}{x}, \underset{\sim}{y}) \simeq w + \tfrac{1}{2}\Sigma w_{rs}\tau_{rs} + \tfrac{1}{2}\Sigma\Lambda_{rst} w_u \tau_{rs}\tau_{tu}, \tag{21}$$

where w, w_u, w_{rs} are values of $w(\underset{\sim}{\xi})$ and its first two partial derivatives evaluated at the posterior mode $\hat{\underset{\sim}{\xi}}$, Λ_{rst} the third partial derivatives of the log posterior at $\hat{\xi}$, and τ_{rs}, the covariance of $\underset{\sim}{\xi}$ approximated by the inverse of the matrix of negative second partials of the log posterior evaluated at $\hat{\underset{\sim}{\xi}}$.

If the posterior distribution of $\underset{\sim}{\xi}$ is normal with mean $\hat{\underset{\sim}{\xi}}$ and covariances τ_{rs}, as it will approximately be for large n, then $\Lambda_{rst} \equiv 0$ and the approximation reduces to

$$E(\theta | \underset{\sim}{x}, \underset{\sim}{y}) \simeq w + \tfrac{1}{2}\Sigma w_{rs}\tau_{rs}. \tag{22}$$

A heuristic justification of this may be seen by first approximating $w(\underset{\sim}{\xi})$ by the Taylor series approximation,

$$\begin{aligned} w(\underset{\sim}{\xi}) &\simeq w + \Sigma w_r(\mu_r - \hat{\mu}_r) + \tfrac{1}{2}\Sigma w_{rs}(\hat{\mu}_r - \hat{\mu}_r)(\mu_s - \hat{\mu}_s) \\ &+ \tfrac{1}{6}\Sigma w_{rst}(\mu_r - \hat{\mu}_r)(\mu_s - \hat{\mu}_s)(\mu_t - \hat{\mu}_t), \end{aligned} \tag{23}$$

where $(\mu, \ldots, \mu_m) = \underset{\sim}{\xi}$, $(\hat{\mu}_1, \ldots, \hat{\mu}_m) = \hat{\underset{\sim}{\xi}}$ and w_{rst} the third partials of $w(\underset{\sim}{\xi})$ evaluated at $\hat{\xi}$. If the posterior is normal as described above, the third moments of $\underset{\sim}{\xi}$ vanish and (22) follows. The posterior variance of θ may be similarly approximated by replacing $w(\underset{\sim}{\xi})$ in (19) by

$$w(\xi) = E\{(\theta - \overline{\theta})^2 | \underset{\sim}{\xi}, \underset{\sim}{x}, \underset{\sim}{y}), \tag{24}$$

where $\overline{\theta}$ is the approximation to $E(\theta | \underset{\sim}{x}, \underset{\sim}{y})$ given by (22). Mosteller and Wallace (1964) refer to this type of approximation as the standard one and caution the reader that it fails to account for the bias of $\hat{\underset{\sim}{\xi}}$ as an approximation to the posterior mean and the skewness of the posterior distribution. More recently Tierney and Kadane (1986) have proposed an approximation to the posterior moment which does not require the evaluation of Λ_{rst}. I will not present the details here, but merely point out what seems to be required for posterior expectation of θ. This expectation can be represented as

$$E(\theta|x, y) = \int E(\theta|x, \xi)P(x, y|\xi)p(\xi)d\xi \Big/ \int P(\underset{\sim}{x}, \underset{\sim}{y} | \underset{\sim}{\xi})p(\underset{\sim}{\xi})d\underset{\sim}{\xi}, \tag{25}$$

where I have used the fact that $E(\theta | \underset{\sim}{x}, \underset{\sim}{y}, \underset{\sim}{\xi}) = E(\theta | \underset{\sim}{x}, \underset{\sim}{\xi})$ due to the conditional independence of $(\underset{\sim}{x}, \theta)$ and $\underset{\sim}{y}$ given $\underset{\sim}{\xi}$.

The Tierney and Kadane approximation requires the separate maximization of

$$P(\underset{\sim}{x}, \underset{\sim}{y} | \underset{\sim}{\xi})p(\underset{\sim}{\xi}) \tag{26}$$

and

$$E(\theta | \underset{\sim}{x}, \underset{\sim}{\xi}) P(\underset{\sim}{x}, \underset{\sim}{y} | \underset{\sim}{\xi}) p(\underset{\sim}{\xi}), \tag{27}$$

with respect to $\underset{\sim}{\xi}$. Although the maximization of (26) may be performed by using the EM algorithm as implemented in Tsutakawa and Lin (1986), the maximization of (26) is quite difficult since it must be done simultaniously with respect to all components of $\underset{\sim}{\xi}$.

Numerical illustration

Now I would like to turn to a numerical example comparing the Bayesian estimate (22) to the more conventional maximum likelihood type estimates. Three data sets are necessary for this illustration. The first set consists of 2,000 cases from a 1981 American College Testing Service (ACT) 40-item math test. This set is used to formulate a prior distribution for $\underset{\sim}{\xi}$. The second set consists of $n = 400$ cases from a 1987 ACT 40-item math test. This set is used for calibration, or the estimation of $\underset{\sim}{\xi}$. The third set consists of 100 additional cases from the 1987 test and is used to estimate 100 θ values.

For the prior distributions, $\xi_1, \ldots, \xi_K$ are assumed mutually independent and identically distributed, according to a distribution induced by a prior distribution on $\underset{\sim}{p} = (p_1, p_2, p_3)$, the height of the unknown 3PL curve at 3 predetermined points, $t_1 < t_2 < t_3$. Thus p_j is the probability of correct response to the item by an individual whose ability θ equals t_j.

In particular, consider the pdf of a constrained ordered Dirichlet distribution (Wilks, 1961) defined by

$$\begin{aligned} &g(p_1, p_2, p_3) \\ &= \text{constant} \times \frac{\Gamma(N)}{\prod_{s=1}^{4} \Gamma(\pi_s N)} p_1^{\pi_1 N - 1} (p_2 - p_1)^{\pi_2 N - 1} (p_3 - p_2)^{\pi_3 N - 1} (1 - p_3)^{\pi_4 N - 1} \end{aligned}$$

if $(p_1, p_2, p_3) \in \mathscr{P}$ and 0 otherwise, where $\pi_j > 0$, $\Sigma_{j=1}^{4} \pi_j = 1$, and $\mathscr{P}$ is defined as the set of all $\underset{\sim}{p} = (p_1, p_2, p_3)$ such that there exists some $\xi \in \Omega$ for which the 3PL curve passes through $\underset{\sim}{p}$. The distribution is uniquely determined by its hyperparameter (π_1, π_2, π_3, N). It can be shown that there exists a 1-1 continuous transformation between the parameter space Ω and $\mathscr{P}$, defined through the relation

$$p_j = c + (1 - c)/1 + \exp\{-a(t_j - b)\},$$

$j = 1, 2, 3$. The value of the hyperparameters were chosen to be identical for all items andspecified by $(\pi_1, \pi_2, \pi_3, N) = (.217, .441, .835, 9.8)$ with

TABLE 1. Summary of posterior distribution of item parameters for $n = 400$.

Item	Item score	Posterior mean b	c	d	Posterior SD b	c	d	Posterior correlation bc	bd	cd
1	327	−1.43	0.01	0.10	0.87	0.53	0.21	0.97	0.81	0.69
2	325	−0.02	0.56	0.69	0.31	0.09	0.28	0.89	0.77	0.69
3	304	−0.77	0.12	0.32	0.53	0.28	0.22	0.96	0.86	0.78
4	304	−0.05	0.41	1.42	0.11	0.06	0.23	0.67	0.62	0.47
5	298	−0.17	0.36	0.62	0.31	0.13	0.25	0.92	0.82	0.76
6	294	−0.61	0.05	0.63	0.30	0.19	0.20	0.94	0.85	0.78
7	279	0.07	0.35	0.96	0.14	0.07	0.20	0.77	0.61	0.53
8	276	−0.34	0.13	0.57	0.27	0.14	0.21	0.93	0.82	0.76
9	274	0.22	0.39	1.16	0.12	0.05	0.22	0.69	0.55	0.49
10	274	−0.31	0.15	0.48	0.31	0.16	0.21	0.94	0.82	0.77
11	265	−0.49	0.00	0.39	0.34	0.19	0.20	0.95	0.84	0.79
12	254	0.74	0.47	0.74	0.17	0.05	0.27	0.71	0.38	0.58
13	253	0.13	0.26	0.57	0.19	0.08	0.20	0.85	0.64	0.63
14	245	0.18	0.24	0.52	0.23	0.10	0.24	0.90	0.74	0.75
15	240	0.27	0.25	0.29	0.33	0.12	0.27	0.93	0.76	0.78
16	234	0.45	0.31	0.98	0.12	0.05	0.22	0.70	0.50	0.56
17	225	0.13	0.11	0.91	0.11	0.05	0.16	0.72	0.55	0.58
18	224	0.12	0.10	1.01	0.11	0.06	0.18	0.73	0.61	0.65
19	216	0.55	0.28	0.87	0.13	0.06	0.23	0.74	0.51	0.63
20	215	0.42	0.22	0.77	0.14	0.06	0.21	0.77	0.56	0.65
21	213	0.42	0.22	0.98	0.11	0.05	0.21	0.70	0.51	0.61
22	211	0.44	0.21	0.65	0.16	0.07	0.22	0.82	0.58	0.70
23	205	0.37	0.16	1.07	0.09	0.04	0.18	0.64	0.42	0.54
24	195	0.70	0.24	0.61	0.15	0.06	0.25	0.75	0.44	0.69
25	195	0.33	0.09	0.60	0.14	0.06	0.18	0.79	0.55	0.68
26	189	1.03	0.28	0.39	0.20	0.06	0.28	0.68	0.25	0.69
27	187	0.46	0.11	0.49	0.15	0.07	0.20	0.81	0.53	0.70
28	184	0.69	0.21	0.85	0.11	0.05	0.22	0.63	0.32	0.59
29	180	0.59	0.16	0.76	0.12	0.05	0.21	0.71	0.42	0.66
30	178	0.83	0.23	0.73	0.13	0.05	0.23	0.62	0.23	0.61
31	174	0.43	0.00	−0.12	0.56	0.21	0.31	0.97	0.83	0.88
32	172	0.52	0.10	0.80	0.10	0.05	0.18	0.66	0.39	0.64
33	168	1.05	0.29	1.24	0.09	0.03	0.24	0.31	−0.11	0.33
34	166	0.60	0.12	1.05	0.08	0.03	0.16	0.44	0.08	0.36
35	160	0.99	0.21	0.67	0.14	0.06	0.30	0.63	0.32	0.77
36	151	1.87	0.30	0.41	0.30	0.04	0.35	0.01	−0.65	0.52
37	149	0.98	0.20	0.99	0.10	0.03	0.23	0.41	−0.04	0.50
38	147	0.68	0.10	1.18	0.07	0.03	0.16	0.32	−0.03	0.29
39	144	0.79	0.12	0.93	0.09	0.03	0.18	0.44	−0.01	0.47
40	133	0.88	0.09	0.72	0.10	0.04	0.20	0.46	0.00	0.61

$(t_1, t_2, t_3) = (-1.28, 0, 1.28)$, the 10, 50 and 90 percentage points of the $N(0, 1)$ distribution. (Details can be found in Tsutakawa, 1988.)

Table 1 gives a summary of the approximate posterior mean and covariances of $\underset{\sim}{\xi}$ using the above prior on $n = 400$ cases. The table does not include the covariances between items which were used in the computation

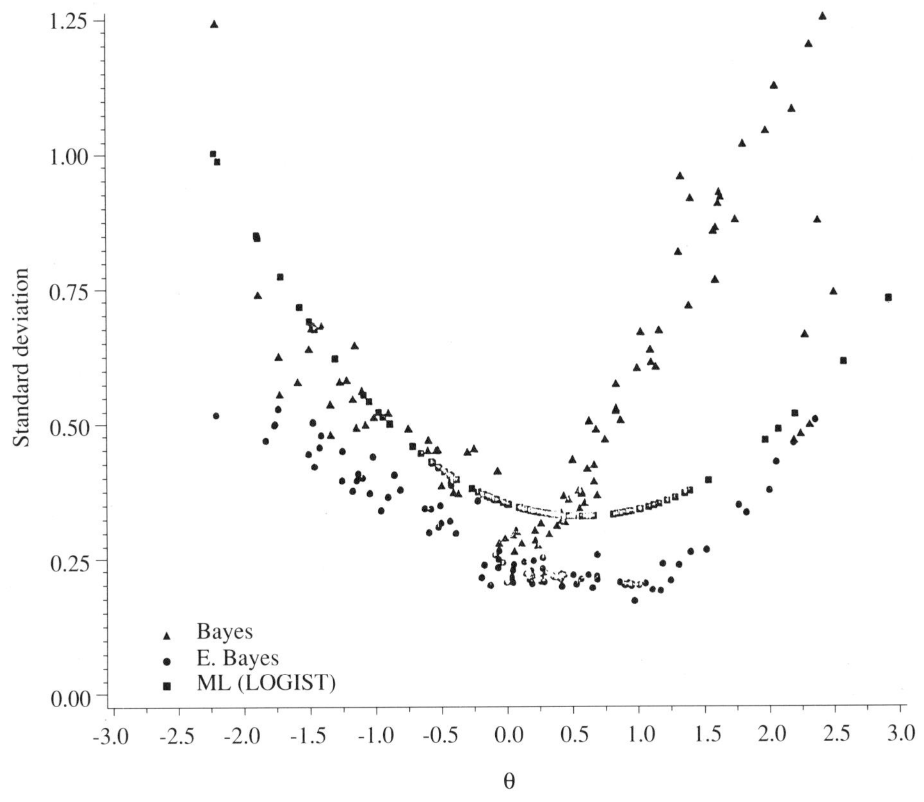

FIGURE 1. Estimated θ and standard deviations under three procedures for $n = 400$.

but too numerous to tabulate. Moreover, the parameter $d = \log(a)$ was used rather than a in order to enhance the asymptotic normality of the posterior.

Figure 1 shows plots of the estimates of the $100\,\theta$'s versus their standard deviation based on maximum likelihood (Wingersky, Barton, and Lord, 1982), empirical Bayes and the Bayesian approximation. The three methods give similiar results for the lower and central values of θ. Note, however, that the standard deviations under Bayes are much larger than those under the other methods for the larger estimated values of θ. Figures 2 and 3, on pp. 84 and 85, compare the maximum likelihood estimate with the Bayes approximation of θ and the empirical Bayes with the Bayes approximation, respectively.

To see the effect of the underestimating the standard deviation on the interval estimate, Figures 4 and 5, on pp. 86 and 87, compare interval estimates of the first 50 of the 100 cases. We note the obvious tendency of the maximum likelihood and empirical Bayes approaches to underestimate the standard deviation. Most of this is due to treating $\underset{\sim}{\xi}$ as if it were known and ignoring its uncertainty.

Additional comparisons including a case using $n = 1000$ for the calibration appear in Tsutakawa and Johnson (1990).

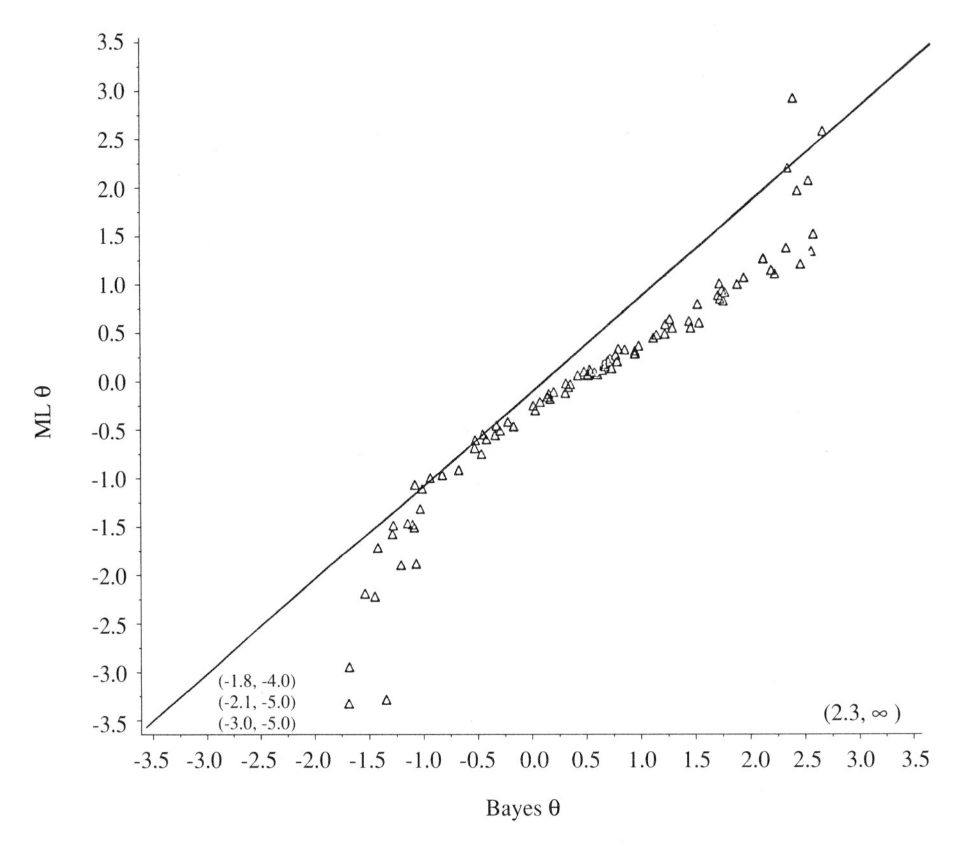

FIGURE 2. Maximum likelihood vs. Bayes estimates of θ for $n = 400$.

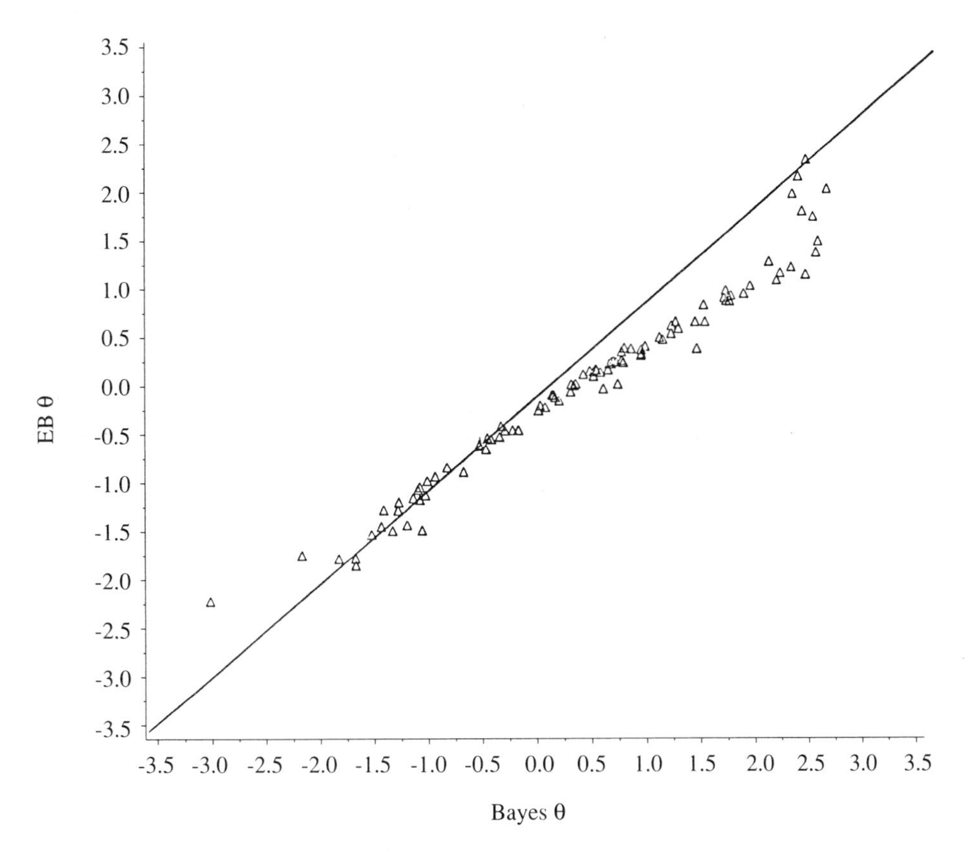

FIGURE 3. Empirical Bayes vs. Bayes estimates of θ for $n = 400$.

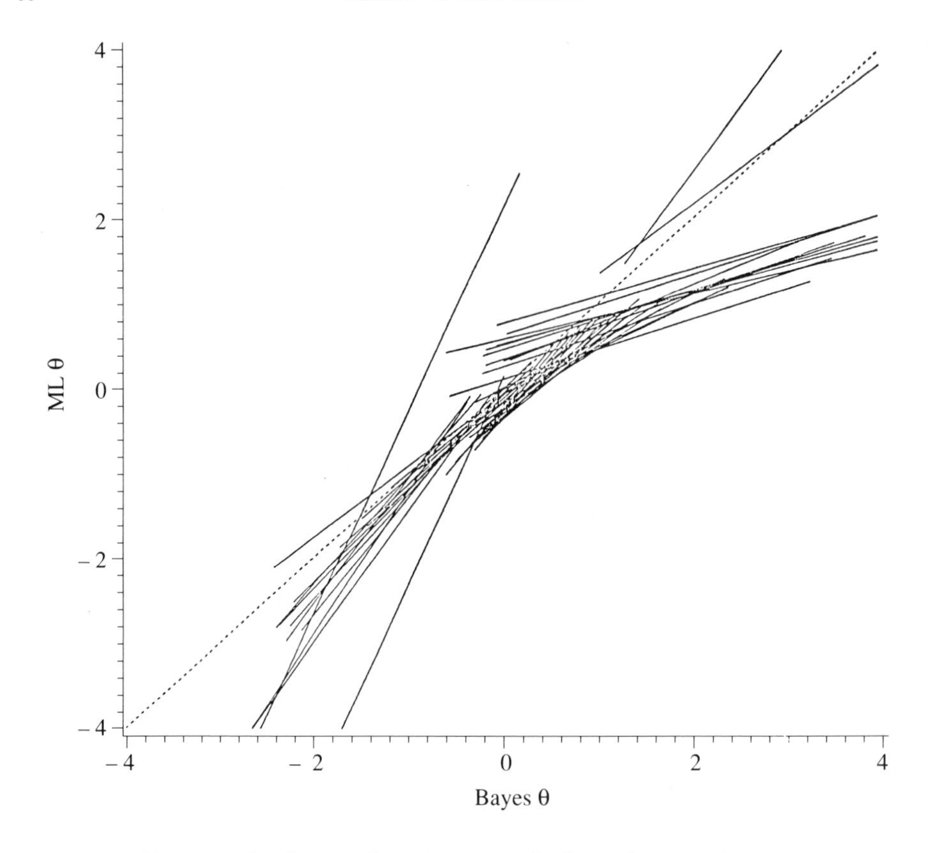

FIGURE 4. Interval estimates of θ under maximum likelihood vs. Bayes for 50 examinees for $n = 400$.

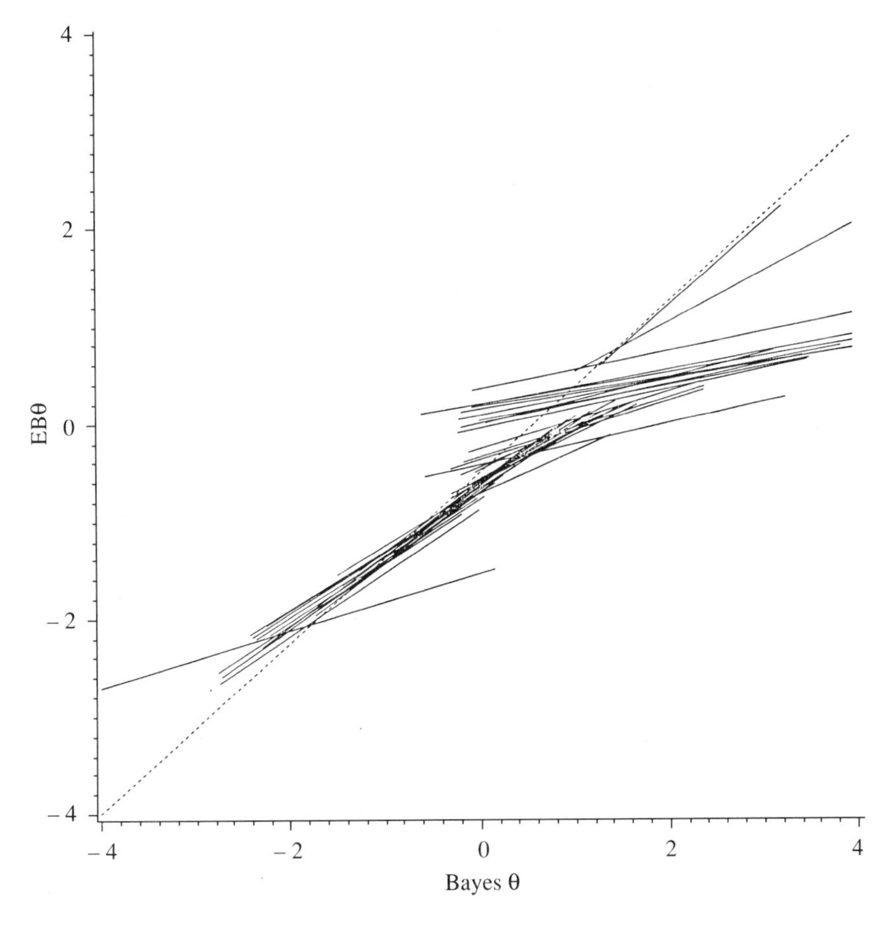

FIGURE 5. Interval estimates of θ under empirical Bayes vs. Bayes for 50 examinees using $n = 400$.

Concluding remarks

I tried to present a real problem with large scales social implications. Due to large random errors that can be expected in measuring abilities, it is important that these errors be properly assessed.

I have pointed out where evaluations of multiple integrals are needed in Bayesian solutions of a psychometric problem. The approximations illustrated here are indeed very crude. We need approximations that are not only more accurate, but more computationally-efficient for use on a routine basis of the type encountered in standarized testing.

References

A. Birnbaum (1968), *Some latent trait models and their use in inferring an examinee's ability*, Statistical Theories of Mental Test Scores (F. Lord and M. Novick, eds.), Addison-Wesley, Reading, MA, pp. 395–479.

R. D. Bock and M. Aitkin (1981), *Marginal maximum likelihood estimation of item parameters: An application of an EM algorithm*, Psychometrika **46**, 443–459.

M. H. DeGroot (1986), *Probability and statistics*, 2nd ed. Addison-Wesley, Reading, MA.

A. P. Dempster, N. M. Laird and D. B. Rubin (1977), *Maximum likelihood from incomplete data via the EM algorithm (with discussion)*, J. Roy. Stat. Soc. Ser. B **39**, 1–38.

D. V. Lindley (1980), *Approximate Bayesian methods*, Trabajos Estadistica **31**, 223–237.

D. V. Lindley, and A. F. M. Smith (1972), *Bayes estimates for the linear model (with discussion)*, J. Roy. Stat. Soc. Ser. B **34**, 1–41.

F. M. Lord (1980), *Applications of item response theory to practical testing problems*, Erlbaun, Hillsdale, NJ.

F. Mosteller and D. L. Wallace (1964), *Inference and disputed authorship: The federalist*, Addison-Wesley, Reading, MA.

L. Tierney and J. B. Kadane (1986), *Accurate approximations for posterior moments and marginal densities*, J. Amer. Stat. Assoc. **81**, 82–86.

R. K. Tsutakawa (1988), *Dirichlet prior in Bayesian estimation of item response curves*, Mathematical Sciences Technical Report no. 143, Department of Statistics, University of Missouri, Columbia, MO.

R. K. Tsutakawa and J. C. Johnson (1990), *The effect of the uncertainty of item parameter estimation on ability estimates*, Psychometrika **55**, 371–390.

R. K. Tsutakawa and H. Y. Lin (1986), *Bayesian estimation of item response curves*. Psychometrika **51**, 251–267.

S. S. Wilks (1962), *Mathematical statistics*, Wiley, New York.

M. S. Wingersky, M. A. Barton, and F. M. Lord (1982), *LOGIST Users Guide*, Educational Testing Service, Princeton, NJ.

Department of Statistics University of Missouri, Columbia, Missouri 65211

Contemporary Mathematics
Volume **115**, 1991

Laplace's Method in Bayesian Analysis

ROBERT E. KASS, LUKE TIERNEY, AND JOSEPH B. KADANE

ABSTRACT. We summarize results on asymptotic approximation of integrals arising in Bayesian statistics. The primary motivation for this work is the need for integration methods that can be applied in interactive Bayesian data analysis. The approximations we discuss can be easily and rapidly computed and are often sufficiently accurate to be useful.

Introduction

Among integration problems in general, many arising in Bayesian statistics have important distinctive features: they involve integrands that are likely to be smooth and to have a single dominant peak in a region that is roughly known. In these cases, it is desirable to employ computational methods that make use of this information, and thereby gain efficiency. Such methods may fail when the integrands in question turn out to be very poorly behaved, but in that situation statistical inference would usually be problematic even if correct values of integrals were available. An additional consideration in choosing appropriate integration strategies is that the results need not be extremely accurate; computational error need only be small relative to the inherent statistical uncertainty that enters the process of drawing inferences from data.

Even for this special class of problems it is unlikely that there would emerge a single best method. One reason is that there is a trade-off between speed and accuracy, so the choice of integration technique will depend on the time one might be willing to wait for an answer and the degree of accuracy required. Our work has been oriented toward creation of a computational environment in which Bayesian data analysis could proceed interactively. By this we mean an environment in which it would be easy for a data analyst to

1980 *Mathematics Subject Classification* (1985 *Revision*). Primary 62F15; Secondary 41A60.
Key words and phrases. Asymptotic expansions, interactive data analysis, posterior distributions.
This work was supported by the National Science Foundation under grant NSF/DMS-8705646.
This paper is in final form and no version of it will be submitted for publication elsewhere.

0271-4132/91 $1.00 + $.25 per page

"fiddle" with the specification of the problem, that is, to consider such things as alterations in the statistical model, transformations of variables, changes in the prior probability distribution, or effects of omitting certain data points. For this purpose it is necessary to have a variety of integration tools available, but because many integrals will need to be computed repetitively after modifications of integrands are made, speed will be important.

In this paper we provide a sketch of efforts made to obtain approximate solutions to Bayesian integration problems through the use of asymptotic expansions. Our purpose is to describe available methodology briefly, rather than comprehensively. In particular, we are omitting many references. Additional references may be found in Kass, Tierney, and Kadane (1988) and Tierney, Kass, and Kadane (1989b), from which much of the material here is taken.

Laplace's method. Assuming h is a smooth function of an m-dimensional parameter θ with $-h$ having a maximum at $\hat{\theta}$, Laplace's method approximates an integral of the form

$$I = \int f(\theta)\exp(-nh(\theta))d\theta \tag{1.1}$$

by expanding h and f about $\hat{\theta}$. Considering first the case in which θ is one-dimensional, the resulting factor $\exp(-nh''(\hat{\theta})(\theta-\hat{\theta})^2)$ in the integrand is proportional to a Normal density; when integrated against this Normal density, the order $O(n^{-1/2})$ terms of the expansion of f and $\exp(-nh)$, which are odd functions of $\theta-\hat{\theta}$, vanish and the integral satisfies

$$I = \hat{I}\cdot\{1+0(n^{-1})\},$$

where

$$\hat{I} = (2\pi/n)^{1/2}\sigma f(\hat{\theta})\exp(-nh(\hat{\theta}))$$

and $\sigma = h''(\hat{\theta})^{-1/2}$. In the m-dimensional case,

$$\hat{I} = (2\pi/n)^{m/2}\det(\Sigma)^{1/2} f(\hat{\theta})\exp(-nh(\hat{\theta}))$$

and $\Sigma^{-1} = D^2h(\hat{\theta})$ (the Hessian of h at $\hat{\theta}$). Higher-order approximations may be derived by retaining higher-order terms in the expansions of f and h. A good general reference on asymptotic expansions is Bleistein and Handelsman (1986).

Applications in Bayesian statistics. Suppose the data vector $Y=(Y_1, Y_2, \ldots, Y_n)$ has a distribution with density $p(y|\theta)$, where $\theta\in\Theta$ is an unknown m-dimensional parameter. The likelihood function is $L(\theta)\propto p(y|\theta)$, where the proportionality constant is arbitrary. In Bayesian statistics, a prior probability density $\pi(\theta)$ is introduced and inferences are based on the posterior probability density, $p(\theta|y)\propto L(\theta)\pi(\theta)$, where the proportionality constant is determined by the requirement that $p(\theta|y)$ integrate to one. One important problem is to compute the posterior expectation of a real-valued function $g(\theta)$,

$$E(g(\theta) \mid y) = \int g(\theta)L(\theta)\pi(\theta)d\theta \bigg/ \int L(\theta)\pi(\theta)d\theta.$$

The likelihood function may be considered a product of n components, corresponding to the n components of the data vector. The loglikelihood function $\ell(\theta) = \log(L(\theta))$ is thus a sum of n components and we may define $h(\theta) = \ell(\theta)\pi(\theta)/n$. This is an abuse of notation because h is not a fixed function: it is a different function for each value of n, but we will ignore that here (see Kass, Tierney, and Kadane, (1990)). We then have $L(\theta)\pi(\theta) = \exp(-nh(\theta))$, and the expectation may be approximated by applying Laplace's method to both the numerator and denominator (with f defined in (1.1) respectively equal to g and 1) to yield the *first-order* expansion

$$E(g(\theta)) = g(\hat{\theta}) \cdot \{1 + 0(n^{-1})\}.$$

This approximation is often called the *modal approximation* because $\hat{\theta}$ is the mode of the posterior density. Modal approximations are commonly used in Bayesian data analysis because they are generally easy to compute, and if they are accurate the interpretation of inferences is simplified. A major motivation for work on integration in Bayesian analysis is to improve on modal approximations.

Our work involves the use of more accurate asymptotic approximations, based on *second-order* expansions. In computing a posterior expectation, second-order expansions have errors of order $O(n^{-2})$. Here it is helpful to obtain a more general form for the integrands by introducing an arbitrary smooth positive function $b(\theta)$, defining h such that $L(\theta)\pi(\theta) = b(\theta)\exp(-nh(\theta))$, and then allowing the definition of h and b used in the numerator and denominator integrals in the expectation to be different. Letting N and D, standing for numerator and denominator, index b and h, the expectation may then be written

$$E(g(\theta) \mid y) = \frac{\int b_N(\theta)\exp(-nh_N(\theta))d\theta}{\int b_D(\theta)\exp(-nh_D(\theta))d\theta}. \tag{1.2}$$

Various alternative choices for b and h in the numerator and denominator may be of interest, and the general form for the second-order expansion is given by Tierney, Kass, and Kadane (1989b).

An example. To illustrate the methods we are discussing, we take results from Tierney, Kass, and Kadane (1989b) concerning a reanalysis of the the Stanford Heart Transplant Data using the Pareto model of Turnbull, Brown,

TABLE 1

δ_t	w	δ_e	s	δ_t	w	δ_e	s	δ_t	w	δ_e	s
0	49	1	0	0	91	0	0	1	0	1	44
0	5	1	0	0	427	0	0	1	1	0	780
0	17	1	0	1	0	1	15	1	20	1	51
0	2	1	0	1	35	1	3	1	35	0	710
0	39	1	0	1	50	1	624	1	82	0	663
0	84	1	0	1	11	1	46	1	31	1	253
0	7	1	0	1	25	1	127	1	40	1	147
0	0	1	0	1	16	1	61	1	9	1	51
0	35	1	0	1	36	1	1350	1	66	0	479
0	36	1	0	1	27	1	312	1	20	1	322
0	1400	0	0	1	19	1	24	1	77	0	442
0	5	1	0	1	17	1	10	1	2	1	65
0	34	1	0	1	7	1	1024	1	26	0	419
0	15	1	0	1	11	1	39	1	32	0	362
0	11	1	0	1	2	1	730	1	13	1	64
0	2	1	0	1	82	1	136	1	56	1	228
0	1	1	0	1	24	0	1379	1	2	1	65
0	39	1	0	1	70	1	1	1	9	0	264
0	8	1	0	1	15	1	836	1	4	1	25
0	101	1	0	1	16	1	60	1	30	0	193
0	2	1	0	1	50	0	1140	1	3	0	196
0	148	1	0	1	22	0	1153	1	26	1	63
0	1	1	0	1	45	1	54	1	4	1	12
0	68	1	0	1	18	1	47	1	45	0	103
0	31	1	0	1	4	1	0	1	25	0	60
0	1	1	0	1	1	1	43	1	5	0	43
0	20	1	0	1	40	0	971				
0	118	0	0	1	57	0	868				

and Hu (1974). The data are listed in Table 1, and the loglikelihood function $\ell(\theta) = \log(L(\theta))$ for this model is

$$\ell(\lambda, \tau, p) = \sum_i \{\delta_e(i)(\log(p) + \delta_t(i)\log(\tau)) + p\log(\lambda) - (p + \delta_e)\log(\lambda + w_i + s_i\tau)\}$$

where $\theta = (\lambda, \tau, p)$ (a three-dimensional parameter vector), $\delta_t(i)$ indicates (is zero or one) whether or not the ith individual was in the transplant group, $\delta_e(i)$ indicates whether or not the ith individual died, and w_i and s_i are the survival times before and after transplant. The prior density is taken to be $\pi(\lambda, \tau, p) = 1$. Here, the functions g are simply $g(\lambda, \tau, p) = \lambda$, $g(\lambda, \tau, p) = \tau$ and $g(\lambda, \tau, p) = p$. Thus, for example, an expectation we wish to compute is

$$E(\lambda \mid y) = \int \lambda L(\lambda, \tau, p) d\lambda d\tau dp \Big/ \int L(\lambda, \tau, p) d\lambda d\tau dp$$

where $L(\lambda, \tau, p) = \exp(\ell(\lambda, \tau, p))$.

The modal approximation and two versions of second-order approximations (described in the next section) are shown in Table 2. It may be seen that the modal (first-order) approximations are rather inaccurate. For instance, the posterior expectation (mean) of λ has an error that is nearly one-third of the exact value, and a substantial fraction of the standard deviation.

TABLE 2

	Mean			SD		
Method	τ	λ	p	τ	λ	p
Exact	1.04	32.5	.500	.470	16.2	.140
1st order	.81	21.9	.434	.332	10.3	.110
Direct	1.04	32.1	.493	.494	16.1	.138
MGF	1.01	30.2	.487	.437	14.0	.131

Summary of results

Regularity conditions. In statistical applications the function h in the exponent of (1.1) depends on n, as it does in (1.2). If we were to write, more appropriately, on $h = h_n$, regularity conditions on h would become conditions on the uniform behavior of h_n. Kass, Tierney, and Kadane (1990) state such conditions. In statistical applications these become conditions on a sequence of loglikelihood functions, and simple requirements on the rest of the integrand; together these insure the validity of expansions based on Laplace's method for a particular infinite sequence of data. In particular, $-nh_n$ must have a uniformly dominant peak and the derivatives of h_n (of appropriate orders) must be uniformly bounded near the maxima of $-h_n$. In that paper, conditions are also given to insure that the sequence of loglikelihood functions is, in the appropriate sense, well-behaved with probability one.

Second-order approximations for expectations. To obtain a second-order approximation, the expansions of f and h in (1.1) must be carried out to include the order $O(n^{-1})$ and $O(n^{-3/2})$ terms, with the latter dropping out on integration. The resulting formula can then be applied to the numerator and denominator (1.2). Assuming $h = h_N = h_D$ and $b = b_N = b_D$, after simplification there is

$$E(g(\theta) \mid y) = g(\hat{\theta}) + n^{-1} \cdot \sum_{i,j} \{h^{ij} g[b(\hat{\theta})^{-1} b_j - (1/2) \sum_{r,s} h^{rs} h_{rsj}] + (1/2) h^{ij} g_{ij}\} + O(n^{-2}) \tag{2.3}$$

where subscripts indicate partial derivatives evaluated at $\hat{\theta}$ and h^{ij} is the (i, j)-component of the inverse of the Hessian matrix $(h_{ij}) = D^2 h(\hat{\theta})$. Expansions of this kind were used by Lindley (1961, 1980) and Mosteller and Wallace (1964).

Now suppose $g(\theta)$ is positive, and in (1.2) let $-nh_N(\theta) = -nh_D(\theta) + \log g(\theta)$, so that $b_N = b_D$. In this case there is cancelation of leading error terms when first-order approximations are applied to the numerator and denominator in (1.2), and there results the second-order expansion

$$E(g(\theta)) = \frac{\det(\Sigma_N)^{1/2} b(\theta_N) \exp(-nh_N(\theta_N))}{\det(\Sigma_D)^{1/2} b(\theta_D) \exp(-nh_D(\theta_D))} \cdot \{1 + 0(n^{-2})\} \tag{2.4}$$

where θ_K is the maximum of $-h_K$ and $\Sigma_K = (D^2 h_K(\theta))^{-1}$ where $K = N, D$. Thus, by applying Laplace's method in this alternative form a remarkable simplification is obtained. This was pointed out by Tierney and Kadane (1986).

We call the form of the approximation given by (2.4) *fully exponential*, and that given by (2.3) *standard*. The fully exponential form converts the evaluation of $m(m+1)(m+2)/6$ third derivatives to a second maximization (of $-h_N$ once $-h_D$ is maximized) and an evaluation of two Hessian determinants. The second maximization, however, is very easily carried out, because θ_D serves as a good initial value for computing θ_N. There is thus a modest improvement in efficiency (which increases with m) and a substantial improvement in programming time when the third derivatives must be derived and correctly coded. An additional advantage of the fully exponential method is that it yields a second-order approximation to the variance of a positive function,

$$V(g(\theta)) = \{\hat{E}(g(\theta)^2) - \hat{E}(g(\theta))^2\} \cdot \{1 + 0(n^{-2})\}$$

where $\hat{E}$ is the expectation approximation given by (2.4). To obtain a second-order variance approximation from the standard form, fourth and fifth derivatives of the loglikelihood would be required.

The fully exponential method just described applies only to positive functions g. A simple way of obtaining the expectation of a nonpositive function $g(\theta)$ is to first approximate the moment generating function $M(s) = E(\exp(sg(\theta)))$ by $\hat{M}(s)$ according to (2.4), which is applicable since $\exp(sg(\theta))$ is positive. Then, the derivative of the approximation at zero

$\hat{M}'(0)$ furnishes an approximation to $E(g(\theta))$. An approximation to the variance is given by $\hat{M}''(0) - \hat{M}'(0)^2$. Each of these may be shown to have multiplicative error of order $O(n^{-2})$. In practice, the derivatives are computed by finite differencing.

These approximations are reexpressed and discussed by Tierney, Kass, and Kadane (1989b). Results of applying the second-order approximations to the Stanford Heart Transplant example are given in Table 2. Since λ, τ, and p are each positive, the direct fully exponential method may be applied. Values based on it, and also values based on the moment generating function (MGF) approximation are given in the table. These provide an important improvement on the first-order approximations, and their accuracy is quite adequate for practical purposes. For instance, the error in the expectation of λ using the first-order approximation is very large, but that using the MGF approximation is less than 15% of the standard deviation of λ (and the direct approximation is even better).

Two further points made by Tierney, Kass, and Kadane (1989b) should be mentioned here. First, the moment generating function approximation may be shown to be equivalent to adding a very large constant to g, performing the fully exponential approximation (2.4), and then subtracting the constant from the result. Second, it turns out that the approximation $\hat{M}'(0)$ is mathematically equivalent to (2.3). Notice, however, that the moment generating function approximation is computed differently, and retains the advantages of the fully exponential method mentioned in the previous section.

Marginal densities. The basic first-order expansion $\widehat{I}$ following (1.1) may be applied immediately to the approximation of marginal posterior densities. Suppose $\Theta = \Theta_1 \times \Theta_2$ and we are interested in the marginal density on θ_1,

$$p(\theta_1 \mid y) \propto \int \exp[\tilde{\ell}(\theta_1, \theta_2)] d\theta_2$$

where $\tilde{\ell} = \ell + \log \pi$ with ℓ being the loglikelihood function based on data y and π being the prior. Taking $h = \tilde{\ell}/n$ and $f = 1$ in (1.1) we obtain,

$$\hat{p}(\theta_1 \mid y) \propto \det(\Sigma(\theta_1))^{1/2} \exp(\tilde{\ell}(\theta_1, \hat{\theta}_2(\theta_1))) \tag{2.5}$$

where $\hat{\theta}_2(\theta_1)$ is the maximum of $\tilde{\ell}(\theta_1, \cdot)$, and $\Sigma(\theta_1)$ is inverse of the Hessian of $-\tilde{\ell}(\theta_1, \cdot)$ at $\hat{\theta}_2(\theta_1)$. The normalizing constant may also be approximated with error of order $O(n^{-1})$, but in practice one may as well ignore the joint normalizing constant, and then renormalize the marginal approximation after it is evaluated at suitably many points θ_1. This approximation appears to have first been suggested by Leonard (1982). As Tierney and Kadane (1986) noted, after the normalization constant is obtained, its relative error is of order $O(n^{-3/2})$ on neighborhoods about the mode that shrink at the rate $n^{-1/2}$. This may be compared to the approximating Normal density, which has error of order $O(n^{-1/2})$ on these shrinking neighborhoods.

Perhaps more importantly, the approximation is of order $O(n^{-1})$ uniformly on any bounded neighborhood of the mode.

It is often convenient to have an expression for the marginal density of a general function $g = g(\theta)$. Assuming the gradient of g is nonzero and writing $\gamma = g(\theta)$ for the argument of the density, Tierney, Kass, and Kadane (1989a) obtained an approximation which they claimed had the properties of (2.5). Unfortunately, there was an error in the derivation. The correct formula, which does have these good properties, is

$$\hat{p}(\gamma \mid y) \propto [\det(\bar{\Omega}(\gamma))/((Dg)^T\bar{\Omega}(\gamma)(Dg))]^{1/2}\exp(\tilde{\ell}(\hat{\theta}(\gamma)))$$

where $\hat{\theta}(\gamma)$ is the maximum of $\tilde{\ell}$ constrained by $g(\theta) = \gamma$, Dg is evaluated at $\hat{\theta}(\gamma)$, and using the Lagrangian $\bar{H} = \tilde{\ell} - \lambda g$, $\bar{\Omega}(\gamma)$ is the inverse of the Hessian of $-\bar{H}$ at $\hat{\theta}(\gamma)$.

Further results

Influence and sensitivity. Assessment of sensitivity of inferences to either a change in prior or the deletion of an observation is amenable to asymptotic approximation. For a new prior q replacing π, we define $\rho(\theta) = \log\{q(\theta)/\pi(\theta)\}$, which we call the *perturbation function.* Writing the likelihood function as $L(\theta) = f(y \mid \theta)$, with the initial prior π the posterior density is proportional to $\tilde{L}(\theta) = L(\theta)\pi(\theta)$. With the new prior q, the posterior density is proportional to $\tilde{L}_{\text{NEW}}(\theta) = \exp[\log\{\tilde{L}(\theta)\} + \rho(\theta)]$. Letting E_π and E_q denote posterior expectations computed with the priors π and q, Kass, Tierney, and Kadane (1989) derived, in several different forms, approximations to $E_q[g(\theta) \mid y] - E_\pi[g(\theta) \mid y]$ having error of order $O(n^{-2})$. The most easily interpreted is based on the joint posterior mode $\tilde{\theta}$ and the modal covariance matrix $\tilde{\Lambda} = \{-D^2\tilde{\ell}(\tilde{\theta})\}^{-1}$, where $\tilde{\ell} = \log(\tilde{L})$. They obtained

$$E_q[g(\theta) \mid y] - E_\pi[g(\theta) \mid y] = (D\rho)^T\tilde{\Lambda}(Dg) + O(n^{-2}) \tag{2.6}$$

where the derivatives $D\rho$ and Dg are evaluated at $\tilde{\theta}$. Thus, the change in the posterior mean is, approximately, the posterior covariance of the perturbation and the function being estimated.

In addition to providing a simple interpretation, this expression may be used to find perturbations that tend to alter inferences about $g(\theta)$, that is, it may be used as a basis for approximate sensitivity calculations. The same methods may be used to assess sensitivity to deletion of an observation from the sample by taking the perturbation function to be $\rho(\theta) = \log\{1/L(\theta)\}$.

Applications to hierarchical models. Kass and Steffey (1989), following earlier work of Deely and Lindley (1981), have discussed approximate Bayesian methods in what they call "conditionally independent hierarchical models," which are sometimes also called "parametric empirical Bayes models". These models are of great importance in statistics, and they provide a challenging class of integration problems. As reviewed by Kass and Steffey (1989), each posterior expectation that needs to be calculated takes the form of an iterated integral with the inner and outer integrals both being multidimensional. There is structure to these integrals that can be used to gain efficiency, but the computations can be demanding in many applications. A comparatively simple, yet interesting example is given by Tsutakawa (1985).

Kass and Steffey (1989) considered and exemplified first-order and second-order approximations in these models, and showed that they can be quite useful. Those authors also gave an explicit sense in which parametric empirical Bayes estimates approximate fully Bayesian posterior means, and showed that first-order approximations to posterior variances provide a simple accounting for estimation of a "hyperparameter" in using approximate (or "empirical") Bayes estimates.

Concluding comments

There are many available strategies to solve integration problems in Bayesian analysis. Developments within statistical contexts have used methods based on importance sampling (Geweke (1987), Stewart (1985), van Dijk and Kloek (1985), Zellner and Rossi (1984)), other sampling-based methods (Tanner and Wong (1987), Gelfand and Smith (1988)), Gauss-Hermite quadrature (Naylor and Smith (1982), Smith, Skene, Shaw, Naylor, and Dansfield (1985)), and quasi-Monte Carlo quadrature (Shaw (1988)). The approximations we have discussed here offer alternative solutions, their chief advantage being that they can be rapidly and easily computed. Our main motivation has been to create tools that could be part of an interactive computing environment for Bayesian analysis (Tierney, Kass, and Kadane (1987), Kass, Tierney, and Kadane (1988), and Tierney (1989)).

Since asymptotic approximations rely on expansions that ignore terms that vanish as the sample size becomes infinite, one might think that they could only be used for large samples. Certainly it is advisable, when possible, to check solutions with nonasymptotic techniques. Examples show, however, that these approximate methods can be sufficiently accurate to be of use for data analysis in many settings. In addition, they can be useful for other reasons, as well. First of all, they can sometimes produce intuitive interpretations (as in the variance approximations discussed by Kass and Steffey (1989)). Secondly, they may be used in conjunction with other methods: both Gauss-Hermite quadrature and importance-function monte carlo require initial centering and scaling which is usually done by first-order approximation. Second-order approximation should be an improvement. We believe there are further opportunities for the use of asymptotic expansions in conjunction with nonasymptotic numerical methods of integration, and hope to see these developed in the near future.

REFERENCES

N. Bleistein and R. A. Handelsman (1986), *Asymptotic Expansions of Integrals,* 2nd ed., Dover, New York.

J. J. Deely and D. V. Lindley (1981), *Bayes empirical Bayes*, J. Amer. Statist. Assoc. **76**, 833-841.

A. E. Gelfand and A. F. M. Smith (1988), *Sampling based approaches to calculating marginal densities*, Technical Report, Nottingham Statistics Group.

J. Geweke (1989), *Bayesian inference in econometric models using Monte Carlo integration*, Econometrica **57**, 1317-1339.

R. E. Kass. and D. Steffey (1989), *Approximate Bayesian inference in conditionally independent hierarchical models (parametric empirical Bayes models)*, J. Amer. Statist. Assoc. **84**, 717-726.

R. E. Kass, L. Tierney, and J. B. Kadane (1989), *Approximate methods for assessing influence and sensitivity in Bayesian analysis*, Biometrika, **76**, 663-674.

——(1990), *The validity of posterior expansions based on Laplace's method*, Bayesian and Likelihood Methods in Statistics and Econometrics: Essays in Honor of George A. Barnard (S. Geisser, J.S. Hodges, S.J. Press, and A. Zellner, eds.), North-Holland, Amsterdam, pp. 473–488.

T. Leonard (1982), *Comment on A simple predictive density function*, J. Amer. Statist. Assoc. **77**, 657-658.

D. V. Lindley (1961), *The use of prior probability distributions in statistical inference and decisions*, Proc. of the 4th Berkeley Symposium **1**, 453-468.

——(1980), *Approximate Bayesian methods, Bayesian Statistics*, Proceedings of the First International Meeting held in Valencia (Spain), May 28-June 2, 1979 (J. M. Bernardo, M. H. DeGroot, D. V. Lindley, and A. M. F. Smith, eds.), University Press, Valencia, Spain, pp. 223-245.

F. Mosteller and D. L. Wallence (1964), *Inference and disputed authorship: the federalist*, reprinted in Applied Bayesian and classical inference: The case of the Federalist papers, 1984, Springer-Verlag, New York.

J. C. Naylor and A. F. M. Smith (1982), *Applications of a method for the efficient computation of posterior distributions*, Appl. Statist. **31**, 214-225.

J. E. H. Shaw (1988), *A quasirandom approach to integration in Bayesian statistics*, Ann. of Statist. **16**, 895-914.

A. F. M. Smith, A. M. Skene, J. E. H. Shaw, J. C. Naylor, and M. Dansfield (1985), *The implementation of the Bayesian paradigm*, Comm. Stat. **A14**, 1079-1102.

L. Stewart (1985), *Multiparameter Bayesian inference using Monte Carlo integration: Some techniques for bivariate analysis*, Bayesian Statistics (J. M. Bernardo, M. H. DeGroot, D. V. Lindley, and A. F. M. Smith, eds.), vol. 2, North-Holland, Amsterdam, pp. 495–510.

M. Tanner and W. Wong (1987), *The calculation of posterior distributions by data augmentation (with discussion)*, J. Amer. Statist. Assoc. **82**, 528-550.

L. Tierney (1989), *Bayesian analysis in New S and Lisp*, Proceedings of the Statistical Computing Section of the American Statistical Association (to appear).

L. Tierney and J. B. Kadane (1986), *Accurate approximations for posterior moments and marginal densities*, J. Amer. Statist. Assoc. **81**, 82-86.

L. Tierney, R.E. Kass, and J. B. Kadane (1989a), *Approximate marginal densities for nonlinear functions*, Biometrika **76**, 425-433. Correction, **78** (to appear).

——(1989b), *Fully exponential Laplace approximations to expectations and variances of nonpositive functions*, J. Amer. Statist. Assoc. **84**, 710-716.

R. K. Tsutakawa (1985), *Estimation of cancer mortality rates: a Bayesian analysis of small frequencies*, Biometrics **41**, 69-79.

B. W. Turnbull, B. H. Brown, and M. Hu (1974), *Survivorship analysis of heart transplant data*, J. Amer. Statist. Assoc. **69**, 74-80.

H. K. Van Dijk and T. Kloek (1985), *Experiments with some alternatives for simple important sampling in Monte Carlo integration*, Bayesian Statistics (J. M. Bernardo, M. H. DeGroot, D. V. Lindley, and A. F. M. Smith, eds.), vol. 2, North-Holland, Amsterdam, pp. 511–530.

A. Zellner and P. J. Rossi (1984). *Bayesian analysis of dichotomous quantal response models*, J. Econometrics **25**, 365-393.

DEPARTMENT OF STATISTICS, CARNEGIE MELLON UNIVERSITY, PITTSBURGH, PENNSYLVANIA 15213

E-mail address, Robert E. Kass: Kass@stat.cmu.edu

E-mail address, Joseph B. Kadane: jk20@stat.cmu.edu

(Luke Tierney) SCHOOL OF STATISTICS, UNIVERSITY OF MINNESOTA, MINNEAPOLIS, MINNESOTA 55645

E-mail address, Luke Tierney: luke@umnstat.stat.umn.edu

Contemporary Mathematics
Volume **115**, 1991

Monte Carlo Integration in Bayesian Statistical Analysis

ROBERT L. WOLPERT

ABSTRACT. This paper presents a review of Monte Carlo methods for approximating the high-dimensional integrals that arise in Bayesian statistical analysis. Emphasis is on the features of many Bayesian applications that make Monte Carlo methods especially appropriate, and on Monte Carlo variance-reduction techniques especially well suited to Bayesian applications. A generalized logistic regression example is used to illustrate the ideas, and high-precision formulas are given for implementing Bayesian Monte Carlo integration.

1. Background: Bayesian analysis

In Bayesian analysis uncertainty about a quantity θ is represented in the form of a probability measure $\pi(d\theta)$ with which one can calculate the probability $\mathbf{P}^{\pi}[\theta \in A]$ that X lies in any measurable set A, or (more generally) the expectation $\mathbf{E}^{\pi} g(\theta)$ of any function of interest that depends on θ. Upon observing the value of some random quantity X whose probability distribution depends on θ, Bayes' theorem gives the rule for calculating the probability distribution $\pi^{\star}(d\theta)$ for uncertainty about θ *after* observing $X = x$ as the ratio

$$\pi^{\star}(d\theta) = f(x|\theta)\,\pi(d\theta) \Big/ \int_{\Theta} f(x|\theta')\,\pi(d\theta'), \tag{1}$$

where $\pi(d\theta)$ is the probability distribution representing uncertainty about θ *before* observing $X{=}x$, and where $f(x|\theta)$ is the probability density function for X (with respect to an arbitrary dominating measure $\mu(dx)$), evaluated at the observed value $X{=}x$. The expectation of some function of interest

1980 *Mathematics Subject Classification* (1985 *Revision*). Primary 62F15; Secondary 65C05.
Key words and phrases. Bayesian analysis, numerical integration, simulation, Monte Carlo method, variance reduction.
This research was supported in part by NSF Grants No. SES-8605867 and SES-8921227.
This paper is in final form and no version of it will be submitted for publication elsewhere.

$g(\theta)$ (possibly a vector) is given by a similar ratio:

$$\begin{aligned} \overline{g} &\equiv \mathbf{E}[g(\theta)|X=x] \\ &= \int_\Theta g(\theta)\,\pi^\star(d\theta) \\ &= \int_\Theta g(\theta) f(x|\theta)\,\pi(d\theta) \Big/ \int_\Theta f(x|\theta)\,\pi(d\theta). \end{aligned} \tag{2}$$

Although we do not require that $\pi(d\theta)$ be properly normalized or even that $\pi(\Theta) < \infty$, we will assume throughout that $\int_\Theta f(x|\theta)\,\pi(d\theta) < \infty$ and $\int_\Theta |g(\theta)|\, f(x|\theta)\,\pi(d\theta) < \infty$ for every x, so both numerator and denominator will be well-defined in (1), (2), and similar formulas to follow.

The posterior covariance matrix $\Sigma \equiv \mathbf{E}[(g(\theta) - \overline{g})(g(\theta) - \overline{g})'|X=x]$ gives one indication of how well $g(\theta)$ is determined by the prior distribution $\pi(d\theta)$ and the observation $X=x$, and so measures the accuracy or precision of the estimate $\overline{g}$ of $g(\theta)$. It also is given by a ratio of integrals

$$\begin{aligned} \Sigma &\equiv \mathbf{E}[(g(\theta) - \overline{g})(g(\theta) - \overline{g})'|X=x] \\ &= \frac{\int (g(\theta) - \overline{g})(g(\theta) - \overline{g})' f(x|\theta)\,\pi(d\theta)}{\int f(x|\theta)\,\pi(d\theta)}. \end{aligned} \tag{3}$$

A more complete representation of the uncertainty attendant $g(\theta)$ following the observation of $X=x$ would be the (joint) *posterior density function* for $g(\theta)$, if it exists. Although $\xi = g(\theta)$ may not have a density function (its distribution will not be absolutely continuous if $\pi(d\theta)$ is supported on a lower-dimensional manifold, for example, or if $g(\theta)$ is constant on a set of positive measure) it can be approximated arbitrarily well by a distribution with a density of the form

$$\begin{aligned} \pi_\varepsilon^\star(\xi) &\equiv \int_\Theta K_\varepsilon(\xi - g(\theta))\,\pi^\star(d\theta) \\ &= \frac{\int_\Theta K_\varepsilon(\xi - g(\theta)) f(x|\theta)\,\pi(d\theta)}{\int_\Theta f(x|\theta)\,\pi(d\theta)} \end{aligned} \tag{4}$$

for any approximate identity $K_\varepsilon(x) = \varepsilon^{-q} K(x/\varepsilon) \geq 0$ satisfying $\int_{R^q} K_\varepsilon(x)\,dx = 1$ with suitably small $\varepsilon > 0$. For plotting marginal posterior densities and for many other purposes it is enough to know $\pi_\varepsilon^\star(\xi)$ at a few hundred points $\{\xi_j\}$, i.e., to find the posterior expectation of a several-hundred-dimensional function $\{K_\varepsilon(\xi_j - g(\theta))\}_{j \in J}$.

If the method of selecting a probability distribution $\pi(d\theta)$ to represent knowledge about θ before the experiment does not depend on the distribution of X, then Bayesian statistical analysis based on $\pi^\star$ is consistent with the Likelihood Principle, i.e., depends on the observation $X=x$ only through the "likelihood function"

$$L(\theta) = L(\theta|X=x) = f(x|\theta).$$

Birnbaum and others (reviewed and extended by Berger and Wolpert (1988)) have shown that any violation of the Likelihood Principle also violates either the Weak Conditionality Principle (which asserts that, if one randomly selects between two experiments, then only the experiment performed is relevant) or the Sufficiency Principle (which asserts that all evidence about θ from observing X is also contained in any sufficient statistic $T(X)$). Since these two principles are widely held, it is hard to justify the use of a statistical procedure inconsistent with the Likelihood Principle.

Although the likelihood function $L(\theta) = f(x|\theta)$ is defined as a probability density function *for* X *given* θ, with respect to some arbitrary reference measure $\mu(dx)$, the Bayesian statistician (in light of Equation (1)) regards it as the probability density function *for* θ *given* $X=x$, with respect to a reference measure proportional to $\pi(d\theta)$. The likelihood function is central for non-Bayesian statistical methods as well: the maximum likelihood estimate (MLE) $\hat{g}$ of $g(\theta)$ is just the function $g(\theta)$ evaluated at a point $\hat{\theta}$ where $L(\theta)$ attains its maximum

$$\hat{g} = \widehat{g(\theta)} = g(\hat{\theta}).$$

For a (nearly) uniform prior distribution $\pi(\theta)$ the MLE $\hat{g}$ is (nearly) the *mode* of the distribution of $g(\theta)$ under the posterior $\pi^\star(d\theta)$, while $\overline{g}$ in (2) is the *mean.*

The difference between a mode and mean doesn't seem so dramatic, and in many cases it is not; the real issues arise in trying to represent the degree of certainty with which $g(\theta)$ is known, following the observation of $X=x$. The Bayesian uses the same measure $\pi^\star$ given in (1) to evaluate the *posterior probabilities* that $g(\theta)$ lies in specified sets (especially *posterior HPD regions*) or to evaluate the *posterior covariance* Σ of $\overline{g}$ as in (3) or the marginal *posterior density* for some components of $g(\theta)$ as in (4), and thus stays faithful to the likelihood principle, while the frequentist constructs *p-values, confidence sets* and *standard errors* by considering what the likelihood function *would have been* if other, near-by values of $X = \tilde{x} \approx x$ had been observed instead of $X=x$, and in so doing leaves the Likelihood Principle behind.

Unfortunately the integrals necessary for calculating the ratios in Equations (2–4) are seldom amenable to analytical methods. Frequently the parameter θ takes values in a high-dimensional space, making the integral resistant to quadrature methods; the simplest quarterly or monthly time-series applications lead to problems in four or twelve dimensions, for example, and problems in five to ten or twenty dimensions and more are common in many fields of application. Tensor-product quadrature methods are unthinkable in such problems.

One recourse is to appeal to large-sample asymptotic normality and approximate the mean $\overline{g}$ in (2) by the mode $\hat{g}$, and the covariance Σ in (3) by the inverse of the information matrix. Another is to approximate the likelihood function $L(\theta) = f(x|\theta)$ and the prior distribution $\pi(\theta)$ by members of some conjugate pair of density families for which the integrals can be evaluated in closed form for a suitable class of functions $g(\theta)$ (e.g., polynomials). With the emergence of fast desk-top computers and appropriate numerical algorithms a third choice has emerged: to approximate the integrals in (2)–(4) through Monte Carlo integration.

2. Background: Monte Carlo integration

The theory of Monte Carlo integration is simple. If $\theta_i \in \Theta$ is a sequence of synthetic random variates, each drawn from some probability distribution $\Pi(d\theta)$ that dominates $\pi(d\theta)$, then let $w(\theta)$ be the Radon–Nikodym derivative

$$w(\theta) = f(x|\theta)\pi(d\theta)/\Pi(d\theta), \tag{5}$$

and consider the weighted averages

$$\overline{g}_n \equiv \sum_{i=1}^{n} g(\theta_i)w(\theta_i) \Big/ \sum_{i=1}^{n} w(\theta_i). \tag{6}$$

It is easy to calculate the expectations of the numerator and denominator in Equation (6), both of which are well defined and finite by the earlier assumptions that both the likelihood function $f(x|\theta)$ and the product $g(\theta)f(x|\theta)$ are integrable with respect to the prior $\pi(d\theta)$, i.e., $\int_\Theta f(x|\theta)\,\pi(d\theta) < \infty$ and $\int_\Theta |g(\theta)|\,f(x|\theta)\,\pi(d\theta) < \infty$. The numerator has expectation

$$\begin{aligned}\mathbf{E}\sum_{i=1}^{n} g(\theta_i)w(\theta_i) &= \sum_{i=1}^{n}\int_\Theta g(\theta_i)f(x|\theta_i)\frac{\pi(d\theta_i)}{\Pi(d\theta_i)}\Pi(d\theta_i)\\ &= n \times \int_\Theta g(\theta)f(x|\theta)\,\pi(d\theta),\end{aligned}$$

whereas the denominator has expectation

$$\begin{aligned}\mathbf{E}\sum_{i=1}^{n} w(\theta_i) &= \sum_{i=1}^{n}\int_\Theta f(x|\theta_i)\frac{\pi(d\theta_i)}{\Pi(d\theta_i)}\Pi(d\theta_i)\\ &= n \times \int_\Theta f(x|\theta)\,\pi(d\theta).\end{aligned}$$

Thus $\overline{g}_n$ is the ratio of unbiased estimates of the numerator and denominator of (2); it doesn't follow that $\overline{g}_n$ is an unbiased estimate of $\overline{g}$, of course, but it suggests that the approach is promising. Notice that the sample size and joint distribution of $\{\theta_i\}_{i\leq n}$ were not specified; we are free to choose n and $\Pi(d\theta)$ in any convenient way, and the $\{\theta_i\}$ need not be stochastically independent. If they are at least *mixing* (e.g., if each θ_i is independent of all but finitely many θ_j) then two applications of the strong ergodic theorem (one to the numerator and one to the denominator) give an immediate proof of the consistency of Monte Carlo estimation, i.e., the almost-sure convergence of $\overline{g}_n$ to $\overline{g}$.

The Monte Carlo estimate of $\Sigma = \mathbf{E}[(g(\theta) - \overline{g})(g(\theta) - \overline{g})'|X{=}x]$ is given by the matrix expression

$$\Sigma_n \equiv \frac{\sum_{i=1}^{n}(g(\theta_i) - \overline{g}_n)(g(\theta_i) - \overline{g}_n)' w(\theta_i)}{\sum_{i=1}^{n} w(\theta_i)} \tag{7}$$

and that of the marginal density $\{\pi_\varepsilon^\star(\xi_j)\}$ by the weighted kernel density estimate

$$\pi_{\varepsilon n}^\star(\xi_j) \equiv \frac{\sum_{i=1}^{n} K_\varepsilon(\xi_j - g(\theta_i)) w(\theta_i)}{\sum_{i=1}^{n} w(\theta_i)}. \tag{8}$$

If $\xi = g(\theta)$ has an absolutely continuous posterior distribution then, for fixed ε, the sequence of estimates $\overline{g}_n$ will not be consistent estimators of the density function $\pi_0^\star(\xi_j)$; in fact, the sequence will converge pointwise almost surely to $\pi_\varepsilon^\star(\xi)$, the convolution of the true density $\pi_0^\star(\xi)$ with the kernel $K_\varepsilon(\xi)$. If a sequence $\varepsilon_n \to 0$ converges to zero sufficiently slowly that $(\varepsilon_n)^q n \to \infty$ (e.g., $\varepsilon_n = n^{-(q+1)^{-1}}$) then one can verify pointwise convergence of $\pi_{\varepsilon_n n}^\star(\xi_j)$ to $\pi_0^\star(\xi_j)$ at all points of continuity ξ_j.

The quantities $\overline{g}_n$, Σ_n, and $\pi_{\varepsilon n}^\star(\xi_j)$ are random variables; we should choose n and the joint sampling distribution of the variates $\{\theta_i\}$ in such a way as to minimize their computation time and some measure of their likely errors in estimating $\overline{g}$, Σ, and $\pi_0^\star(\xi_j)$.

The mean square error $\sqrt{\mathbf{E}|\overline{g}_n - \overline{g}|^2}$ in Monte Carlo importance sampling falls off as $\sigma/\sqrt{n}$ for some constant $\sigma > 0$ if w, $gw \in \mathscr{L}^2(\Theta, \Pi(d\theta))$. For independent, identically-distributed (iid) sampling it is easy to show that the constant is approximately

$$\sigma \approx \frac{\sqrt{\int (g(\theta) - \overline{g})^2 w(\theta)^2 \Pi(d\theta)}}{\int w(\theta)\Pi(d\theta)},$$

but *variance reduction techniques* can often lead to much smaller constants. In d dimensions, the errors of tensor-product quadrature rules of local order m fall off as $n^{-m/d}$; the commonly used fourth-order Runge-Kutta scheme, for example, has errors that fall as $n^{-4/d}$, faster than those of iid Monte Carlo methods in dimensions $d \leq 7$. For this reason Monte Carlo methods were widely discounted in the 1950's, until the discovery by Hammersley and Morton of "antithetic accelertion" (or, at least, its popularization; see Tukey (1957)), the most important of several methods of reducing the constant σ above. For small d Monte Carlo integration may not be efficient *asymptotically*, in the limit as $n \to \infty$, but it can be quite efficient for achieving moderate precision with modest n if σ is made sufficiently small. Popular acceleration methods for reducing σ enough to make Monte Carlo practical for Bayesian integration include:

1. Importance sampling, the use of a sampling distribution $\Pi(d\theta)$ lending relatively little weight to the "unimportant" regions to which

$f(x|\theta)\,\pi(d\theta)$ gives little weight and instead concentrating on those areas to which $f(x|\theta)\,\pi(d\theta)$ gives great weight (or, roughly, choosing $\Pi(d\theta)$ to insure that $w(\theta)$ is bounded and nearly constant). See Curtiss, et al. (1951) for an account of the early development of importance sampling by Fermi, von Neumann, and Ulam. Specific cases or generalizations include *Russian roulette, splitting, stratified sampling,* and *conditional Monte Carlo.*

2. Control variables, the indirect use of Monte Carlo methodology to estimate the difference $\int (g(\theta)-h(\theta))w(\theta)\Pi(d\theta)/\int w(\theta)\Pi(d\theta) = \overline{g}-\overline{h}$ for some function $h(\theta) \approx g(\theta)$ with known expectation $\overline{h}$ (see Fieller and Hartley (1954)). A generalization of this method is the *regression* method.
3. Antithetic variates, the use of non–iid variates (especially negatively-correlated pairs). Generalizations of this include *random quadrature* methods. See Hammersley and Morton (1956).

These and other methods (e.g., orthogonal polynomials) are discussed in Hammersley and Handscomb, Chapter 5, (1964), Rubinstein (1981, 1986), and Wilson (1984).

A renewed interest in Monte Carlo methods accompanied the appearance in the early 1980's of widely-available minicomputers and low-cost microcomputers. Kloek and van Dijk (1978) introduced *adaptive* importance sampling, the dynamic adjustment of the sampling distribution $\Pi(d\theta)$ to improve incrementally the algorithm's efficiency, while Naylor and Smith (1982), Smith et al. (1985), and others at the University of Nottingham developed adaptive orthogonalization techniques improving the efficiency of both quadrature and Monte Carlo techniques. Geman and Geman (1984) in a study of Bayesian methods for image reconstruction developed and applied the theory of Gibbs sampling schemes, later recognized by Gelfand and Smith (1988) to be a broadly useful tool in high-dimensional Bayesian analysis. Tanner and Wong (1987) introduced a related technique (substitution sampling). Geweke (1988) proved a central limit theorem for Monte Carlo sampling schemes in Bayesian analysis. Many other recent contributions deserve mention in this active field.

One of the reasons these methods are so successful in Bayesian statistics is that statisticians seldom must integrate pathological functions with great local variation; indeed, likelihood functions and their products with prior densities can often be approximated strikingly well (sometimes after a nonlinear transformation removing positivity or monotonicity constraints) by simple elliptically symmetric functions such as multivariate Student t or the Gaussian forms

$$f(x|\theta)\pi(\theta) \approx c e^{-\frac{1}{2}(\theta-\hat{\theta})'\Lambda(\theta-\hat{\theta})} \tag{9}$$

for some vector $\hat{\theta}$ and Hermitian form Λ. Important problems with non-unimodal integrands do arise (e.g., in estimating the central tendency in problems with broad-tailed distributions, such as Cauchy or Student t), and some problems exhibit sufficient skewness to require novel, asymetric sampling distributions, but Bayesian statistical methods are now more practical than ever before precisely because the integrands encountered in Equations (2)–(4) above are commonly well enough behaved that Monte Carlo methods work well, even in high-dimensional problems. In the example below the parameters in a seven-dimensional variation on a logistic regression model are fit to about two decimals of precision in several minutes' computation time on a desktop workstation by Monte Carlo methods, while a five-parameter submodel could not be fit to $\pm 10\%$ precision in a week-end of computation using fourth-order Runge-Kutta. Asymptotically the quadrature method must win out, of course, since eventually $c_1 n^{-4/7} < c_2 n^{-1/2}$, but the asymptotics are not terribly relevant to the practical problem of finding an approximate solution to the problem at hand. The solution can be found using Monte Carlo methods, and (apparently) not using quadrature methods.

It is worth noting that nowhere in the discussion of the Monte Carlo method and methods of accelerating its convergence has it been necessary to consider the dimension of the parameter space Θ. Quadrature methods calling for the evaluation of an integrand at points of a lattice require an amount of computation which increases exponentially in the dimension, while Monte Carlo does not.

Of course this point can be emphasized too strongly. In many high-dimensional problems the matrix Λ in (9) above is badly conditioned, so the measure $f(x|\theta)\,\pi(d\theta)$ is supported in a neighborhood of some lower-dimensional (and possibly curved) space; a failure to detect this situation or imprecision in identifying the space can lead to inefficient sampling schemes and even to gross undetected estimation errors. If the curvature is significant then nonlinear reparametrizations may be required. Graphical methods for exploring likelihood contours can be helpful in revealing this and similar problems and in suggesting nonlinear reparametrizations to correct them.

3. Special features of Bayesian Monte Carlo analysis

Implementing Bayesian statistical analysis calls for the numerical estimation of several integrals such as those in Equations (2)–(4) which share certain features (illustrated in the example below) that make some Monte Carlo techniques seem especially appropriate. Each of the integrands in the numerators and denominators of equations (2)–(4) is a product of three terms: a multi-dimensional function of interest like $g(\theta)$, $(g(\theta)-\overline{g})(g(\theta)-\overline{g})'$, $K_\varepsilon(\xi_j - g(\theta))$, or simply 1; a nonnegative likelihood function $f(x|\theta)$; and a nonnegative prior density function, $\pi(\theta)$. In many applications:

1. The integrand is high dimensional. The function $g(\theta)$ often has hundreds of components (e.g., in kernel density estimation) each of which is easy to compute (e.g., components might include $[\theta]_j$, $[\theta]_j[\theta]_k$,

$1_{A_j}(\theta)$, or $K_\varepsilon(\xi_j - g(\theta))$). Using Monte Carlo techniques, all the terms in Equations (4)–(6) can be calculated simultaneously using the same sequence of random deviates $\{\theta_i\}$.

2. The likelihood function $L(\theta) = f(x|\theta)$ in each of the required integrals is often slow and expensive to compute. When no sufficient statistics are available (e.g., when exponential families are inappropriate), calculating $f(x|\theta)$ may require a loop through the entire dataset for each distinct value of θ. The computational burden of generating random deviates $\theta_i \sim \Pi(d\theta)$ is usually negligible when compared to that of calculating $f(x|\theta_i)$, so there is little reward for using especially efficient random number generators; conversely, variance-reduction techniques are quite important to limit the number of points where $f(x|\theta_i)$ must be evaluated.
3. The prior density function $\pi(\theta)$ in each of the required integrals is also comparatively expensive to compute if subjective prior distributions are used (summarized in tables, or using density functions drawn with a "mouse.") Some prior densities intended to be "noninformative" are especially expensive computationally, e.g., the *reference priors* of Berger and Bernardo (1989) or even Jeffreys' priors (1960) in multi-dimensional models which are not exponential families.
4. For many problems the product of the likelihood function and prior density is unimodal and "bell-shaped," i.e., the negative logarithm $\ell(\theta) \equiv -\log[f(x|\theta)\pi(\theta)]$ is well approximated by a quadratic form in a neighborhood of its minimum.
5. High precision isn't important; $\pm 5\%$ or $\pm 1\%$ is usually quite adequate. Other uncertainties and approximations arising in the modeling process usually make it inappropriate to seek machine accuracy (6-16 decimals) in statistical calculations.
6. Integrals are moderately high dimensional. In applications to structured, hierarchical Bayesian models and in routine economic time series Θ often range from 4–24 dimensions or more, while latent variable models and nonparametric survival or density estimation lead to problems with hundreds of variables.
7. As insights and experience are gained the function of interest $g(\theta)$ often changes; as new data become available the likelihood function $f(x|\theta)$ changes; even the prior density $\pi(\theta)$ may change. "Optimal" methods tailored to a particular integrand aren't helpful unless the tailoring can be done and revised almost instantly.
8. Speed is important in some applications and not in others. For interactive elicitation of prior densities in hierarchical models it would be useful to calculate predictive distributions and prior marginal densities within seconds. This is not yet possible for any but the simplest of models.

4. A bioassay example

In Wolpert and Warren-Hicks (1990) details are given of a hierarchical Bayesian analysis combining laboratory data and field observations to study the effects on fish survival of three features often associated with so-called "acid rain" (low pH and high concentrations of calcium and monomeric aluminum). Multicollinearity in some observational datasets makes it difficult or impossible to use all three quantities to good advantage in a model selected through the use of field observations alone; hierarchical Bayesian models provide a coherent logical structure for combining field data and bioassay data, despite uncertainties about the differences between field and laboratory settings, and offer a way to circumvent the multicollinearity.

Threat and tolerance. Denote by $\mathbf{X}_i$ the vector of explanatory variables $\mathbf{X}_i = (\text{pH}_i, \log[\text{Al}]_i, \log[\text{Ca}]_i)$ associated with the water chemistry of some lake (indexed by i). With each such vector is associated a "threat" ζ_i, but the correspondence $\mathbf{X}_i \mapsto \zeta_i = \zeta(\mathbf{X}_i)$ is uncertain. Initially take the association to be linear

$$\zeta_i = \mathbf{X}_i\beta = X_{i1} + X_{i2}\beta_2 + X_{i3}\beta_3,$$

for an uncertain 3-vector β, normalized by the constraint $\beta_1 = 1$. With each lake in the field dataset associate an uncertain "tolerance" τ_i representing unrecorded and uncontrolled environmental factors (food supply, water temperature, etc.) which may affect the lake's ability to support a brook trout population, with the understanding that the lake *will* support brook trout if $\tau_i \geq \zeta_i$ and it *will not* support trout if $\tau_i < \zeta_i$. If tolerances are taken to be random, drawn independently of $\mathbf{X}_i$ from a specified location-scale family with standardized CDF $\Psi(x)$ (say, the logistic) and uncertain location and scale parameters α_F and σ_F, then the probability that a lake with the explanatory variables $\mathbf{X}_i$ would be viable for brook trout would be

$$\mathbf{P}[\tau_i \geq \zeta_i] = 1 - \Psi\left(\frac{\zeta_i - \alpha_F}{\sigma_F}\right) = \left(1 + e^{(\mathbf{X}_i\beta - \alpha_F)/\sigma_F}\right)^{-1}.$$

In the laboratory bioassay experiments fish are presumed to be endowed with unobservable logistically distributed tolerances τ_j and are presumed to experience a constant hazard of $h_j = 0$ if $\tau_j \geq \mathbf{X}_j\beta$, and $h_j = (\mathbf{X}_j\beta - \tau_j)c$ if $\tau_j < \mathbf{X}_j\beta$; it follows that the probability p_j of surviving for at least the duration t_j of the bioassay is

$$p_j = Z_j + \frac{Z_j(1 - Z_j)}{1 + c\sigma_L t_j}\, {}_2F_1(1, 2; 1 + c\sigma_L t_j; 1 - Z_j),$$

where $Z_j = \left(1 + e^{(\mathbf{X}_j\beta - \alpha_L)/\sigma_L}\right)^{-1}$ is the probability $\mathbf{P}[\tau_j \geq \mathbf{X}_j\beta]$ and where ${}_2F_1(a, b; c; z)$ is Gauss' hypergeometric function. The second term in this expression is the probability of a right-censored death time. The likelihood function for the seven-dimensional parameter $\theta = (\beta_2, \beta_3, \alpha_F, \sigma_F, \alpha_L, \sigma_L, c)$ on the basis of 177 field observations ($P_i = 1$ for presence, $P_i = 0$ for absence of fish) and 164 bioassay observations (S_j surviving and D_j dead fish) is

$$L(\theta) = \prod_{i=1}^{177} \left(1 + e^{(\mathbf{X}_i\beta - \alpha_F)/\sigma_F}\right)^{-P_i} \left(1 + e^{-(\mathbf{X}_i\beta - \alpha_F)/\sigma_F}\right)^{P_i - 1}$$

$$\times \prod_{j=1}^{164} \left(Z_j + \frac{Z_j(1 - Z_j)}{1 + c\sigma_L t_j} \, {}_2F_1(1, 2; 1 + c\sigma_L t_j; 1 - Z_j)\right)^{S_j}$$

$$\times \prod_{j=1}^{164} \left(1 - Z_j - \frac{Z_j(1 - Z_j)}{1 + c\sigma_L t_j} \, {}_2F_1(1, 2; 1 + c\sigma_L t_j; 1 - Z_j)\right)^{D_j} .$$

The functions of interest include the marginal densities for each of the four parameters $(\beta_2, \beta_3, \alpha_F, \sigma_F)$ governing field observations, the posterior mean and covariance matrix for these parameters, and the predictive distributions for the probability of fish presence $\left(1 + e^{(\mathbf{X}\beta - \alpha_F)/\sigma_F}\right)^{-1}$ for several specified chemistries X (all of which are reported in Wolpert and Warren-Hicks (1990)). Features of this model include:

1. High-dimensional integrand (several hundred dimensions for the density estimation).
2. Slow computation for the likelihood function; transcendental and even special functions must be calculated for each observation at every point θ.
3. Slow computation of the prior density, if the Jeffreys prior is used; again transcendental functions must be calculated inside a loop, and now a 7×7 determinant must be evaluated as well.
4. A unimodal and "bell-like" likelihood function (after logarithmic transformations for σ_F and σ_L), with nearly elliptical contours for each pair of parameters.
5. Satisfactory and attainable precision of $\pm 1\%$ with several minutes' computation on a desktop Unix workstation.
6. Modest dimensionality (7) of the parameter space Θ. We also consider models with quadratic terms in the uncertain dependence of threat ζ upon the three explanatory variables, increasing the dimension to 13; the dimension increases quickly as higher order terms or more explanatory variables are added.
7. Changing *functions of interest* (predictive distributions were added), *likelihood function* (observations were added, and probit models were considered), and *prior density* (both uniform and Jeffreys priors were studied).
8. Badly conditioned linear approximations in the original coordinate system; the elliptical likelihood contours are highly eccentric, and

the condition numbers for the information matrices at the posterior modes are high.

Wolpert and Warren-Hicks (1990) found approximations to the predictive survival distributions and posterior parameter distributions for this model using Monte Carlo methods with antithetic variates drawn from multivariate Student t sampling distributions. A two-step adaptive importance sampling scheme was used with initial location vectors and dispersion matrices suggested by an analysis of the likelihood function's behaviour near its maximum, later improved using Monte Carlo estimates of the mean and covariance. The degrees-of-freedom parameters were chosen to match the tail behaviour of the likelihood function.

5. Appendix: implementation details

Elementary statistics textbooks sometimes give "shortcuts" for calculating the sample mean and variance from running totals $\mathrm{S}_X \equiv \sum X_i$ and $\mathrm{S}_{XX} \equiv \sum X_i^2$; unfortunately these widely-implemented formulas represent S_n^2 as the small difference of two large numbers, and so entail a great loss of precision. To find $S_3^2 = 2/3$ correctly to d decimals using this technique for the data set $\{2999, 3000, 3001\}$, for example, requires that all intermediate calculations be carried out accurately to about $7+d$ decimals; even for this simple problem *zero* decimals are available in Fortran single-precision! The expected number of bits of precision lost is about $\log_2\left(1+(\mu/\sigma)^2\right)$ in general, or 23 bits for the example.

One remedy for this unnecessary loss of precision is to use the defining relation for S_n^2: first compute $\overline{X}_n$, then sum the squared deviations $(X_i - \overline{X}_n)^2$. This is unattractive because it requires that all n observations be stored while the usual but imprecise formulas require only the three summary statistics N, S_X, and S_{XX}. An alternative is to initialize $\overline{X}_0 \equiv 0$ and $S_0^2 \equiv 0$, and then for $n \geq 1$ use the recursive formulas:

$$\Delta_n \equiv [X_n - \overline{X}_{n-1}],$$

$$\overline{X}_n \equiv \overline{X}_{n-1} + \frac{1}{n}\Delta_n,$$

$$S_n^2 \equiv \left[S_{n-1}^2 + \frac{1}{n}\Delta_n^2\right]\left(1 - \frac{1}{n}\right),$$

which lead to full-precision values for $\overline{X}_n$ and S_n^2.

The same precision problem arises in calculating the weighted mean $\overline{g}_n$ in Equation (6) and especially in estimating the precision (7); the use of antithetic variates to reduce σ can exacerbate the problem, since $[1+(\mu/\sigma)^2]$ is then so large. The algorithm presented below gives high-precision recursive estimates $\overline{g}_n$ of the mean vector $\overline{g}$, Σ_n of the covariance matrix Σ, and MSE_n of the estimation error matrix $\mathbf{E}(\overline{g}_n - \overline{g})(\overline{g}_n - \overline{g})'$, for any antithetic sampling scheme.

Recursive estimation formulas. Let $\pi(d\theta)$ be a prior measure and $L(\theta) = f(x|\theta)$ a likelihood function on some measure-space $(\Theta, \mathscr{F}, d\theta)$ and let $g(\theta)$ be an $\mathbb{R}^q$ valued measurable function on Θ. The problem at hand is to find a sequence of estimates

$$\overline{g}_n \approx \overline{g} \equiv \frac{\int g(\theta)L(\theta)\,\pi(d\theta)}{\int L(\theta)\,\pi(d\theta)},$$

$$\Sigma_n \approx \Sigma \equiv \frac{\int \left(g(\theta)-\overline{g}\right)\left(g(\theta)-\overline{g}\right)' L(\theta)\,\pi(d\theta)}{\int L(\theta)\,\pi(d\theta)},$$

$$\mathrm{MSE}_n \approx \mathbf{E}\left[\left(\overline{g}_n - \overline{g}\right)\left(\overline{g}_n - \overline{g}\right)' \mid X{=}x\right].$$

For each n choose an integer m_n and draw an antithetic series $\theta_1^{(n)}, \ldots, \theta_{m_n}^{(n)}$ consisting of m_n not-necessarily-independent draws $\theta_j^{(n)}$, each with marginal distribution $\Pi(d\theta)$. The number m_n of draws need not be constant as n varies. Within each antithetic series estimates will be calculated of $\overline{g}$ and Σ, each an average weighted by $\mathrm{w}_j^{(n)} = w(\theta_j^{(n)})$ at the observed points $\theta_j^{(n)}$.

Within each antithetic series. For each n set $\mathrm{w}_0^{(n)} \equiv 0$, $\overline{g}_0^{(n)} \equiv 0$, $\mathrm{msq}_0^{(n)} \equiv 0$. For $1 \le j \le m_n$, set

$$\begin{aligned}
\omega_j^{(n)} &\equiv w(\theta_j^{(n)}) && = L(\theta_j^{(n)})\,\pi(d\theta_j^{(n)})/\Pi(d\theta_j^{(n)}),\\
\mathrm{w}_j^{(n)} &\equiv \mathrm{w}_{j-1}^{(n)} + \omega_j^{(n)} && = \sum_{i\le j} \omega_i^{(n)},\\
h_j^{(n)} &\equiv \omega_j^{(n)}/\mathrm{w}_j^{(n)} && \\
g_j^{(n)} &\equiv g(\theta_j^{(n)}) && \\
\delta_j^{(n)} &\equiv g_j^{(n)} - \overline{g}_{j-1}^{(n)} && \\
\overline{g}_j^{(n)} &\equiv \overline{g}_{j-1}^{(n)} + h_j^{(n)}\delta_j^{(n)} && = \sum_{i\le j} \omega_i^{(n)} g_i^{(n)} \Big/ \sum_{i\le j} \omega_i^{(n)},\\
\mathrm{msq}_j^{(n)} &\equiv (1-h_j^{(n)})\left[\mathrm{msq}_{j-1}^{(n)} + h_j^{(n)}(\delta_j^{(n)})(\delta_j^{(n)})'\right] && \\
& && = \sum_{i\le j} \omega_i^{(n)}(g_i^{(n)}-\overline{g}_j^{(n)})(g_i^{(n)}-\overline{g}_j^{(n)})' \Big/ \sum_{i\le j}\omega_i^{(n)}.
\end{aligned}$$

The four summary statistics from the nth antithetic series are:

m_n	The number of function evaluations in the series;
$\mathrm{w}_n \equiv \mathrm{w}_{m_n}^{(n)}$	The total weight for the series;
$g_n \equiv \overline{g}_{m_n}^{(n)}$	The $w(\theta_j^{(n)})$-weighted average of the vectors $g(\theta_j^{(n)})$;
$\mathrm{msq}_n \equiv \mathrm{msq}_{m_n}^{(n)}$	The mean-square variation, i.e., the $w(\theta_j^{(n)})$-weighted average of the matrices $(g_i^{(n)}-g_n)(g_i^{(n)}-g_n)'$.

There is no information in these "within" statistics about the variability arising from the importance sampling because of the allowed dependence among the $\theta_j^{(n)}$ for a given n. This variability will be reflected in a "between" mean-square summarizing the variability of the vector quantities g_n across repeated independent replicates.

If samples were drawn directly from the posterior density, the function $w(\theta)$ would be constant and g_n and msq_n would be the unweighted average and sample variance of the $\{g_i\}_{i \le m_n}$, respectively. In the common case $m_n = 2$ the formulas above reduce to

$$
\begin{aligned}
m_n &= 2, \\
\mathrm{w}_n &= \omega_1^{(n)} + \omega_2^{(n)}, \\
g_n &= \Big(\frac{\omega_1^{(n)}}{\omega_1^{(n)} + \omega_2^{(n)}}\Big) g_1^{(n)} + \Big(\frac{\omega_2^{(n)}}{\omega_1^{(n)} + \omega_2^{(n)}}\Big) g_2^{(n)}, \\
\mathrm{msq}_n &= \Big(\frac{\omega_1^{(n)}}{\omega_1^{(n)} + \omega_2^{(n)}}\Big)\Big(\frac{\omega_2^{(n)}}{\omega_1^{(n)} + \omega_2^{(n)}}\Big)(g_1^{(n)} - \overline{g}_2^{(n)})(g_1^{(n)} - \overline{g}_2^{(n)})',
\end{aligned}
$$

or simply $w_n = 2$, $\overline{g} = (g_1^{(n)} + g_2^{(n)})/2$ and $\mathrm{msq}_n = (g_1^{(n)} - \overline{g}_2^{(n)})(g_1^{(n)} - \overline{g}_2^{(n)})'/4$ for direct sampling from the posterior.

Combining antithetic series. For estimating mean-square estimation errors it is useful to have both w-weighted and w^2-weighted averages; the latter we distinguish with asterisk superscripts. Initialize $N_n \equiv 0$, $W_0 \equiv 0$, $W_0^* \equiv 0$, $\overline{g}_0 \equiv 0$, $\overline{g}_0^* \equiv 0$, $\Sigma_0 \equiv 0$, $\mathrm{MSB}_0 \equiv 0$, and $\mathrm{MSB}_0^* \equiv 0$; with each succeeding n set

$$
\begin{aligned}
N_n \equiv N_{n-1} + m_n &= \sum_{j \le n} m_j, \\
\mathrm{w}_n \equiv \mathrm{w}_{m_n}^{(n)} &= \sum_{i \le m_n} \omega_i^{(n)}, \\
g_n \equiv \overline{g}_{m_n}^{(n)} &= \sum_{i \le m_n} \omega_i^{(n)} g_i^{(n)} / \mathrm{w}_n, \\
\mathrm{msq}_n \equiv \mathrm{msq}_{m_n}^{(n)} &= \sum_{i \le m_n} \omega_i^{(n)} (g_i^{(n)} - g_n)(g_i^{(n)} - g_n)' / \mathrm{w}_n,
\end{aligned}
$$

$$
\begin{aligned}
W_n &\equiv W_{n-1} + \mathrm{w}_n = \sum_{j \le n} \mathrm{w}_j, & W_n^* &\equiv W_{n-1}^* + (\mathrm{w}_n)^2 = \sum_{j \le n} (\mathrm{w}_j)^2, \\
H_n &\equiv \mathrm{w}_n / W_n, & H_n^* &\equiv (\mathrm{w}_n)^2 / W_n^*, \\
\Delta_n &\equiv [g_n - \overline{g}_{n-1}], & \Delta_n^* &\equiv [g_n - \overline{g}_{n-1}^*], \\
\overline{g}_n &\equiv \overline{g}_{n-1} + H_n \Delta_n = \sum_{j \le n} \mathrm{w}_j g_j / W_n, & \overline{g}_n^* &\equiv \overline{g}_{n-1}^* + H_n^* \Delta_n^* = \sum_{j \le n} (\mathrm{w}_j)^2 g_j / W_n^*,
\end{aligned}
$$

$$
\begin{aligned}
\mathrm{MSW}_n &\equiv (1-H_n)\mathrm{MSW}_{n-1} + H_n\,\mathrm{msq}_n \\
&= \frac{\sum_{j\le n} \mathrm{w}_j\,\mathrm{msq}_j}{\sum_{j\le n}\mathrm{w}_j}, \\
&= \frac{\sum_{j\le n}\sum_{i\le m_j} \omega_i^{(j)}(g_i^{(j)}-g_j)(g_i^{(j)}-g_j)'}{\sum_{j\le n}\sum_{i\le m_j}\omega_i^{(j)}}, \\
\mathrm{MSB}_n &\equiv (1-H_n)\left[\mathrm{MSB}_{n-1} + H_n(\Delta_n)(\Delta_n)'\right] \\
&= \frac{\sum_{j\le n}\mathrm{w}_j(g_j-\overline{g}_n)(g_j-\overline{g}_n)'}{\sum_{j\le n}\mathrm{w}_j}, \\
\mathrm{MSB}_n^* &\equiv (1-H_n^*)\left[\mathrm{MSB}_{n-1}^* + H_n^*(\Delta_n^*)(\Delta_n^*)'\right] \\
&= \frac{\sum_{j\le n}(\mathrm{w}_j)^2(g_j-\overline{g}_n^*)(g_j-\overline{g}_n^*)'}{\sum_{j\le n}(\mathrm{w}_j)^2}, \\
\Sigma_n &\equiv \mathrm{MSW}_n + \mathrm{MSB}_n \\
&= H_n\,\mathrm{msq}_n + (1-H_n)\left[\Sigma_{n-1} + H_n(\Delta_n)(\Delta_n)'\right], \\
&= \frac{\sum_{j\le n}\sum_{i\le m_j}\omega_i^{(j)}(g_i^{(j)}-\overline{g}_n)(g_i^{(j)}-\overline{g}_n)'}{\sum_{j\le n}\sum_{i\le m_j}\omega_i^{(j)}}.
\end{aligned}
$$

The estimates. With these in hand we can now estimate $\overline{g}$, the posterior expectation of $g(\theta)$, by

$$
\begin{aligned}
\overline{g}_n &= \sum\sum\omega_i^{(j)}g_i^{(j)}/\sum\sum\omega_i^{(j)} \\
&\approx \int w(\theta)g(\theta)\Pi(d\theta)\Big/\int w(\theta)\Pi(d\theta) \\
&= \int g(\theta)L(\theta)\,\pi(d\theta)\Big/\int w(\theta)\Pi(d\theta), \\
&= \overline{g},
\end{aligned}
$$

and the posterior variance by

$$
\begin{aligned}
\Sigma_n &\equiv \mathrm{MSW}_n + \mathrm{MSB}_n \\
&= \sum\sum\omega_i^{(j)}(g_i^{(j)}-\overline{g}_n)(g_i^{(j)}-\overline{g}_n)'/\sum\sum\omega_i^{(j)}, \\
&\approx \int w(\theta)(g(\theta)-\overline{g})(g(\theta)-\overline{g})'\Pi(d\theta)\Big/\int w(\theta)\Pi(d\theta) \\
&= \int (g(\theta)-\overline{g})(g(\theta)-\overline{g})'L(\theta)\,\pi(d\theta)\Big/\int L(\theta)\,\pi(d\theta), \\
&= \Sigma.
\end{aligned}
$$

The mean-square error of estimation is more interesting. A second-order Taylor-series expansion of the function $f(x, y, z) = xy/z^2$ about the means $\mu_X=0$, $\mu_Y=0$, and $\mu_Z\neq 0$ of three real-valued random variables X, Y, and

Z (not necessarily independent) reveals that $\mathbf{E}(XY/Z^2) \approx (\mathbf{E}XY)/(\mathbf{E}Z)^2$. Thus the mean-square estimation error is

$$\begin{aligned}
\mathbf{E}(\overline{g}_n - \overline{g})(\overline{g}_n - \overline{g})' &= \mathbf{E}\left(\frac{\sum_{j\le n} \mathrm{w}_j(g_j - \overline{g})}{\sum_{j\le n} \mathrm{w}_j}\right)\left(\frac{\sum_{j\le n} \mathrm{w}_j(g_j - \overline{g})}{\sum_{j\le n} \mathrm{w}_j}\right)', \\
&\approx \frac{\mathbf{E}\left(\sum_{j\le n} \mathrm{w}_j(g_j - \overline{g})\right)\left(\sum_{j\le n} \mathrm{w}_j(g_j - \overline{g})\right)'}{\left(\mathbf{E}\sum_{j\le n} \mathrm{w}_j\right)^2} \\
&= \frac{\mathbf{E}\sum_{j\le n} (\mathrm{w}_j)^2 (g_j - \overline{g})(g_j - \overline{g})'}{\left(\mathbf{E}\sum_{j\le n} \mathrm{w}_j\right)^2}, \\
&\approx \frac{\sum_{j\le n} (\mathrm{w}_j)^2 (g_j - \overline{g}_n)(g_j - \overline{g}_n)'}{\left(\sum_{j\le n} \mathrm{w}_j\right)^2} \\
&= \frac{1}{(W_n)^2} \sum_{j\le n} (\mathrm{w}_j)^2 \left(g_j - \overline{g}_n^* + \overline{g}_n^* - \overline{g}_n\right)\left(g_j - \overline{g}_n^* + \overline{g}_n^* - \overline{g}_n\right)', \\
&= \frac{1}{(W_n)^2} \sum_{j\le n} (\mathrm{w}_j)^2 [(g_j - \overline{g}_n^*)(g_j - \overline{g}_n^*)', \\
&\qquad\qquad + (\overline{g}_n^* - \overline{g}_n)(\overline{g}_n^* - \overline{g}_n)'] \\
= \mathrm{MSE}_n &\equiv \frac{W_n^*}{(W_n)^2} [\mathrm{MSB}_n^* + (\overline{g}_n^* - \overline{g}_n)(\overline{g}_n^* - \overline{g}_n)'].
\end{aligned}$$

This error estimate has two components: MSB_n^*, a measure of how much the g_j differ, and $(\overline{g}_n^* - \overline{g}_n)(\overline{g}_n^* - \overline{g}_n)'$, a measure of how much the w_j differ. The first component can be made small by capturing as much variability as possible *within* each antithetic series (in MSW_n) and leaving as little as possible in $\mathrm{MSB}_n \approx \mathrm{MSB}_n^*$, while the second component can be made small by sampling from an importance function similar to the posterior distribution so the weights w_j will be nearly constant and hence the (w_j)-weighted and $(w_j)^2$-weighted means $\overline{g}_n$ and $\overline{g}_n^*$ will be nearly equal.

With iid sampling from the posterior density function the mean-square error in estimating the ith component of $\overline{g}$ would have been

$$\begin{aligned}
\frac{1}{N_n}\boldsymbol{\Sigma}^{ii} &\approx \frac{1}{N_n}\boldsymbol{\Sigma}_n^{ii} \\
&= \mathrm{RE}_n^i \times \mathrm{MSE}_n^{ii}
\end{aligned}$$

where, following Hammersley and Handscomb, the "relative efficiency" for the ith component is defined to be

$$\begin{aligned}\mathrm{RE}_n^i &\equiv \left(\frac{1}{N_n}\Sigma_n^{ii}\right)/\mathrm{MSE}_n^{ii}\\ &= \frac{(W_n)^2\left[\mathrm{MSW}_n^{ii}+\mathrm{MSB}_n^{ii}\right]}{N_n W_n^*\left[\mathrm{MSB}_n^{*ii}+|(\overline{g}_n^*-\overline{g}_n)_i|^2\right]}.\end{aligned}$$

The RE indicates the efficiency of a given importance-sampling scheme, relative to the benchmark of iid sampling directly from the posterior distribution. The given procedure with n function evaluations attains the same precision as would iid sampling with $\mathrm{RE}\times n$ function evaluations. For some problems relative efficiencies of well over 100% are possible with well-chosen antithetic schemes.

References

J. O. Berger and R. L. Wolpert (1988), *The likelihood principle* (2nd Edition), Institute of Mathematical Statistics Press, Hayward.

J. O. Berger and J. M. Bernardo (1989), *Estimating a product of means: Bayesian analysis with reference priors*, J. Amer. Statist. Assoc. **84**, 200–207.

J. H. Curtiss et al. (1951), *Monte Carlo method*, National Bureau of Standards Applied Mathematics Series, **12**.

E. C. Fieller and H. O. Hartley (1954), *Sampling with control variables*, Biometrika **41**, 494–501.

A. E. Gelfand and A. F. M. Smith (1990), *Sampling based approaches to calculating marginal densities*, J. Amer. Statist. Assoc. **85**, 398–409.

S. Geman and D. Geman (1984), *Stochastic relaxation, Gibbs distributions, and the Bayesian restoration of images*, IEEE Trans. Pattern Analysis and Machine Intelligence **6**, 721–741.

J. Geweke (1988), *Antithetic acceleration of Monte Carlo integration in Bayesian inference*, J. Economics **38**, 73–89.

J. M. Hammersley and K. W. Morton (1956), *A new Monte Carlo technique: antithetic variates*, Proc. Cambridge Philos. Soc. **52**, 449–475.

J. M. Hammersley and D. C. Handscomb (1964), *Monte Carlo methods*, Chapman and Hall, London.

H. Jeffreys (1960), *Theory of probability* (3rd edition), Oxford University Press, London.

T. Kloek and H. J. van Dijk (1978), *Bayesian estimates of equation system parameters: an application of integration by Monte Carlo*, Econometrica **46**, 1–19.

J. C. Naylor and A. F. M. Smith (1982), *Applications of a method for the efficient computation of posterior distributions*, Appl. Stat. **31**, 214–225.

R. Y. Rubinstein (1981), *Simulation and the Monte Carlo method*, John Wiley and Sons, New York.

——(1986), *Monte Carlo optimization, simulation, and sensitivity of queueing networks*, John Wiley and Sons, New York.

A. F. M. Smith, A. M. Skene, J. E. H. Shaw, and M. Dransfield (1985), *The implementation of the Bayesian paradigm*, Comm. Stat. **A14**, 1079–1102.

M. Tanner and W. Wong (1987), *The calculation of posterior distributions by data augmentation (with discussion)*, J. Amer. Statist. Assoc. **82**, 528–550.

J. W. Tukey (1957), *Antithesis or regression?* Proc. Cambridge Philos. Soc. **53**, 923–924.

J. R. Wilson (1984), *Variance reduction techniques for digital simulation*, Amer. J. Math. Manage. Sci. **4**, 277–313.

R. L. Wolpert and W. J. Warren-Hicks (1990), *Bayesian hierarchical logistic models for combining field and laboratory data*, in preparation.

Institute of Statistics and Decision Sciences, Duke University, Durham, North Carolina 27706

E-mail address: rlw@isds.duke.edu

Contemporary Mathematics
Volume **115**, 1991

Generic, Algorithmic Approaches to Monte Carlo Integration in Bayesian Inference

JOHN GEWEKE

ABSTRACT. A program of research in generic, algorithmic approaches to Monte Carlo integration in Bayesian inference is summarized. The goal of this program is the development of a widely applicable family of solutions of Bayesian multiple integration problems, that obviate the need for case-by-case treatment of arcane problems in numerical analysis. The essentials of the Bayesian inference problem, with some reference to econometric applications, are set forth. Fundamental results in Monte Carlo integration are derived and their current implementation in software is described. Potential directions for fruitful new research are outlined.

1. Introduction

The central technical problem in Bayesian inference is multiple integration, often in high dimensions. The central technical problem in non-Bayesian inference is optimization, often in high dimensions. The latter problem has been solved with some degree of generality so that (for example) a statistician developing a new model may with confidence use a Newton–Raphson ascent algorithm to find maximum likelihood estimates, and approximate the sampling properties of the estimator from the Hessian of the likelihood function; and a consulting statistician has at his disposal software for the application of existing models to a new data set. The former problem has not been solved in any such degree of generality, because multiple integration is technically more demanding than is optimization. Problems in Bayesian inference have historically been solved on a case-by-case basis, with the

1980 *Mathematics Subject Classification* (1985 *Revision*). Primary G2F15, G5C05.

Key words and phrases. Antithetic acceleration, forecasting, importance sampling, signal extraction.

Financial support from NSF Grant SES-8908365 is gratefully acknowledged.

This paper is based on a talk by the author at the 1989 Joint Summer Research Conference on Statistical Multiple Integration, Humboldt State University, Arcata, California, June 18–22, 1989. This is an overview of research previously published elsewhere by the author.

solution of one problem providing little to help with the solution of the next. The statistician who must get on with the problem at hand usually has no practical Bayesian options. Given the desirability of Bayesian inference in principle, this is unfortunate. A necessary condition for the routine use of Bayesian inference in real (as opposed to illustrative) problems is the establishment of generic, algorithmic approaches to Bayesian multiple integration problems, that can be taken in new situations and are embodied in reliable software.

Bayesian multiple integration problems have several characteristics which distinguish them from those that have received the attention of numerical analysts.

(1) The dimension of integration is the number of unknown parameters. Serious, as opposed to illustrative, statistical models have at least six unknown parameters in most cases, and often many more: a few dozen is not uncommon.

(2) The functional form to be integrated is specific to the model, and to the information combined with the data to which the model is applied. There are perhaps a few dozen types of models which together account for the bulk of applied statistical work, but there are many minor variations on each and the information to be combined with the data is application-specific. The problem is not one of finding very efficient solutions to the integration of a handful of specific functions.

(3) The functional form to be integrated is also specific to the use to which the model is put. For example, in a forecasting application a few dozen functions of perhaps a score of parameters must be integrated each period, and these functions change frequently as the data are updated. The applied Bayesian statistician therefore confronts a multitude of high-dimensional integration problems that must be solved with negligible human time and reasonable computational expense per problem.

(4) Whereas each integration problem is unique, the integrand can often be approximated roughly as a multivariate normal distribution. Moreover, from classical statistics the situations in which this approximation is likely to be good or poor are well understood. Indeed, the reliability of the only generic non-Bayesian alternatives to Bayesian inference depend on the adequacy of this assumption.

(5) Most statisticians have no formal training and very limited experience in numerical analysis, and they do not work in groups that include this expertise. An acceptable solution to Bayesian multiple integration problems cannot require the intervention of a numerical analyst before it is brought to application.

(6) Contemporary innovations in statistical inference are widely applied only after incorporation in high-level statistical software. An attrac-

tive solution of the Bayesian multiple integration problem is one that can become part of currently widespread statistical software packages.

These characteristics of the problem imply some necessary characteristics of the solution.

(1) The solution must provide reliable answers in real time for *most* applied problems. Whereas faster algorithms and software are better than slow, the design of highly efficient or optimal procedures per se cannot be motivated by the demands of application.
(2) The solution must provide reliable indicators of reliability for *all* applied problems. It is not necessary, and probably is not possible, to design methods that work well in virtually all cases. But the statistician (who is not a numerical analyst) must known when the solution is not reliable.
(3) The solution must be adaptable to unforeseen uses. This includes not only models yet to be created, but also the functions of parameters of interest in applications. Algorithmic solutions are especially attractive because they lead to software libraries that can be applied in these unforeseen cases.
(4) The solutions must be incorporated in high-level statistical software.

This paper summarizes aspects of a program of research in generic, algorithmic approaches to Monte Carlo integration in Bayesian inference, whose goal is a solution of Bayesian multiple integration problems with these desirable characteristics. The next section sets forth the mathematical essentials of the Bayesian inference problem, with some reference to econometric applications. Section 3 derives some fundamental results in Monte Carlo integration, and some of the current implementations of Monte Carlo integration are described in §4. The final section is devoted to a summary and some speculations on the agenda for future research.

2. The Bayesian inference problem

2.1. The essentials. Statistical models are usually expressed in terms of probability density function $f(x|\theta)$ for a vector of observed data x which is known up to a vector of unknown parameters θ. Information, Q, independent of the data provides a prior density $\pi(\theta|Q)$ for the parameters. Conditional on the data, x, and the prior information, Q, the probability density for the unknown parameter θ is

$$p(\theta|Q, x) = f(x|\theta)\pi(\theta|Q)/h(x) \propto f(x|\theta)\pi(\theta|Q),$$

where $h(x)$ is the probability density function of x. The integral $\int_\Theta f(x|\theta)\pi(\theta|Q)\,d\theta$ typically converges, and in all that follows we shall assume this to be the case. The function $p(\theta|Q, x)$ is the posterior density for the parameters. Since its integral is one,

$$p(\theta|Q, x) = f(x|\theta)\pi(\theta|Q) \Big/ \int_{\Theta} f(x|\theta)\, \pi(\theta|Q)\, d\theta,$$

where Θ is the domain of θ. The function $L(\theta) \propto f(x|\theta)$ is known as the likelihood function; with this notation, and suppressing dependence on prior information Q and data x, we have

$$p(\theta) = L(\theta)\pi(\theta) \Big/ \int_{\Theta} L(\theta)\pi(\theta)\, d\theta.$$

Bayesian inference problems can be expressed as the evaluation of the expectation of a function of interest $g(\theta)$ with respect to the posterior probability density function,

$$\text{(1)} \quad E[g(\theta)] = \int_{\Theta} g(\theta)p(\theta)\, d\theta = \int_{\Theta} g(\theta)L(\theta)\pi(\theta)\, d\theta \Big/ \int_{\Theta} L(\theta)\pi(\theta)\, d\theta.$$

Of course, the integral in the numerator of this expression must converge. Establishing this convergence is the essential first step in any approach to the evaluation of (1), and in all that follows we shall assume that this has been done. The evaluation of (1) is the generic Bayesian multiple integration problem.

2.2. Some typical likelihood functions. A few examples will indicate the variety of functions $L(\theta)$ that arise in practice.

Multiple linear regression model. One of the simpler forms arises from the normal multiple linear regression model,

$$x_{ip} = \sum_{j=1}^{p-1} \beta_j x_{ij} + \varepsilon_i, \qquad \varepsilon_i \sim IIDN(0, \theta_p^2)$$

for observations $i = 1, \dots, n$. Standard textbook derivations lead to a log likelihood function

$$\text{(2)} \qquad \log[L(\theta)] = -(n/2)\log(\theta_p^2) - (1/2\theta_p^2)(\theta^* - \hat{\theta}^*)'C(\theta^* - \hat{\theta}^*),$$

where $\theta^* \equiv (\theta_1, \dots, \theta_{p-1})'$, and $\hat{\theta}^*$ and C are trivial functions of the data. Conditional on θ_p^2 this function is quadratic in θ^*, and so conditional on θ_p^2 the posterior probability density function of θ^* is multivariate normal. If $g(\theta)$ is a linear function of θ the Bayesian multiple integration problem can be solved analytically (e.g. Zellner (1971, ch. 3)). Typically $g(\theta)$ is not linear (§2.4) and there is no analytical solution. The number of parameters p varies greatly from one application to another: the range of 10 to 20 is common, and $p > 50$ is not rare (Geweke (1986)).

Markov chain models. The data x_{rt} take on the integer values $1, \dots, m$, indicating one of m discrete states occupied by individual r at time t;

$$P[x_{rt} = j | x_{r,t-1} = i;\ x_{r,t-2}, x_{r,t-3}, \dots] = \theta_{ij},$$

where $r = 1, \ldots, n$; $t = 1, \ldots, T \geq 2$; $i = 1, \ldots, m$; $j = 1, \ldots, m$; and $\sum_{j=1}^{m} \theta_{ij} = 1$. Then

$$\log[L(\theta)] = \sum_{i=1}^{m} \sum_{j=1}^{m} m_{ij} \log(\theta_{ij}), \tag{3}$$

where m_{ij} denotes the number of observations for which $x_{r,t-1} = i$ and $x_{rt} = j$. If the data actually arise from the specified model with all $\theta_{ij} > 0$, then as the number of observations n becomes large, $\log[L(\theta)]$ becomes quadratic (Wald (1943)). Thus, for sufficiently large n, the posterior density of the θ_{ij} will be approximately normal. If the prior density is from the beta distribution family and the functions of interest are the parameters themselves, then the Bayesian multiple integration problem can be solved analytically (Zellner (1971, §2.13) and Geweke (1989b, §5)). This is a very special case. Applications of this model with scores of parameters are common (Geweke, Marshall, and Zarkin (1986)).

Nonlinear, multivariate, multiple regression models. A commonly occurring model in econometrics is

$$x_{ij} = f_j(x_i^*, \theta_1) + \varepsilon_{ij}, \quad (j = 1, \ldots, m; \ i = 1, \ldots, n);$$
$$(\varepsilon_{i1}, \ldots, \varepsilon_{im})' \sim IIDN(0, \textstyle\sum(\theta_2)) \quad (i = 1, \ldots, n).$$

Typically the variance matrix $\sum(\theta_2)$ is restricted only to be positive definite, but the functional forms of the m equations are nonlinear in the vector of parameters θ_1 which appears in all of them. For example, the widely applied neoclassical econometric production and consumption models are of this form (Deaton (1986) and Jorgenson (1986)). The vector θ_2 contains the $m(m+1)/2$ elements of the Choleski factorization L of the inverse of the variance matrix $\sum$. The log likelihood function is

$$-n\sum_{i=1}^{m} \log(L_{ii}) - \left(\frac{1}{2}\right) \sum_{i=1}^{n} \sum_{r=1}^{m} \sum_{s=1}^{m} \sum_{u=1}^{r} \sum_{v=1}^{s} L_{ru} L_{sv} [x_{ir} - f_r(x_i^*, \theta_1)][x_{is} - f_s(x_i^*, \theta_1)].$$

The Bayesian multiple integration problem generally cannot be solved analytically, for any function of interest or priors.

2.3. Some typical prior densities. Prior densities often fall into one of three categories.

Noninformative prior densities. In many applications of Bayesian inference very little is known about the parameters beyond the information in the data. In these cases, the prior density may be taken to be the limit of a sequence of prior densities which place an increasingly diffuse distribution on θ. This limit need not itself be a probability density function. As an example, suppose that the prior density $\pi(\theta)$ for a single parameter θ is from the gamma family,

$$\pi(\theta) = [\beta^{\alpha+1}/\Gamma(\alpha+1)]\theta^{\alpha} \exp(-\theta\beta), \qquad \theta > 0,$$

which has mean β/α and variance $(\alpha+1)/\beta^2$. Fixing the mean and allowing the variance to increase without bound, $\lim[\pi(\theta)] = \theta^{-1}$ for all θ; since the integral of this function diverges on $\theta > 0$, the limit is not a proper probability density function. However, the limit of (1) will still be found by replacing $\pi(\theta)$ with θ^{-1}.

Noninformative priors are often chosen to render the Bayesian multiple integration problem analytically more tractable, rather than to reflect a particular form of prior knowledge; Zellner (1971) provides many examples. For the objectives of this paper it suffices to note that for noninformative or improper priors, the shape of $p(\theta)$ is essentially the shape of $L(\theta)$.

Restrictions on Θ. In many instances the prior will specify that $\theta \in \Theta^* \in \Theta$ by setting $\pi(\theta) = 0$ on $\Theta - \Theta^*$, and perhaps imposing a diffuse prior elsewhere. The imposition of linear restrictions in the multiple linear regression model is a very commonly arising instance (Geweke, (1986)). In many circumstances the boundaries of Θ cannot be expressed in closed form, although for any given θ it is straightforward to determine whether or not $\theta \in \Theta$. Examples include the imposition of stability on stochastic difference equations in the multiple linear regression model (Geweke, (1988a)) and the imposition of embeddability in the Markov chain model (Geweke, Marshall, and Zarkin, (1986)). Restrictions like these are very common in applications, and in their presence the Bayesian multiple integration problem rarely has an analytical solution.

Informative prior densities. When the prior density is informative, the posterior density may be quite different from the likelihood function. In particular, while the likelihood function may be well approximated by a multivariate normal probability density function, the posterior density need not be.

2.4. Functions of interest. In most applications the parameter vector θ is a technically useful device for indexing a family of probability density functions, but the parameters themselves are not of any inherent interest. In a typical application there will be several, or many, functions of interest. As noted, when these functions are linear, the likelihood function is simple, and the prior density is carefully chosen, an analytical solution of the Bayesian multiple integration problem may be attainable. Generally functions of interest are not linear. Three classes of nonlinear functions of interest are worth noting.

Closed form nonlinear expressions. Statistical models are often used to evaluate conditional propositions, e.g., the rate of change of an element in x_1 with respect to the rate of change of an element in x_2 in the nonlinear multivariate multiple regression model, which involves derivatives of the f_j. These derivatives, in turn, are nonlinear functions of the parameters. By choosing $g(\theta)$ to be an indicator function, 1 on Θ^* and 0 elsewhere, (1) becomes the posterior probability $P[\theta \in \Theta^*]$. This, in turn, becomes the basis for construction of histograms corresponding to the posterior density of a function; i.e., by selecting $g(\theta) = 1$ if $h(\theta) \in H^*$ and 0 elsewhere, a histogram for the posterior probability density function of $h(\theta)$ may be formed. Observe that for indicator functions, the convergence of the integral in the numerator of (1) is immediate if the integral in the denominator converges.

Implicit functions of interest. A function of interest may be implicit. Examples include evaluation of the posterior probability of the concavity of a function at a point, which reduces to determining the largest eigenvalue of the matrix of derivatives (Barnett, Geweke, and Yue (1990)), or the polar representation of the roots of a stochastic difference equation (Geweke, (1988a)). It would at best be extremely awkward to parameterize the models directly in terms of such functions of interest.

Integrals. In forecasting or signal extraction applications, the objective is to determine $P[x^* \in X]$, where x^* is a vector observed in the future, or a vector of lantent variables. Typically $g(\theta) = P[x^* \in X|\theta]$ can be formed, sometimes as a closed form expression; but more often this is an integration problem itself. (In either case, since $0 \leq g(\theta) \leq 1$, the convergence of the integral in the numerator of (1) is not at issue, given convergence of the integral in the denominator of (1).) The results described in §3 can be applied to this integration problem (Geweke (1989a)). In many instances the evaluation of this probability amounts to simulation of the model with the parameter vector fixed through the conditioning on θ. Such simulation is often much more straightforward than any attempt to provide the probability density function for x^* explicitly. Such problems constitute much of the work of applied statisticians, and it is interesting to note that they require the solution of two multiple integration problems.

3. Monte Carlo integration

3.1. Simple Monte Carlo integration. The principle of Monte Carlo integration is evident in an important class of simple cases. This class is defined by the ability to generate a sequence of synthetic, independent and identically distributed random vectors $\{\theta_i\}$, from the probability density function

$$L(\theta)\pi(\theta) \Big/ \int_\Theta L(\theta)\pi(\theta)\,d\theta, \quad \text{or} \quad L(\theta) \Big/ \int_\Theta L(\theta)\,d\theta.$$

For example, in the multiple linear regression model if the variance parameter $\theta_p^2 \sim s^2\chi^2(n-p)$ (distributed as chi-square with $n-p$ degrees of freedom, scaled by a function of the data s^2), and conditional on θ_p^2, $\theta^* \sim N(\hat{\theta}, \theta_p^2 C^{-1})$ (distributed as multivariate normal with mean $\hat{\theta}$ and variance matrix $\theta_p^2 C^{-1}$) then the logarithm of their joint probability density function is (2). In the Markov chain model, if $z_{ij} \sim \chi_2(2m_{ij}+2)$, all the z_{ij} are independent, and $\theta_{ij} = z_{ji} / \sum_{j=1}^{m} z_{ij}$, then the logarithm of the joint probability density function of the θ_j is given by (3), up to an additive constant. In these cases,

$$\tilde{E}[g(\theta_i)\pi(\theta_i)] = \int_\Theta g(\theta)L(\theta)\pi(\theta)\,d\theta \Big/ \int_\Theta L(\theta)\pi(\theta)\,d\theta = E[g(\theta)] \ .$$

(Here, and in what follows, the tilde above expectation operators denotes integration against the distribution generating the random vectors θ_i. Unmodified expectation operators continue to denote expectation with respect to the posterior density.) If $\{\theta_i\}_{i=1}^{n}$ is a sequence of independent drawings with probability density function $L(\theta)/\int_\Theta L(\theta)\,d\theta$, then by the strong law of large numbers

$$\overline{g}_n \equiv \sum_{i=1}^{n} g(\theta_i)\pi(\theta_i) \Big/ \sum_{i=1}^{n} \pi(\theta_i) \to \tilde{E}[g(\theta_i)\pi(\theta_i)] = E[g(\theta)] \equiv \overline{g},$$

where "$\to$" denotes almost sure convergence in n, the number of Monte Carlo replications. If $\tilde{\text{var}}[g(\theta)\pi(\theta)]$ exists, then by the central limit theorem,

$$n^{1/2}(\overline{g}_n - \overline{g}) \Rightarrow N(0, \tilde{\text{var}}[g(\theta)\pi(\theta)]),$$

where "$\Rightarrow$" denotes convergence in distribution. We therefore take $\overline{g}_n$ to be a Monte Carlo numerical approximation to $\overline{g}$ based on n replications, and

$$\left\{ [n(n-1)]^{-1} \sum_{i=1}^{n} [g(\theta_i)\pi(\theta_i) - \overline{g}_n]^2 \right\}^{1/2}$$

to be the standard error of the distribution of that approximation.

Similarly one can sometimes generate synthetic, independent and identically distributed random vectors from the posterior density $L(\theta)\pi(\theta)/\int_\Theta L(\theta)\pi(\theta)\,d\theta$. In the multiple linear regression model with the improper prior $(\theta_p^2)^{-1/2}$ on θ^2, the posterior density is characterized by $\theta_p^2 \sim s^2\chi^2(n-p+1)$ (where s^2 is the same function of the data as previously), and conditional on θ_p^2, $\theta^* \sim N(\hat{\theta}, \theta_p^2 C^{-1})$. In the Markov chain model, if the logarithm of the prior density is of the form $\sum_{i=1}^{m}\sum_{j=1}^{m} \alpha_{ij}\log(\theta_{ij})$ up to an additive constant, then θ_{ij} may be drawn from the posterior density as before, if $z_{ij} \sim \chi^2(2m_{ij} + 2\alpha_{ij} + 2)$. In these cases,

$$\tilde{E}[g(\theta_i)] = E[g(\theta)],$$

$$\overline{g}_n \equiv \sum_{i=1}^{n} g(\theta_i) \to \tilde{E}[g(\theta_i)] = E[(g(\theta)] \equiv \overline{g},$$

and if the posterior variance of $g(\theta)$ exists, then

$$n^{1/2}(\overline{g}_n - \overline{g}) \Rightarrow N(0, \text{var}[g(\theta)]) .$$

This procedure possesses several of the desirable features of a solution of the Bayesian multiple integration problem listed in §1. It provides an indication of the reliability of the approximation, through the computation of the standard errors

$$\left\{[n(n-1)]^{-1}\sum_{i=1}^{n}[g(\theta_i)\pi(\theta_i)-\overline{g}_n]^2\right\}^{1/2}$$

or

$$\left\{[n(n-1)]^{-1}\sum_{i=1}^{n}[g(\theta_i)-\overline{g}_n]^2\right\}^{1/2}.$$

It is easy to adapt the method to different functions of interest, $g(\,)$. Observe that the accuracy of approximation does not depend in any direct way on the dimensionality of the integration problem. For large values of n, it is only $\tilde{\text{var}}[g(\theta)\pi(\theta)]$ or $\text{var}[g(\theta)]$, and the size n, that matter. For smaller values of n, nonnormality of $\overline{g}_n$ may be important in determining the adequacy of the limiting approximation. What matters is the dispersion of $g(\theta)\pi(\theta)$ relative to $L(\theta)$, or of $g(\theta)$ relative to $L(\theta)\pi(\theta)$. In the former case, this means that sampling from the likelihood function will be adequate if the behavior of $g(\theta)\pi(\theta)L(\theta)$ is driven by the behavior of $L(\theta)$ when $L(\theta)$ is small; this precludes informative priors and functions of interest with very high second moments under the posterior. In the latter case, it means that sampling from the posterior will be adequate if the behavior of $g(\theta)\pi(\theta)L(\theta)$ is dominated by the behavior of $\pi(\theta)L(\theta)$ when $\pi(\theta)L(\theta)$ is small. *It is not necessary for the statistician to establish these properties: their failure will be indicated by an unsatisfactory standard error of numerical approximation.*

The chief limitation of simple Monte Carlo integration is its requirement that one be able to generate synthetic, independent and identically distributed random vectors whose probability density function matches that of the likelihood function or posterior density. The multiple linear regression model and the Markov chain model are special cases: in general, this cannot be done.

3.2. Monte Carlo integration with importance sampling. More generally, suppose that the posterior probability density function of the θ_i is $I(\theta)$, termed the *importance sampling density.* The Monte Carlo approximation to $\overline{g} = E[g(\theta)]$ is then

$$\overline{g}_n \equiv \sum_{i=1}^{n}[g(\theta_i)L(\theta_i)\pi(\theta_i)/I(\theta_i)] \Big/ \sum_{i=1}^{n}[L(\theta_i)\,\pi(\theta_i)/I(\theta_i)].$$

To simplify the notation and highlight an important aspect of this approximation, define the *weight function*:

$$w(\theta) = \pi(\theta)L(\theta)/I(\theta).$$

With this notation,

$$\overline{g}_n \equiv \sum_{i=1}^{n} g(\theta_i)w(\theta_i) \Big/ \sum_{i=1}^{n} w(\theta_i).$$

Simple Monte Carlo integration is either of the special cases $I(\theta_i) \propto L(\theta_i)$, $w(\theta_i) = k_1\pi(\theta_i)$, or $I(\theta_i) \propto L(\theta_i)\pi(\theta_i)$, $w(\theta_i) = k_2$.

So long as the support of $I(\theta)$ includes the support of the posterior density [equivalently, the support of $L(\theta)\pi(\theta)$]—i.e., so long as sampling from the importance sampling density does not systematically exclude regions of Θ relevant to the Bayesian multiple integration problem, $\overline{g}_n \to \overline{g} = E[g(\theta)]$ (Geweke (1989b, Theorem 1)). The assumption about the support of $I(\theta)$ is quite weak, but so is the result: nothing is asserted about the rate of convergence. What is needed is an appropriate central limit theorem.

THEOREM (Geweke (1989b, Theorem 2)). *Suppose that* $E[w(\theta)] < \infty$, $E[g(\theta)^2 w(\theta)] < \infty$. *Let*

$$\sigma^2 = E\{[g(\theta) - \overline{g}]^2 w(\theta)\}, \tag{4}$$

$$\hat{\sigma}_n^2 = \sum_{i=1}^{n}[g(\theta_i) - \overline{g}_n]^2 w(\theta_i)^2 \Big/ \left[\sum_{i=1}^{n} w(\theta_i)\right]^2 .$$

Then

$$n^{1/2}(\overline{g}_n - \overline{g}) \Rightarrow N(0, \sigma^2]),$$

$$n\hat{\sigma}_n^2 \to \sigma^2 .$$

We shall refer to $\hat{\sigma}_n \equiv (\hat{\sigma}_n^2)^{1/2}$ as the numerical standard error of $\overline{g}_n$. The conditions of this theorem must be verified, analytically, if the computed numerical standard error is to be used to assess the accuracy of $\overline{g}_n$ as an approximation of $E[g(\theta)]$. These conditions typically can be established by showing either

$$w(\theta) < \overline{w} < \infty \ \forall\theta \in \Theta, \quad \text{and} \quad \text{var}[g(\theta)] < \infty; \tag{5}$$

or,

$$\Theta \text{ is compact}, \quad p(\theta) < \overline{p} < \infty, \quad I(\theta) > \varepsilon > 0 \ \forall\theta \in \Theta. \tag{6}$$

Demonstration of (5) involves comparison of the tail behaviors of $L(\theta)\pi(\theta)$ and $I(\theta)$. Demonstration of (6) is generally simple when Θ is compact. Meeting these conditions does not establish the reliability of $\overline{g}_n$ and $\hat{\sigma}_n$ in any practical sense: σ^2 may still be so large that $\overline{g}_n$ is unreliable even for very large n. Rather, these conditions provide a starting point for the algorithmic construction of $I(\theta)$, to which we return in §4.

3.3. Appraising the adequacy of the importance sampling density. The expression for σ^2 indicates that the numerical standard error is adversely affected by large $\text{var}[g(\theta)]$, and by large relative values of the weight function. The former is inherent in the function of interest and in the posterior density, but the latter can in principle be controlled through the choice of the importance sampling density. A simple benchmark for comparing the adequacy of importance sampling densities is the numerical standard error that would result if the importance sampling density were the posterior density itself, i.e., $I(\theta) \propto \pi(\theta)L(\theta)$. In this case $\sigma^2 = \text{var}[g(\theta)]$ and the number of replications controls the numerical standard error relative to the posterior standard deviation of the function of interest—e.g., $n = 10,000$ implies the former will be one percent of the latter. This is a very appealing metric for numerical accuracy.

As we have noted, only in particular cases is it possible to sample directly from the posterior density. But since $\text{var}[g(\theta)]$ can be approximated numerically as a routine byproduct of Monte Carlo integration, it is possible to see what the numerical variance would have been, had this been possible. Define the relative numerical efficiency of the importance sampling density for the function of interest $g(\theta)$,

$$\text{RNE} \equiv \text{var}[g(\theta)]/\sigma^2 .$$

The inverse of RNE is the ratio of the number of replications required to achieve any specified numerical standard error using the importance sampling density $I(\theta)$, to the number required using the posterior density as the importance sampling density. The numerical standard error for $\overline{g}_n$ is the fraction $(\text{RNE}\cdot n)^{-1/2}$ of the posterior standard deviation.

Low values of RNE indicate that there exists an importance sampling density (namely, the posterior density itself) that does not have to be tailored specifically to the function of interest, and provides substantially greater numerical efficiency. Thus they alert the statistician to the possibility that more efficient, yet practical, importance sampling densities might be found.

3.4. Acceleration methods. Improvements on Monte Carlo integration with importance sampling may be achieved by drawing an m-tuple $\theta_i^1, \ldots, \theta_i^m$ on each Monte Carlo replication rather than a single drawing θ_i. Each θ_i^j is drawn from the same importance sampling density $I(\theta)$, with θ_i^j independent of θ_k^h if $i \neq k$, but θ_i^j and θ_i^h are in general dependent. The value of $\overline{g}_n = E[g(\theta)]$ is numerically approximated by

$$\overline{g}_{n,m} \equiv \sum_{i=1}^{n}\sum_{j=1}^{m} g(\theta_i^j)w(\theta_i^j) \Big/ \sum_{i=1}^{n}\sum_{j=1}^{m} w(\theta_i^j) .$$

Under the conditions of the theorem in §3.2,

$$n^{1/2}(\overline{g}_{n,m} - \overline{g}) \Rightarrow N(0, \sigma_m^2]),$$

as $n \to \infty$ with m fixed; and defining

$$\hat{\sigma}_{n,m}^2 = \sum_{i=1}^{n}\left\{\left[\sum_{j=1}^{m} w(\theta_i^j)g(\theta_i^j) \Big/ \sum_{j=1}^{m} w(\theta_i^j) - \overline{g}_{n,m}\right]^2\right\} \Big/ \left[\sum_{i=1}^{n}\sum_{j=1}^{m} w(\theta_i^j)\right]^2 ,$$

then

$$n\hat{\sigma}_{n,m}^2 \to \sigma^2 \quad \text{(Geweke, (1988b))}.$$

The accuracy of any scheme for choosing an m-tuple at each Monte Carlo replication may be assessed through its numerical standard error $(\hat{\sigma}^2_{n,m})^{1/2}$. The efficiency of this procedure stems from the fact that in many situations there are very simple schemes for which $\sigma^2_m \ll \sigma^2$. We note two methods here.

Antithetic acceleration. A simple, generic method is based on the techniques of antithetic variants introduced by Hammersley and Morton (1956). A pair of identically distributed but negatively correlated vectors, θ^1_i and θ^2_i, are drawn from $I(\ )$. If the function of interest $g(\ )$ and weight function $w(\ )$ are smooth, then $w(\theta^1_i)g(\theta^1_i)$ and $w(\theta^2_i)g(\theta^2_i)$ will tend to be negatively correlated as well; the more negative the correlation the smaller is σ^2_m, and as the correlation approaches -1, σ^2_m approaches 0. Geweke (1988c) has demonstrated that this must happen as sample size increases in a special but interesting class of statistical models, and that σ^2_m/σ^2 is then inversely proportional to sample size. Experience with this method (Geweke (1988b, 1988c, 1989a)) suggests that this relationship is valid much more generally. The ratio $\operatorname{var}[g(\theta)]/\sigma^2_m$ is the RNE of the antithetically accelerated importance sampler: examples in Geweke (1988c) show RNE's of over 100.

Grid acceleration. A feasible method for problems of k dimensions, where k is small, proceeds as follows. Let $u^* = (u^*_1, \dots, u^*_k)'$ be chosen at random from the unit hypercube in R^k, and define an h-grid in R^k by all points of the form $(u_1, \dots, u_k)'$, $u_j = u^*_j + i/h$ $(i = 0, \dots, h-1)$ $(j = 1, \dots, k)$ modulo 1. Map this grid into h^k points in Θ via the inverse cumulative distribution function of the importance sampling density $I(\theta)$. This method suffers from the same curse of dimensionality as do quadrature methods. (Although both methods can cope with high dimensional problems in which the integration of all but k dimensions may be carried out analytically, and k is small.) However, it has the important advantage that its accuracy may be assessed as a trivial byproduct of the computations through $\hat{\sigma}^2_{n,m}$, rather than by increasing computation time several fold to the next order of approximation and loosely appealing to a Cauchy convergence notion as is typically done with quadrature. RNE's of over 1,000 are demonstrated for this method in a real problem worked in Geweke (1988b).

4. Algorithmic implementations

4.1. The problem. The appropriate choice of the importance sampling density is critical to the success of Monte Carlo integration with importance sampling. As discussed in §1, however, it is not practical to require validation of the conditions set out in §3.2 for every application of Bayesian inference. The strategy taken in algorithmic implementations is, rather, to define a class of importance sampling densities $I(\theta, \alpha)$, indexed by the vector α. The conditions of the theorem in §3.2 should be met for at least some values of α; the idea is to devise an algorithm which will determine a value of α within this set which produces a good RNE (relative to that produced by other values of α) for functions of interest of the kind typically employed. The process of choosing α can be transparent to the investigator. The RNE for each function of interest can be computed routinely, and the investigator may use the RNE and $\hat{\sigma}_n$ to identify any cases in which the algorithm for choosing α has produced an insufficiently accurate numerical approximation to $E[g(\theta)]$.

The construction of families of importance sampling densities is based on the behavior of likelihood functions for very wide classes of statistical models, whose study goes back over fifty years. To review the essentials of this theory, let $\hat{\theta}$ denote the mode of a log-likelihood function that is at least thrice-differentiable with respect to the parameter vector θ. Let H denote the Hessian of the log-likelihood function at $\hat{\theta}$. Suppose furthermore that the data were actually generated by the hypothesized model, with $\theta = \theta^*$. Then, as sample size increases, the difference between the likelihood function and a multivariate normal distribution with mean $\hat{\theta}$ and variance H^{-1} becomes small. Especially for a diffuse prior, this suggests that a multivariate normal importance sampling density would be a good approximation to the posterior density.

This suggestion is misplaced if taken literally. The multivariate normal approximation to the posterior density is good in the sense that confidence intervals of fixed size come into increasingly close agreement as sample size increases. But it is poor in the sense of the theorem in §3.2—in fact, as discussed in Geweke (1989b), the conditions of this theorem will never be met using the multivariate normal approximation, no matter how large the sample size, in most applications. The difficulty is that the tails of posterior densities typically decay algebraically whereas those of the multivariate normal decay exponentially, leading to an unbounded weight function.

4.2. The multivariate split Student importance sampling density. The rate of decay of the likelihood function, in its tails, can generally be identified from inspection of the log-likelihood function. Occasionally this rate is exponential, and in these cases the multivariate normal is an appropriate family from which to choose an importance sampling density. Much more often, the rate of decay matches that of a multivariate Student distribution with degrees of freedom that are readily determined. A candidate importance sampling density is then the multivariate Student with indicated degrees of freedom, mean equal to the posterior mode, and variance equal to minus the inverse of the Hessian of the log-posterior evaluated at its mode.

In practice this importance sampling density can be poor, because the Hessian poorly predicts (especially if it underpredicts) the behavior of the posterior density away from the mode, because the posterior density is substantially asymmetric, or both. To cope with them, we modify the multivariate normal or Student, comparing the posterior density and importance sampling density in each direction along each of the axes indicated by a Choleski decomposition of minus the inverse of the Hessian, in each case adjusting the importance sampling density as required to keep the weight function bounded. With k dimensions, there are $2k$ adjustments, so α is a $2k \times 1$ vector.

A little more formally, let $\operatorname{sgn}^+(\,)$ be the indicator function for nonnegative real numbers, and let $\operatorname{sgn}^-(\,) = 1 - \operatorname{sgn}^+(\,)$. The k-variate split normal density $N^*(\mu, T, q, r)$ is described by construction of a member x of its population:

$$\varepsilon \sim N(0, I_k);$$
$$\eta_i = [q_i \operatorname{sgn}^+(\varepsilon_i) + r_i \operatorname{sgn}^-(\varepsilon_i)]\varepsilon_i \qquad (i = 1, \ldots, k);$$
$$x = \mu + T\eta.$$

The log-probability density function is (up to an additive constant)

$$-\sum_{i=1}^{n}[\log(q_i)\operatorname{sgn}^+(\varepsilon_i) + \log(r_i)\operatorname{sgn}^-(\varepsilon_i)] - (1/2)\varepsilon'\varepsilon.$$

In our application, μ is the posterior mode, and T is a factorization such that the inverse of TT' is the negative of the Hessian of the log posterior density evaluated at its mode. A variate from the multivariate split Student density $t^*(\mu, T, q, r, n)$ is constructed the same way, except that

$$\eta_i = [q_i \operatorname{sgn}^+(\varepsilon_i) + r_i \operatorname{sgn}^-(\varepsilon_i)]\varepsilon_i(\zeta/\nu)^{-1/2},$$

with $\zeta \sim \chi^2(\nu)$. The log-probability density function is (up to an additive constant)

$$-\sum_{i=1}^{n}[\log(q_i)\operatorname{sgn}^+(\varepsilon_i) + \log(r_i)\operatorname{sgn}^-(\varepsilon_i)] - [(\nu + k)/2]\log(1 + \nu^{-1}\varepsilon'\varepsilon).$$

To select appropriate values for q and r, we explore each axis in each direction, to find that slowest rate of decline in the posterior density relative to a univariate normal (or Student) density, and then choose the variance of the normal (or Student) density to match that slowest rate of decline. A little more formally, let $e(i)$ be a $k \times 1$ indicator vector, $e_i^{(i)} = e^{(i)\prime}e^{(i)} = 1$. For the split normal define $f_i(\delta)$ according to

$$p(\hat{\theta} + \delta T e^{(i)})/p(\hat{\theta}) = \exp[-\delta^2/2f_i(\delta)^2]$$
$$\Rightarrow f_i(\delta) = |\delta|\{2[\log(p(\hat{\theta})) - \log(p(\hat{\theta} + \delta T e^{(i)}))]\}^{-1/2}$$

Then take $q_i = \sup_{\delta>0} f_i(\delta)$ and $r_i = \sup_{\delta<0} f_i(\delta)$. For the split Student the procedure is the same except that

$$f_i(\delta) = \nu^{-1/2}|\delta|\{[p(\hat{\theta})/p(\hat{\theta} + \delta T e^{(i)})]^{2/(\nu+k)} - 1]\}^{-1/2}.$$

In practice, carrying out the evaluation for $\delta = 1/2, 1, \ldots, 6$, seems to be satisfactory.

This algorithm has been incorporated in software, which in turn may be applied to many statistical models. The determination of the posterior mode, and the Hessian of the log-posterior at the mode, are standard computations in non-Bayesian inference. Thus, the algorithmic construction of the importance sampling density for the split Student family is a step following the determination of the mode, and it is as transparent to the user of the software as is mode determination itself.

4.3. Mixed importance sampling and acceptance–rejection methods. The RNE attainable from the split Student family, or from any importance sampling density family, depends on the regularity of the posterior density. For irregular posterior densities—i.e., for those that are not well approximated by any member of the family of potential importance sampling densities, RNE will be small. In these cases, a high fraction of the total weight $\sum_{i=1}^{n} w(\theta_i)$ is contributed by just a small fraction of the θ_i. If the computation time required for evaluation of the functions of interest $g(\theta_i)$ is small relative to the computation of $w(\theta_i)$, little is lost by proceeding with the evaluation of the functions of interest. But if this time is substantial, then the bulk of the computing time may well be devoted to evaluation of functions of interest of θ_i for which $w(\theta_i)$ is so small that the contribution to $\overline{g}_n$ is negligible.

In such cases, it will be more efficient to revise the weight function from $w(\theta)$ to $\tilde{w}(\theta)$,

$$\tilde{w}(0) = w(\theta) \text{ if } w(\theta) > w^*$$

$$\tilde{w}(\theta) = \begin{cases} w^* & \text{with probability } w(\theta)/w^* \\ 0 & \text{with probability } 1 - w(\theta)/w^* \end{cases} \quad \text{if } w(\theta) \leq w^*.$$

In those cases for which $\tilde{w}(\theta) = 0$, the functions of interest need not be evaluated. One can easily verify the results of §3.2 for these modified weight functions, which were first employed in Barnett, Geweke, and Yue (1990). Given the relative computation times for the weight function, on the one hand, and the functions of interest, on the other, w^* can be chosen so as to minimize the computation time required for specified numerical standard error (Müller, this volume, pp. 145–163).

5. Summary and directions for future research

5.1. Advantages of the Monte Carlo method. The Monte Carlo approach to the solution of the Bayesian multiple integration problem has three, key advantages relative to other approaches. The first of these is shared by other approaches, but the second and third are distinguishing features.

Monte Carlo integration with importance sampling is an inherently generic method. The importance sampling densities described in §4.2 are based on the established common structure of all regular likelihood functions—no more, and no less. Modifications for particular problems are entirely algorithmic, and require no problem-specific interventions. One must bring to these procedures precisely what is required in classical likelihood-based non-Bayesian approaches: the log-likelihood function and its first two derivatives.

The accuracy of the numerical approximation is a simple byproduct of the computations. Monte Carlo integration may be viewed as a controlled experiment conducted to obtain evidence on the value of posterior moments. When the number of Monte Carlo replications is large, the summary evidence ($\overline{g}_n$, in our notation) is normally distributed about the posterior moments ($\overline{g}$, in our notation). The variance of this distribution can be estimated with negligible further computations.

Monte Carlo integration is an inherently distributive algorithm. It is immediate from the method that Monte Carlo integration can be carried out in a distributed computing environment. In many applications, most or all of the computations at each Monte Carlo replication will be the same, and in these cases the method will also be well suited to vector or parallel processors.

5.2. Disadvantages of the Monte Carlo method. The Monte Carlo approach to the solution of the Bayesian multiple integration problem has two disadvantages.

Monte Carlo integration is inefficient relative to other methods in specific cases. The point is trivial in some circumstances: e.g., if a closed-form analytical solution is at hand computations are essentially instantaneous. A leading example is the mean of the posterior density, under a standard conjugate prior, for regression coefficients in the multiple linear regress ion model, where the usual least squares formulas provide the answer. The interesting cases are less clear cut: competing approaches like the Laplace expansion (Tierney and Kadane (1986)) can be computed rapidly, but have a fixed and unknown inaccuracy and require additional analytical work. Always, one must assess the tradeoff between computation time and human effort.

Monte Carlo integration can fail completely in irregular problems. The methods described in this paper are inextricably tied to the assumption that a satisfactory importance sampling density can be crafted beginning with a multivariate Student density. Adjustments of the form described in §4.2 will fail to bring about an importance sampling density with a practical relative numerical efficiency, or perhaps even an importance sampling density that satisfies the assumptions of the theorem in §3.2, if the posterior density is sufficiently irregular. For example, multimodal posterior densities are not likely to be handled well by these methods. (Other implementations could be tied to other initial approximations; the problem would still remain.) However, posterior densities with noninformative priors that cannot be handled well by the methods of §4 correspond to likelihood functions for which classical non-Bayesian methods fail completely. Such problems will be extremely difficult to approach by any method, and are very likely not robust with respect to changes in prior information or minor changes in model specification.

Taken together, these two disadvantages imply that the Monte Carlo approach to the Bayesian numerical integration problem is an attractive method for problems that lie between two extremes. On the one hand, for certain (usually simple) problems there will be specific features that can be exploited (with a reasonable amount of analytical effort) to provide very efficient solutions. At the other extreme, there are problems for which posterior densities are sufficiently irregular that a standardized, algorithmic approach to the construction of the importance sampling density will fail. Most of the problems that academic and consulting statisticians confront lie in between.

5.3. The research agenda. The Monte Carlo approach to the Bayesian numerical integration problem has matured to the point where it may be taken in specific applications, with considerable confidence that a workable solution will be obtained. As this is done, valuable experience and ideas for the improvement of the Monte Carlo approach will emerge. At this point, it is evident that there are five fronts on which substantial progress can be made that would benefit all applications.

(1) *Alternative methods for the evaluation of numerical accuracy.* The computation and interpretation of $\hat{\sigma}_n$, the standard error of the numerical approximation, are based on a central limit theorem. While the number of Monte Carlo replications, n, should be large and can be controlled, variants on a simple appeal to asymptotic theory are desirable here as in other contexts. Jackknife and bootstrap methods are possibilities. Another, which is particularly appealing if computations are distributed over processors, is to partition the Monte Carlo replications, carry the computations through to completion on each partition, and then compute the variance over partitions. Indeed, it would be useful to have available variance estimates computed by a variety of such methods, all being asymptotically equivalent.

(2) *More robust and intelligent algorithms for importance sampling density construction.* The algorithmic construction of an importance sampling density is one of learning and adaptation. It is important to do this in an efficient and sensible way, and this requires considerable care. For example, Monte Carlo methods are a very poor way to learn about the characteristics that theory (§3.2) indicates are important for Monte Carlo integration with importance sampling. Yet that is what adaptive methods—those that modify the importance sampling density in the process of the Monte Carlo iterations themselves—suggest. (For examples of this method see Barnett, Geweke, and Yue (1990) and Evans (this volume, pp. 137–143).) The procedures outlined in §4.2 are much more efficient, but better methods are easy to conceive. The procedures of §4.2 concentrate on the axes defined by a Choleski decomposition of the inverse of the Hessian, and ignore other parts of the parameter space. As an alternative, one could apply steepest ascent or

other hill-climbing methods to the log-weight function $w(\theta)$, which would seek out a maximum—exactly those points that the theory (§3.2) indicate can vitiate the efficiency of Monte Carlo integration and the reliability of $\hat{\sigma}_n$. As a byproduct, these methods would suggest adjustments to the importance sampling density to produce a more suitable weight function.

(3) *Further possibilities for acceleration.* This paper has summarized some adaptations of antithetic acceleration and quadrature methods, that (from the perspective of Monte Carlo integration) increase the efficiency of the computations dramatically in some cases, or (from the perspective of these other methods) provide an indication of numerical accuracy that can be computed very quickly. Control variates are another possibility for acceleration, and such variates can be furnished by the Laplace expansion (Tierney and Kadane (1986)).

(4) *Nonstandard Bayesian problems.* This paper, and the Monte Carlo approach generally, has concentrated on Bayesian inference problems that can be cast as the computation of the expectation (under the posterior density) of a function of interest. While formal problems are almost always stated this way, informal, often highly informative treatments are not: graphics and the computation of quantiles are cases in point. The problem of the graphical representation of the posterior probability density function of one or two functions of interest is one that has not yet yielded to a good solution. (For the parameters themselves, substantial results exist, e.g., Smith et al. (1987).) Conveying the numerical accuracy of a graphical representation is an especially important and difficult problem; a related simpler but unsolved problem is obtaining the numerical accuracy of quantiles computed in the obvious way (Geweke (1989b)) as a byproduct of Monte Carlo integration.

(5) *Bayesian inference in nonstandard models.* Increasingly, the statistical models used to study complex phenomena are implicit, in the terminology of Diggle and Gratton (1984): the model is defined in terms of a generating stochastic mechanism, and is not expressable as the parametric specification of the distribution of a random vector. Likelihood functions for such models are implicitly defined. The adaption of Bayesian methods to such models is necessary to provide scientific, as opposed to ad hoc, inference.

REFERENCES

W. Barnett, J. Geweke, and P. Yue (1990), *Semiparametric estimation of the asymptotically ideal model: the Alm demand system*, Nonparametric and Semiparametric Methods in Econometrics and Statistics (W. Burnett, J. Powell, and G. Tauchen, eds.), Cambridge University Press, Cambridge.

A. Deaton (1986), *Demand analysis*, Handbook of Econometrics (Z. Griliches and M. D. Intriligator, eds.), vol. 3, North-Holland, Amsterdam.

P. J. Diggle and R. J. Gratton (1984), *Monte Carlo methods of inference for implicit statistical models*, J. Roy. Statist. Soc. Ser. B, **46**, 193–227.

M. Evans, *Adaptive importance sampling and chaining*, this volume, pp. 137–143.

J. Geweke (1986), *Exact inference in the inequality constrained normal linear regression model*, J. Appl. Econometrics **1**, 127–142.

____(1988a), *The secular and cyclical behavior of real GDP in nineteen OECD countries*, 1957–1983, J. Business and Econ. Stats. **6**, 479–486.

____(1988b), *Acceleration methods for Monte Carlo integration in Bayesian inference*, Proceedings of the 20th Symposium on the Interface: Computationally Intensive Methods in Computing Science and Statistics (E. J. Wegman, D. T. Gantz, and J. J. Miller, eds.), Amer. Statist. Assoc., Alexandria, VA, pp. 587–592.

____(1988c), *Antithetic acceleration of Monte Carlo integration in Bayesian inference*, J. Econometrics **38**, 73–89.

____(1989a), *Exact predictive densities in linear models with ARCH disturbances*, J. Econometrics **40**, 63–86.

____(1989b), *Bayesian inference in econometric models using Monte Carlo integration*, Econometrica **57**, 1317–1340.

J. Geweke, R. C. Marshall, and G. Zarkin (1986), *Exact inference for continuous time Markov chains*, Review of Economic Studies **53**, 653–669.

J. M. Hammersely and K. W. Morton (1956), *A new Monte Carlo technique: antithetic variates*, Proc. Cambridge Philos. Soc. **63**, 449–475.

D. Jorgenson (1986), *Econometric methods for modeling producer behavior*, Handbook of Econometrics (Z. Griliches and M. D. Intriligator, eds.), vol. 3, North-Holland, Amsterdam.

P. Müeller, *Monte Carlo integration in general dynamic models*, this volume, pp. 145–163.

A. F. M. Smith, A. M. Skene, J. E. H. Shaw, and J. C. Naylor (1987), *Progress with numerical and graphical methods for practical Bayesian statistics*, The Statistician **36**, 75–82.

L. Tierney and J. B. Kadane (1986), *Accurate approximations for posterior moments and marginal densities*, J. Amer. Statist. Assoc. **81**, 82–86.

A. Wald (1943), *Tests of statistical hypotheses concerning several parameters when the number of observations is large*, Ann. of Math. Stats. **14**, 426–482.

A. Zellner (1971), *An introduction to Bayesian inference in econometrics*, Wiley, New York.

Department of Economics, University of Minnesota, Minneapolis, Minnesota 55455
E-mail address: geweke@atlas.sosci.umn.edu

Contemporary Mathematics
Volume **115**, 1991

Adaptive Importance Sampling and Chaining

MICHAEL EVANS

ABSTRACT. The problem of numerically evaluating high-dimensioal integrals arises frequently in statistics and in many other fields. There is a need for flexible algorithms which can be used for a wide variety of problems and which can give reasonable accuracy within a modest amount of computing time. One possible approach to this problem is adaptive importance sampling together with chaining. These techniques and a specific implementation are discussed in this paper.

1. Introduction

The necessity of computing high-dimensional integrals arises with considerable frequency in applications of statistics. For example the problem of computing orthant probabilities for the multivariate normal has a long and extensive literature; for some discussion see Evans and Swartz (1988a). Many substantial examples also arise from the need to implement Bayesian analyses.

There are a wide variety of methods available for approximating such integrals; e.g. multiple quadrature, importance sampling, asymptotic techniques, etc. Each of these has contexts where they are of proven value but none can claim to be a universal solution to the problem. For example, importance sampling can be useful in problems where a fairly low level of accuracy is required and computation times are not of serious concern. On the other hand it can be very difficult to obtain an importance sampling algorithm that satisfies even these minimal requirements.

Our motivation here is to address the kinds of problems that arise in Bayesian inference. In particular we are interested in approximating ratios

---

1980 *Mathematics Subject Classification* (1985 *Revision*). Primary 65D30, 65C05.

Key words and phrases. High-dimensional integration, Monte Carlo, adaptive importance sampling, chaining.

This research was supported in part by Grant A3120 of the Natural Sciences and Engineering Research Council of Canada.

This paper is in final form and no version of it will appear for publication elsewhere.

of the form

$$\int_\Omega g(\theta)f(\theta)\,d\theta \Big/ \int_\Omega f(\theta)\,d\theta \tag{1}$$

for many different real-valued g on Ω, where Ω is an open subset of R^k and f is the product of the likelihood and the prior. The g functions could be monomials in the coordinates of θ or indicator functions for various subregions of Ω.

Our computational approach will be based on an extension of the importance sampling technique. For importance sampling we must specify a density h on Ω such that the support of h contains the support of f, generate a sample $\theta_1, \ldots, \theta_N$ from h and then approximate (1) by

$$\sum_{i=1}^N g(\theta_i)w(\theta_i) \Big/ \sum_{i=1}^N w(\theta_i), \tag{2}$$

where $w(\theta) = f(\theta)/h(\theta)$. The Strong Law of Large Numbers gives the convergence of (2) to (1) as N grows. If a poor choice is made for h, however, this approach may not provide an adequate approximation within reasonable computing times. Hence some care is needed in choosing h. The basic principle is to choose h so that it mimics f as closely as possible, as h proportional to f implies that the estimate of the denominator of (2) has variance 0. Provided $g(\theta)f(\theta)$ is not wildly different than $f(\theta)$ this h will then be a good choice for approximating (1).

Typically we have available a family $I = \{h_\lambda | \lambda \in \Lambda\}$ of densities, some elements of which will be useful for importance sampling, and we must find $h \in I$ which is in some sense closest to the normalized f. A natural way of choosing h is to compute characteristics of f and then find h in the family with similar values for these attributes; e.g. Geweke (1988) suggests a family I and a method of matching based on a preliminary computation of varoius characteristics of f including its maximum. In §2 we discuss an approach where some or all of these characteristics are computed via Monte Carlo.

2. Adaptive importance sampling

A natural approach to selecting $h \in I$ is to begin with an $h_1 \in I$, generate a sample from h_1, using this sample estimate characteristics of the posterior, based on these estimates choose a new $h_2 = I$, generate a sample from h_2, etc. We refer to this process as adaptive importance sampling. More formally let $\mathbf{m}_*$ be a vector of characteristics of the posterior each of which can be computed as an expectation; e.g. moments, probability contents. Let $\mathbf{m}(\lambda)$ be the vector of corresponding characteristics of h_λ. Then the algorithm is:

1. choose $h_{\lambda_1} \in I$;
2. generate $\theta_1, \ldots, \theta_N$ from h_{λ_1} and compute estimate $\hat{\mathbf{m}}_1$ of $\mathbf{m}_*$ using (2);
3. find $\lambda_2 \in \Lambda$ such that $\|\mathbf{m}(\lambda) - \hat{\mathbf{m}}_1\|$ is minimized;

4. generate $\theta_{N+1}, \ldots, \theta_{2N}$ from h_{λ_2} and combine with $\theta_1, \ldots, \theta_N$, using the appropriate w function with θ_i in (2), to get $\hat{\mathbf{m}}_2$;
5. iterate until $\hat{\mathbf{m}}_i$ is stable;
6. use the final h_λ for computation of other integrals of interest.

We note that many integrals of interest are often included as elements of $\mathbf{m}_*$ so that at the end of step 5 we will have approximations of these quantities. Further when compared with straight importance sampling the adaptive approach has the advantage that it encodes information learned about the posterior in sampling so that this information is available to improve the efficiency of later computations.

The above is really a rough sketch of the algorithm and a number of aspects must be examined further to ensure success of the approach. For example we must choose I so that it contains a reasonable approximation to the posterior, under the restriction that we have an efficient algorithm for generating from each $h \in I$. The elements of $\mathbf{m}_*$ must be chosen so that they accurately reflect the prominent characteristics of the posterior, can be efficiently estimated, and so that the elements of $\mathbf{m}(\lambda)$ can be relatively easily computed. Perhaps most important is the choice of h_{λ_1}. If h_{λ_1} is too different from the posterior then the process can fail to converge within reasonable computing times. This effect becomes more pronounced as dimension rises. Connected with this is the problem of assessing convergence. While standard errors can be computed, experience suggests that they can give misleading indications of accuracy and cannot be viewed as a complete solution to this problem.

While no definitive answers to the above problems currently exist, there are various examples where adaptive importance sampling has proven to be useful. For some of these and for further discussion see Kloek and van Dijk (1978), Smith, Skene, Shaw, and Naylor (1987), Evans (1988a), Oh and Berger (1989), and Evans, Gilula, and Guttman (1989). Now we discuss the technique of chaining which is designed in part to address the problems of choosing h_{λ_1} and assessing convergence.

3. Chaining

First consider the problem of evaluating the denominator of (1); i.e. the norming constant, via adaptive importance sampling. We consider this problem as a member of a class of integration problems

$$\int_\Omega f_\nu(\theta)\, d\theta \tag{3}$$

indexed by the parameter $\nu \in V$, where V is an open subset of R^m, $f_{\nu_0} = f$ for some $\nu_0 \in V$ and such that (3) is a continuous function of ν. Further we suppose that the family I provides adequate approximations, as importance samplers, to the functions f_ν and for $\nu_1 \in V$ we know a good starting choice for $h_{\lambda_1} \in I$; i.e. we can solve the integration problem given by $\nu = \nu_1$ by starting the algorithm at h_{λ_1}. The process of chaining then proceeds as follows. We set $\nu = \nu_1$, start the adaptive importance sampling at h_{λ_1}, and find the best-fitting $h_{\lambda_2} \in I$ for this problem; i.e. h_{λ_2} is the output from step 6. Given the above description of the ingredients of the problem we know that this step will not fail. Then we make a small change in ν, from ν_1 to ν_2, towards ν_0, and start the adaptive importance sampling for the problem given by ν_2 at h_{λ_2} finding the best-fitting h_{λ_3} for this problem. If the change from ν_1 was small enough then we can feel somewhat confident that the sampling was successful. In this way we construct a chain $(\nu_1, \lambda_1), \ldots, (\nu_n, \lambda_n)$ where $\nu_n = \nu_0$ and find the best-fitting $h_{\lambda_{n+1}}$ for the problem of interest.

The problem still remains of not having absolute confidence in the accuracy of our final computed results but it seems clear that chaining can be used to at least increase our confidence over straight importance sampling or adaptive importance sampling alone. Of course we pay the price of increased computation times. The accuracy of a result produced by chaining can be assessed by running another chain with smaller step-sizes than the old one.

There are typically many different ways of implementing chaining and some will be more efficient and useful in a given context than others. A very general method is discussed in Evans (1988b) which, at least formally, applies to any problem. This involves putting $\nu = (t, u) \in (0, \infty)^2$ and $f_\nu(\theta) = f^{1/t}(\theta) h_{\lambda_1}^{1/u}(\theta)$ for some choice of λ_1. Then for large values of t and for $u = 1$ we have that $f_\nu \sim h_{\lambda_1}$ and we know the correct starting value for λ in the adaptive importance sampling is $\lambda = \lambda_1$. Hence we start the chain with t large and $u = 1$. As t is lowered with $u = 1$ the influence of f in f_ν is increased. When we reach $t = 1$ in the chain we fix t and begin raising u and hence remove the effect of h_{λ_1} until f_ν is effectively f. In Evans (1988b) this method is discussed and it is shown that this approach to chaining can also be used to solve global optimization problems. In the next section we discuss an alternative method of implementing chaining in a specific problem.

An idea similar to chaining has been used to develop algorithms for solving other numerical problems. These are often called homotopy methods and chaining could be called a homotopy method for integration. For a discussion of these methods in the context of root-finding algorithms see, for example, Zulehner (1988) and Watson, Billyes, and Morgan (1989).

4. Example and conclusions

We consider the problem of computing the integral of the function

$$\exp\{c_1\nu + \mathbf{1}' \log f_\tau(k_\tau^{-1} e^{c_2\nu}(c_3 Q\alpha + c_4\mathbf{d}))\}, \tag{4}$$

where $c_1, c_2, c_3, c_4, \tau, k_\tau$ are constants, $Q \in R^{n\times k}$ is a fixed column orthonormal matrix, $\mathbf{d}$ is a unit vector orthogonal to $L(\mathbf{1})$, $f_\tau(x) = (1 + x^2)^{-(\tau+1)/2}$, we interpret $\log f_\tau$ as acting componentwise on a vector argument and $\alpha \in R^k$, $\nu \in R$ are the variables of integration. This integral arises in the Bayesian analysis of the linear model when we are using Student errors and Jeffrey's prior and is effectively the norming constant. A more complete description of this function is given in Evans (1988a). It can be taken as fairly representative of the kinds of complex integration problems which arise in statistics. An algorithm that works well with one set of the constants can fail disastrously with another choice. Hence the need for adaptation. More generally we want the posterior distribution of $\theta = (\alpha, \nu)$.

Now we consider how to evaluate the integral of (4) via adaptive importance sampling for various choices of $\tau \in (0, \infty)$. It is well known that the posterior distribution of α has longer tails than the normal distribution and also exhibits a degree of skewness. Accordingly we developed a family I, which includes the multivariate normal densities, but also allows for these more general behaviours. The family I is parametrized by $(\mu, \Delta, \lambda_1, \lambda_2)$ where we generate from this density via

$$\theta = \mu + \Delta \mathbf{e}, \tag{5}$$

where, $\mu \in R^{k+1}$, Δ is a $(k+1) \times (k+1)$ lower triangular matrix with positive diagonal elements, $\lambda_1, \lambda_2 \in (2, \infty)^{k+1}$ and $e_1, \ldots, e_{k+1}$ are statistically independent with $e_i \sim \mathrm{GS}(\lambda_{i_1}, \lambda_{i_2})$. By $\mathrm{GS}(\lambda_1, \lambda_2)$ we mean a generalized Student distribution with density $g_{\lambda_1, \lambda_2}(e) = \sigma_1(1+\sigma_1^2(e+\eta)^2)^{-(\lambda_1+1)/2}$ when $e \le -\eta$, $g_{\lambda_1, \lambda_2}(e) = \sigma_2(1+\sigma_2^2(e+\eta)^2)^{-(\lambda_2+1)/2}$ otherwise. The quantities η, σ_1, σ_2 are completely determined by the requirements that g_{λ_1, λ_2} be a continuous density with mean 0 and variance 1 and exact formulas can be given for these quantities using the gamma function. For these expressions and a simple algorithm for generating from a GS distribution see Evans (1988a). A GS density is essentially arrived at by gluing together two Student densities at the origin and then normalizing and standardizing to have mean 0 and variance 1. Since we require the GS densities to have means and variances we must restrict the λ_{ij} to be bigger than 2. If some other method of standardization is chosen this restriction can be relaxed. The GS densities are typically skewed with differing tail lengths in each tail. The family includes the standardized Student densities and is similar to one presented in Geweke (1986).

The first $k+1$ elements of $\mathbf{m}_*$ are taken to be the means of the elements of θ and the next $(k+1)(k+2)/2$ elements are taken to be the variances and covariances of these quantities. Hence μ is the posterior mean of θ and Δ is the Cholesky factor of the posterior variance matrix of θ. For fitting the GS densities we took the next $2(k+1)$ coordinates of $\mathbf{m}_*$ to be quantities which measured the skewness and tail-length of the posterior distribution of the coordinates of $\mathbf{e} = \Delta^{-1}(\theta - \mu)$. For the ith coordinate we used the values of the probability content of $[0, \infty)$ and $E[|e_i|^{1/2}]$. These quantities were tabulated for the GS densities and the table was searched at each fitting step.

For further discussion concerning this choice of elements of $\mathbf{m}_*$ see Evans (1988a).

The adaptive importance sampling then proceeds by choosing starting values for $(\mu, \Delta, \lambda_1, \lambda_2)$, a value for N and estimating iteratively the mean, variance matrix and the additional quantities specified above for the posterior distribution of θ. Typically it is better to start initially by estimating the mean alone then, when a reasonable estimate is obtained, begin estimating the variance and then finally the additional quantities.

We are then faced with the problem of choosing a starting value for $(\mu, \Delta, \lambda_1, \lambda_2)$. Experience with this problem indicates that the success of adaptive importance sampling is highly sensitive to this choice, particularily when τ is small and in these circumstances the algorithm can fail. On the other hand when τ is large we know quite a bit about the integrand (4) and, in fact, know that taking $\mu = 0$, Δ equal to the identity matrix, and Student densities for the distributions of the e_i, will provide good approximations to the true posterior distributions of the coordinates of θ. The specific Student densities used depend on additional aspects of the problem which we have not specified here. This is based on exact distribution theory results and asymptotics when $\tau = \infty$. In fact the unspecified constants in (4) are chosen so that this is correct. Hence when we want the integral of (4) for τ small, we chain on τ, starting with large values, and slowly lowering its value as when τ is large we can construct a good starting importance sampler in the family we have described above. Further we also know the values of some of the elements of $\mathbf{m}_*$ to a fair degree of accuracy when τ is large. Hence at the first step of the chain we have an accurate idea of the absolute errors in our estimates of some of the fitting quantities. Thus if we take small steps in the chain and we maintain the sample size N we can hope to maintain or even improve on this. Additional details and a numerical example, when $k = 10$ and $\tau = 1$, are given in Evans (1988a).

We are not necessarily recommending adaptive importance sampling with chaining as the preferred method of carrying out a posterior analysis of linear models. For example the methods of Kass, Tierney, and Kadane (1988) and Geweke (1986) are also applicable here. It does seem reasonable to suggest, however, that these methods be part of the collection of computational techniques that could be used in a given problem. Further work is required, particularly in the development of additional families I. One such family under current development replaces the generalized Students in (5) by distributions whose densities are positive polynomials times normal densities. For some results concerning such a family see Evans and Swartz (1988b) and Geweke (1989).

REFERENCES

M. Evans (1988a), *Monte Carlo computation of marginal posterior quantiles*, Technical Report 11, Department of Statistics, University of Toronoto.

____(1988b), *Chaining via annealing*, Technical Report 14, Department of Statistics, University of Toronto and in press Annals of Statistics.

M. Evans and T. Swartz (1988a), *Monte Carlo computation of some multivariate normal probabilities*, J. Statistical Computation and Simulation **30**, 117–128.

____(1988b), *Sampling from Gauss rules*, SIAM J. Scientific and Statistical Computing **9**, 950–961.

M. Evans, Z. Gilula, and I. Guttman (1989), *Latent class analysis of two-way contingency tables by Bayesian methods*, Biometrika **76**, 557–563.

J. Geweke (1986), *Bayesian inference in econometric models using Monte Carlo integration*, DP 87-02, Department of Economics, Duke University.

____(1989), *Modeling with normal polynomial expansions*, Economic Complexity. Chaos, Sunspots and Nonlinearity (W. A. Bennett, J. Geweke, and K. Shell, eds.), Cambridge University Press, Cambridge, U.K.

R. E. Kass, L. Tierney, and J. B. Kadane (1988), *Asymptotics in Bayesian computations (with discussion)*, Bayesian Statistics 3 (J. M. Bernardo, M. H. DeGroot, D. V. Lindley, and A. F. M. Smith, eds.), Oxford University Press, Oxford, U.K.

T. Kloek and H. K. Dijk (1978), *Bayesian estimates of equation system parameters: an application of integration by Monte Carlo*, Econometrica **46**, 1–19.

M. S. Oh and J. O. Berger (1989), *Adaptive importance sampling in Monte Carlo integration*, Technical Report 89-19C, Department of Statistics, Purdue University, West Lafayette, IN.

A. F. M. Smith, A. M. Skene, J. E. H. Shaw, and J. C. Naylor (1987), *Progress with numerical and graphical methods for practical Bayesian statistics*, The Statistician **36**, 75–82.

L. T. Watson, S. C. Billyes, and A. P. Morgan (1989), *Hompack: A suite of codes for globally convergent homotopy algorithms*, ACM Transactions in Mathematical Software **15**, 99.

W. Zulehner (1988), *On the solution to polynomial systems obtained by homotopy methods*, Numerische Mathematik **54**, 303–317.

DEPARTMENT OF STATISTICS, UNIVERSITY OF TORONTO, TORONTO, ONTARIO, CANADA M5S 1A1

E-mail address: mevans@utstat.toronto.edu

Contemporary Mathematics
Volume **115**, 1991

Monte Carlo Integration in General Dynamic Models

PETER MÜLLER

ABSTRACT. This paper suggests the use of Monte Carlo integration to estimate the required integrals in Bayesian analysis of general dynamic models. The necessary prior and posterior samples are obtained by simulating the model with a sample from the initial prior distribution. The suggested simulation approach does not require the global distributional assumptions and restrictions on the functional form of the model that are required for available analytical and numerical algorithms. The main problem in simulating a general dynamic model is to transform a given sample from the posterior distribution at time $t-1$ to a sample from the posterior distribution at time t, i.e. essentially the simulation of Bayes' theorem. This is solved by an accept/reject like procedure combined with a preceding expansion of the sample size. In an application example the proposed algorithm is compared with an analysis as generalized linear model.

1. Introduction

The framework of time series modeling calls for an analysis that treats the model parameters as evolving, nonconstant quantities. As the dynamic model evolves through time, inferences about the parameters are updated in response to the incoming data and possibly other external information. This can be accomplished most naturally via a Bayesian approach.

Following the notation of Pole and West (1988) the general dynamic model can be defined as:

$$\text{Observation equation: } y_t = f_t(\theta_t) + \nu_t, \tag{1}$$

$$\text{Evolution equation: } \theta_t = g_t(\theta_{t-1} + w_t). \tag{2}$$

Here f_t and g_t are known, nonlinear functions and ν_t and w_t are random variables with arbitrary specified distribution.

1980 *Mathematics Subject Classification* (1985 *Revision*). Primary 65D30, 65C05, 62M20.

Key words and phrases. Dynamic nonlinear models, Bayesian forecasting, numerical integration, Monte Carlo integration.

This research was supported by NSF Grants DMS-8717799 and DMS-8702620 at Purdue University.

The final version of this paper will be submitted for publication elsewhere.

This paper proposes an algorithm for the numerical evaluation of the integrals required in the analysis of the general dynamic model. In §2.1 these integrals are stated and it is argued why it is in general impossible to exactly evaluate them and why even numerical integration is difficult. Sections 2.2 and 2.3 review some restricted versions of the dynamic model for which an exact analysis is feasible. Section 3 states the proposed Monte Carlo integration algorithm. The sample points required for the Monte Carlo integrals are obtained by simulating the dynamic model starting with a sample generated from the initial prior distribution. Section 3.2 gives the algorithm used for the simulation of the dynamic model, leaving the implementation details for the simulation of the observation step for §3.3. Section 4 contains an application of the algorithm to a model of advertising awareness taken from West and Harrison (1989).

2. The analysis of the dynamic model

2.1. The general dynamic model. The following notation will help to describe the analysis of the general dynamic model stated in (1) and (2): Let D_t denote the information set at time t ($t = 0, 1, \dots$), i.e. all information relevant to forming beliefs on the parameter vector θ_t. Suppose that the model is closed to inputs of external information, so that $D_{t+1} = \{y_t\} \cup D_t$. It will be assumed that all continuous probability distributions have a density defined with respect to Lebesgue measure, and $p(X)$ will be used generically to denote the density for a random variable X, and $p(X|D)$ the density of X given D. Expressions of the form $(X|D) \sim N(\mu, \sigma^2)$ refer to the conditional distribution of X given D.

Much of the Bayesian analysis in the dynamic model can be done in terms of integrals with respect to the distributions $p(\theta_t|D_{t-1})$ and $p(\theta_t|D_t)$ at time $t = 1, 2, \dots T$. The distribution $p(\theta_t|D_{t-1})$ reflects the beliefs about θ_t prior to observing y_t. The density $p(\theta_t|D_t)$ summarizes the updated beliefs after observing the data y_t. Since in terms of the isolated experiment at time t $p(\theta_t|D_{t-1})$ represents the prior distribution on θ_t and $p(\theta_t|D_t)$ the posterior distribution, they will be referred to as the prior at time t and the posterior at time t in the following.

Given the initial prior $p(\theta_1|D_0)$, the unknown densities $p(\theta_t|D_t)$ and $p(\theta_t|D_{t-1})$ are recursively determined by

$$p(\theta_t|D_t) \propto p(\theta_t|D_{t-1})p(y_t|\theta_t), \tag{3}$$

$$p(\theta_t|D_{t-1}) = \int p(\theta_t|\theta_{t-1})p(\theta_{t-1}|D_{t-1})d\theta_{t-1}, \tag{4}$$

where $p(y_t|\theta_t)$ is the known likelihood function implied by the sampling model (1) and $p(\theta_t|\theta_{t-1})$ is determined by $g_t(\ )$ and the distribution of ω_t in the system equation (2).

Inference typically requires integration of the prior and/or posterior densities. Examples include point estimation of the parameter vector based on the information available at time t; namely,

$$\hat{\theta}_t = E(\theta_t|D_t) = \int \theta_t p(\theta_t|D_t)d\theta_t,$$

and one-step ahead forecast distributions,

$$p(y_{t+1}|D_t) = \int p(y_{t+1}|\theta_{t+1})p(\theta_{t+1}|D_t)d\theta_{t+1}.$$

K-step ahead forecast densities $p(Y_{t+k}|D_t)$ are also easily obtainable in the framework of the suggested algorithm.

In general, we are interested in estimating integrals of the form

$$E(f(\theta_t)|D_t) = \int f(\theta_t)p(\theta_t|D_t)d\theta_t, \tag{5}$$

$$E(f(\theta_t)|D_{t-1}) = \int f(\theta_t)p(\theta_t|D_{t-1})d\theta_t. \tag{6}$$

Here the integrands involve the unknown posterior density (3) or the prior density (4). The absence of conjugate prior/likelihood pairs and the possibly nonlinear form of g_t and f_t in (1) and (2) make it in general impossible to analytically derive expressions for the prior distribution (4) and the posterior distribution (3) at time t. Therefore an exact evaluation of the integrals (5) and (6) is impossible, and even numerical integration is very difficult, since most schemes require evaluation of the integrand at specific points

Another feature which distinguishes the integration problem in the general dynamic model from other applications is the sequential nature of the problem, requiring estimation of a sequence of integrals over very closely related densities.

2.2. The normal dynamic linear model. An exact analysis of the dynamic model is only possible if some restrictions are imposed on the general dynamic model. The Bayesian analysis of dynamic models becomes most simple in the framework of the Kalman filter with the linear evolution and observation equation, which together with normal distributed noise terms and conjugate initial prior lead to a sequence of conjugate prior/likelihood pairs, making the computational burden of evaluating integrals of the form (5) or (6) trivial. This linear dynamic Bayesian model was described and further developed by e.g. Ho and Lee (1964) and Harrison and Stevens (1976).

Using the notation of Pole, West, and Harrison (1988) the basic normal dynamic linear model (NDLM) can be stated as:

$$\begin{aligned} \text{Observation Equation: } & y_t = F_t'\theta_t + \nu_t \\ & \nu_t \sim N(0, V_t), \\ \text{Evolution Equation: } & \theta_t = G_t'\theta_{t-1} + \omega_t, \\ & \omega_t \sim N(0, \Omega_t), \\ & t = 1, 2, \ldots . \end{aligned}$$

Here y_t is the observation vector, F_t a vector of regression variables, θ_t the parameter vector, G_t the state transition matrix, V_t the observation variance, and Ω_t the system variance. The quadruple $M = \{F_t, G_t, V_t, \Omega_t\}$ fully specifies the model.

With conjugate initial prior $(\theta_1|D_0) \sim N(a_1, R_1)$ the prior $p(\theta_t|D_{t-1})$ and posterior $p(\theta_t|D_t)$ are normal with means and variances, which are derived as follows by induction. Assume the prior at time t is already given by $(\theta_t|D_{t-1}) \sim N(a_t, R_t)$. Then the posterior becomes

$$(\theta_t|D_t) \sim N(m_t, C_t),$$

with $m_t = a_t + A_t e_t$, $C_t = R_t - A_t Q_t A_t'$, $Q_t = F_t' R_t' F t + V_t$, $A_t = R_t F_t Q_t^{-1}$, and $e_t = y_t - F_t' a_t$. The prior for the subsequent period is then given by

$$(\theta_{t+1}|D_t) \sim N(a_{t+1}, R_{t+1}),$$

where $a_t = G_t m_{t-1}$ and $R_t = G_t C_{t-1} G_t' + \Omega_t$. See West and Harrison (1989) for a proof and more discussion. The NDLM covers a wide range of models, allowing incorporation of trend and seasonal components in G_t, a regression term in F_t and possible intervention, just to name a few of the modeling tools developed in West and Harrison (1989).

2.3. Extensions of the NDLM. Pole, West, and Harrison (1988) extended the basic NDLM to allow for nonlinearities in the observation equation by introducing a "guide relationship" and assuming a normal prior at each step. The guide relationship is used as an aid in determining the first two moments of the posterior, which together with the normality assumption and the linear evolution suffices to specify the next prior. Another extension of the NDLM is achieved by replacing the normality assumption by a wider class of conjugate priors. Sorenson and Alspach (1971) propose a discrete mixture of normals to represent observation and system noise term.

West, Harrison, and Migon (1985) developed an algorithm to analyze dynamic generalized linear models (DGLM). The DGLM is an extension of the NDLM, allowing any density from an exponential family as sampling distribution for the observation y_t and specifying the distribution of the evolution noise term only by its first and second order moments. Let θ_t denote the state vector at time t and η_t the natural parameter of the exponential family sampling distribution $p(y_t|\eta_t)$. Then the observation equation is given by:

$$\text{(7)} \qquad \text{Observation model: } p(y_t|\eta_t) \propto \exp\{V_t^{-1}[y_t\eta_t - a(\eta_t)]\},$$

$$\text{(8)} \qquad g(\eta_t) = \lambda_t = F_t'\theta_t,$$

where V_t is a known scale parameter, $a(\eta)$ is a known function, F_t is a vector of regression variables, and $g(.)$ is a known mapping between the parameter η_t and the linear regression λ_t. The evolution equation specifies the distribution of the evolution noise term only by its first and second moments:

$$\text{Evolution equation: } \theta_{t+1} = G_t\theta_t + \omega_t,$$
$$E(\omega_t) = 0,$$
$$\text{Var}(\omega_t) = W_t.$$

The analysis proceeds by determining, at any time, only the first and second moments of the prior and posterior for the state vector. The distribution of $(\eta_t|D_{t-1})$ is assumed in a form conjugate to the likelihood $p(y_t|\eta_t)$; namely,

$$p(\eta_t|D_{t-1}) \propto \exp\{s_t[x_t\eta_t - a(\eta_t)]\},$$

for some parameters s_t and x_t. Then this leads to a sequence of simple updating equations for the moments of the prior $p(\theta_t|D_{t-1})$ and posterior $p(\theta_t|D_t)$, that would correspond to an analysis with fully specified distributions, if the prior and posterior distributions were actually normal and the forecast distribution $p(\eta_t|D_{t-1})$ was actually conjugate to the sampling distribution. See West and Harrison (1989), §14.3 for a complete description of this approach.

The accessible range of models can be even further widened by abandoning the restriction to analytically tractable models and using numerical integration methods. Pole and West (1988) propose an algorithm combining the computational convenience of conjugate models and the flexibility of using numerical integration. They deal with models which are NDLM conditional on a subset of parameters β:

$$M(\beta) = \{F_t(\beta), G_t(\beta), V_t(\beta), \Omega_t(\beta)\}.$$

The model is analyzed by Gaussian quadrature with respect to β, using the analytical solution for $\{F_t(\beta), G_t(\beta), V_t(\beta), \Omega_t(\beta)\}$ for fixed β's. Gaussian quadrature involves at each stage maintaining an optimal grid $K_t = \{\beta_{t,i}\}$ which is used to estimate the posterior integrals and forecast densities. When updating the grid K_{t-1} to K_t, the $p(\beta_{t,i}D_t)$ and the moments of the NDLM parameters conditional on the β's need to be determined on the new grid points. Pole and West (1988) suggest using spline interpolation. Linear interpolation is used to maintain required positive definiteness of $C_t(\beta) = Var(\theta_t|\beta_t, D_t)$.

3. The algorithm

3.1. Monte Carlo integration in the general dynamic model. In §1.1. the general dynamic model was stated as

$$\text{(9)} \qquad \text{Observation equation: } y_t = f_t(\theta_t) + \nu_t,$$
$$\text{(10)} \qquad \text{Evolution equation: } \theta_t = g_t(\theta_{t-1} + w_t),$$

with f_t and g_t known, nonlinear functions and ν_t and w_t random variables with arbitrary specified distribution. As argued in §2.1, under this model it is in general impossible to exactly evaluate the posterior integrals (5) and prior integrals (6). Even numerical integration is difficult since the posterior (3) and the prior (4) cannot be evaluated. The algorithms mentioned in §§2.2 and 2.3 avoid this problem by either making distributional assumptions or by estimating density values, e.g. by spline interpolation or linear interpolation.

One numerical integration approach that fits naturally into the dynamic model framework is Monte Carlo integration (with importance sampling) as described, e.g., in Rubinstein (1981), van Dijk (1984), or Geweke (1989), with the sample points for the Monte Carlo integration coming from a simulation of the dynamic model. A sample from the original prior $p(\theta_1|D_0)$ is propagated step by step through the time series, thereby always maintaining a sample from the prior and posterior at each time step; these can be used for Monte Carlo estimates of prior and posterior integrals as desired. The task of evaluating the posterior and prior is replaced by the problem of simulating the observation and evolution steps by an appropriate process applied to the sample points.

No restrictions on the functional form of the observation equation (1) or on the system equation (2) are required, i.e., f_t and g_t can be any, not necessarily linear, functions. Although no global assumptions about the involved distributions will be necessary, the simulation of the observation step will require that the prior $p(\theta_t|D_{t-1})$ be approximable by a normal density over certain regions.

3.2. Simulating the dynamic model. In the following, let π_t denote the prior $p(\theta_t|D_{t-1})$, p_t the posterior $p(\theta_t|D_t)$, and l_t the likelihood function $p(y_t|\theta_t)$ as a function of θ_t. The simulation of the evolution step is simple: Assume a sample $\{\eta_i, i = 1, \dots, n\}$ from the posterior p_{t-1} is available. Then these sample points are easily mapped into a sample $\{\theta_i, i = 1, \dots, n\}$ from π_t by $\theta_i := g_t(\eta_i + \omega_i)$, where the ω_i are generated from the distribution given by the system equation (10).

The simulation of the observation step, i.e. the transformation of the prior sample from π_t to a posterior sample from p_t, would be conceptually straightforward by an accept/reject procedure: Bayes' theorem, $p_t(\theta_i) \propto \pi_t(\theta_i)l_t(\theta_i)$, implies that the prior sample could be transformed to a posterior sample by deleting and retaining sample points in an accept/reject like way with probabilities $P(\theta_i$ is accepted$) \propto l_t(\theta_i)$. This could be implemented by generating $u_i \sim \text{Uniform}(0, 1)$ and comparing u_i with $P(\theta_i$ is accepted): if $u_i > P(\theta_i$ is accepted) then delete θ_i from the sample, otherwise keep θ_i in the sample. This only requires the evaluation of the likelihood function l_t, which is analytically available; the difficult evaluation of the densities π_t and p_t is avoided. Unfortunately, each accept/reject step will typically lead to a decrease in the sample size, making an unmodified implementation of this approach impractical for simulation of a many-step dynamic model.

The proposed algorithm solves this problem by replacing the accept/reject step by a two step procedure. First the prior sample is transformed into a sample from an envelope density I, which is then in a second step used for the accept/reject procedure. The envelope density as specified in Algorithms 1 and 2 is chosen such that

- a sample from I can be generated by expanding the available prior sample by stratified additional sampling from π_t, restricted to certain regions L_m,
- the accept/reject weights are functions of l_t only, and
- the expected sample size after the accept/reject step is equal to the original prior sample size.

The sets L_m are chosen as regions of high likelihood in a way formalized in Algorithm 2.

The following specific algorithm is proposed.

ALGORITHM 1: SIMULATION OF THE GENERAL DYNAMIC MODEL.

1. *Initial Prior. Draw a sample from the initial prior:* $\theta_i \sim \pi_1$, $i = 1, \dots, n_0$.

Set Time. $t := 1$.

2. *Observation step. The details of the implementation of this step will be explained in Algorithm* 2.

2.1. *Extension of sample size. By stratified additional sampling from* π_t *the sample* $\{\theta_i, i = 1, \dots n_0\}$ *is extended to a sample from a density of the form*:

$$I \in \mathcal{G} = \left\{ f : f = \sum_{m=1}^{M} \gamma_m \pi_t|_{Lm}, \sum \gamma_m = 1 \right\},$$

where $\pi_t(x)|_{Lm} := \begin{cases} \pi_t(x)/\pi_t(L_m) & \text{if } x \in L_m, \\ 0 & \text{otherwise}, \end{cases}$ *denotes* π_t *restricted to* L_m. *The regions* L_m *will be specified in equation* (11) *in Algorithm* 2.

2.2. *Accept/Reject. The density* I *is now used as envelope density for an accept/reject step which, results in a sample from the posterior* p_t:

$$P(\theta_i \text{ is deleted}) \propto 1 - \frac{p_t(\theta_i)}{I(\theta_i)}.$$

2.3. *Posterior sample. The posterior sample* $\{\theta_i, i = 1, \dots, n_t\}$ *can now be used for Monte Carlo estimates of posterior integrals. By choosing* $I \in \mathcal{G}$ *appropriately some minimum expected sample size for the posterior sample can be guaranteed.*

3. *Evolution step. Generate* ω_i, $i = 1, \dots, n_t$ *and set* $\eta_i := g_t(\theta_i) + \omega_i$. *The set* $\{\eta_i, i = 1, \dots, n_t\}$ *now forms a sample from* π_{t+1}, *which can be used to estimate prior integrals.*

4. *Iteration. Increment* $t := t + 1$ *and simulate the next observation step.*

The estimates of the posterior integrals can be improved by using Monte Carlo with importance sampling with the sample from the envelope density

I. The accept/reject weights p_t/I would become the importance sampling weights. This point will not be further explored here, since this is a refinement of the algorithm that is not essential to the main idea.

3.3. Simulating Bayes' theorem. To simulate the observation step by Algorithm 1, it is still necessary to specify the choice of $I \in \mathcal{G}$. In the following, π will denote the prior $p(\theta_t|D_{t-1})$ at time t, l the likelihood function $p(y_t|\theta_t)$, and p the posterior $p(\theta_t|D_t)$.

The envelope density of Algorithm 1, $I = \sum \gamma_m \pi|_{Lm}$, can alternatively be written recursively as

$$I_0 = \pi, \dots, I_m = \alpha_m I_{m-1} + (1-\alpha_m)\pi|_{Lm}, \dots, I = I_M.$$

This suggests that the available sample from π can be transformed into a sample from I_M by iterative expansion of the sample size by drawing additional sample points from $\pi|_{Lm}$.

Simulation of the observation step, i.e. Bayes' theorem, can then be done by the following algorithm.

ALGORITHM 2: SIMULATION OF BAYES' THEOREM.

Assume that a sample $\{\theta_i, i = 1, \dots, n\}$ from the current prior π is available.

1. *Initialization. Estimate $\bar{l} = E_\pi l$ by $\frac{1}{n}\sum_1^n l(\theta_i)$ and set $m := 1$. Set the initial accept/reject weights $w_0(\theta_i) = p/I_0 = p/\pi$ to:*

$$w_0(\theta_i) = l(\theta_i).$$

2. *Iteration. The following steps iteratively expand the original prior sample to a sample from the envelope density I_M.*

2.1. *Additional sampling. Increase the sample size by a factor $1/\alpha_m$ by additional sampling from $\pi|_{Lm}$, where*

$$L_m = \{\theta : w_{m-1}(\theta) > \bar{l}\}. \tag{11}$$

This expands the current sample to a sample from I_m.

2.2. *Weights. The new weights p/I_m are given by:*

$$\frac{p}{I_m} \propto w_m := \begin{cases} w_{m-1} & \text{if } \theta \notin L_m, \\ q_m w_{m-1} & \text{if } \theta \in L_m, \end{cases} \tag{12}$$

where

$$q_m := \left(1 + \frac{1-\alpha_m}{\alpha_m}\frac{1}{I_{m-1}(L_m)}\right)^{-1}.$$

2.3. *Stopping rule. Set $m := m+1$ and continue iterating until $m = M$, where*

$$M = \min\{m : \prod_{k=1}^{m} q_k \le \frac{\bar{l}}{l^*}\}. \tag{13}$$

Here $l^* = \sup(l(\theta))$.

3. *Accept/Reject. Reject sample points with probability*

$$P(\theta_i \text{ is deleted}) \propto 1 - \frac{p_t(\theta_i)}{I(\theta_i)} = 1 - w_M.$$

4. *Posterior sample. The posterior sample* $\{\theta_i, i = 1, \dots, n\}$ *can now be used for Monte Carlo estimates of posterior integrals.*

To derive the expression (12) for w_m, observe the following: The regions L_m are such that $L_M \subset \cdots \subset L_1$, implying $\pi|_{Lm} = I_{m-1}|_{L_m}$. With this the expression for I_n can be written as:

$$\begin{aligned} I_m &= \alpha_m I_{m-1} + (1-\alpha_m) I_{m-1}|_{L_m} \\ &\propto \begin{cases} I_{m-1} & \text{if } \theta \notin L_m, \\ I_{m-1}\left(1 + \frac{1-\alpha_m}{\alpha_m}\frac{1}{I_{m-1}(L_m)}\right) = I_{m-1}/q_m & \text{if } \theta \in L_m. \end{cases} \end{aligned}$$

From this, (12) follows.

The number of iterations (13) is chosen such that the expected sample size of the posterior sample after the accept/reject step is greater than or equal to the original prior sample size. This leads to the stopping rule (13) (see appendix). If the α_m are chosen such that $q_m \leq \frac{1}{2}$ then obviously

$$M \approx \log(2) \log l^* / \bar{l}.$$

There remains the problem of generating from $\pi|_{Lm}$. This is the point at which the simulation algorithm requires some approximation; in particular it will be assumed that π can, on the regions L_m, be approximated by a multivariate normal distribution restricted to L_m. Compared with the global distributional assumptions and linearity restrictions on the functional form of system and evolution equation that are required for available analytical and numerical algorithms, this local approximation of the prior density $\pi|_{Lm}$ seems quite weak. The bias which this approximation introduces into the Monte Carlo integral estimates does not accumulate over time, since adding the noise term ω_t in each evolution step amounts to "discounting" the past history at a certain rate, so that typically only the most recent observations have a strong influence on the prior and posterior at any given point in time.

The following simple example illustrates the suggested algorithm.

EXAMPLE. Assume $X \sim N(\theta, 1)$, where θ is unknown. Let the prior $\pi(\theta)$ be $N(2, 1)$. Then, if $x = -1$ is observed, the likelihood is $l(\theta) = \phi(\theta - (-1))$, where $\phi(.)$ denotes the standard normal density, giving the posterior $p(\theta)$ as $(\theta|x) \sim N(0.5, 0.5)$.

Following Algorithm 2, we first drew a sample from the prior: $\theta_i \sim N(2, 1), i = 1, \dots, 100$. In the first step of the iteration $L_1 = \{l(\theta_i) > \bar{l}\}$ was the interval $(-\infty, 1.29]$, where 1.29 was obtained as $\max\{\theta_i : l(\theta_i) > \bar{l}, i = 1, \dots, 100\}$. Since the likelihood function is analytically available, it would obviously be possible to obtain L_1 exactly. But the only penalty for not accurately identifying L_m is that the expected final posterior sample size

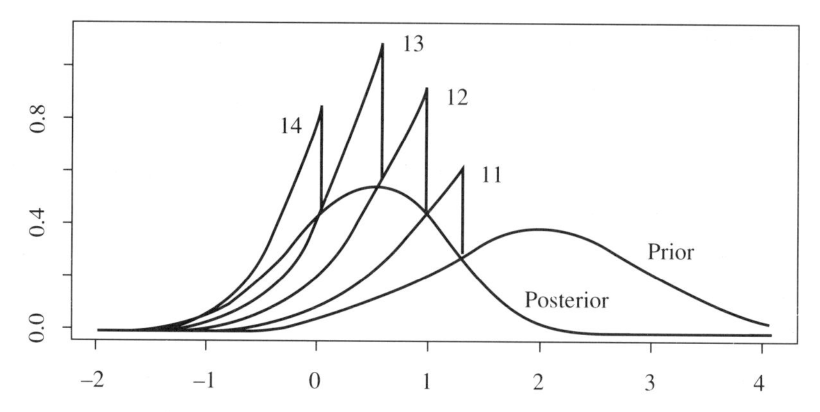

FIGURE 1. Prior π, envelope densities I_1, I_2, I_3, I_4, and posterior p.

will not exactly match the initial prior sample size. The parameter α_1 was (rather arbitrarily) chosen such that q_1 evaluates to $\frac{1}{2}$. With all parameters set, the additional sample was taken from $\pi|L_1$, increasing the total sample size to 122. The new points were added to the sample, and the weights updated to $w_1(\theta) := w_0(\theta)/2 = l(\theta)/2$, if $\theta \in L_1$, respectively left unchanged $w_1(\theta) := w_0(\theta)$ otherwise.

The stopping rule was determined by the ration $\bar{l}/l^*$, which was estimated to be .076, based on the likelihood values $l(\theta_i)$ of the sampled points. Although $l^* = \sup(l(\theta))$ could have been obtained analytically here, this was not done, since underestimation of l^* by taking the sample maximum will only affect the posterior sample size.

The stopping criterion was not met yet. So the iteration continued with $L_2 = \{w_1 > \bar{l}\} = (-\infty, 0.97]$. In the second iteration the sample size was increased to 151. After the forth cycle the stopping criterion was met, and a sample of 191 points resulted from the last envelope density I_4. The final accept/reject step reduced the sample size back to $n = 106$.

The graph in Figure 1 shows the prior density, the four envelope densities and the posterior density, plotted proportional to the sample size; i.e. , if one curve is above another, then a sample corresponding to the first curve can always be transformed by accept/reject to a sample corresponding to the second. The algorithm continues to expand the sample size until the curve corresponding to the final envelope density I_4 is everywhere above the posterior density; i.e., the envelope density sample can by accept/reject be transformed to a posterior sample with expected sample size 100.

4. A model of advertising awareness

4.1. The model. This nonlinear, nonnormal example is taken from West and Harrison (1989). The number of individuals in a survey who are "aware" of certain TV commercials is modeled. Awareness is measured by the proportion of the TV viewers who have seen these TV commercials during a recent, fixed time interval. The data was collected in weekly intervals by questioning sampled members of the TV viewing population as to whether or not they have seen the TV commercials in question. The survey was taken over a period of 75 weeks with a constant sample size of $n = 66$.

The variables involved are:

Y_t: number of positive responses out of the sample size n
X_t: extent of advertising (measured in units called "TVR").

The time series for Y_t and X_t are plotted in Figures 2 and 3.

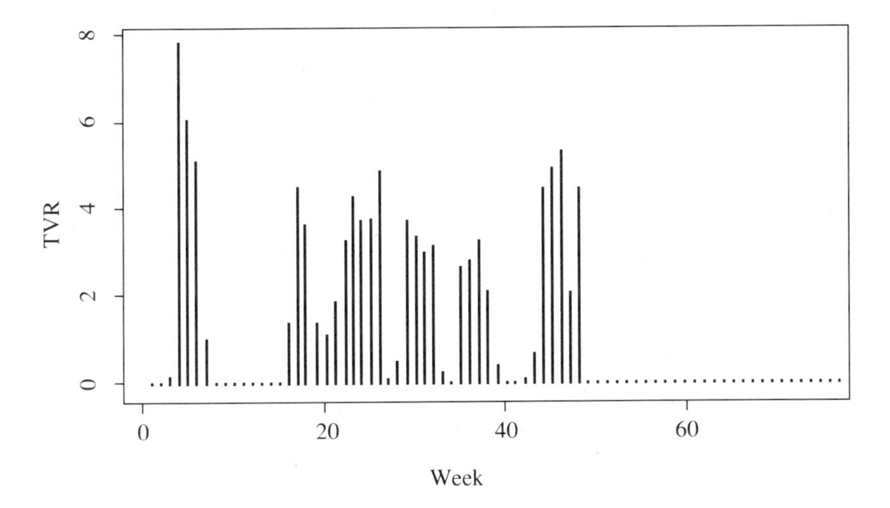

FIGURE 2. Extent of advertising in TVR.

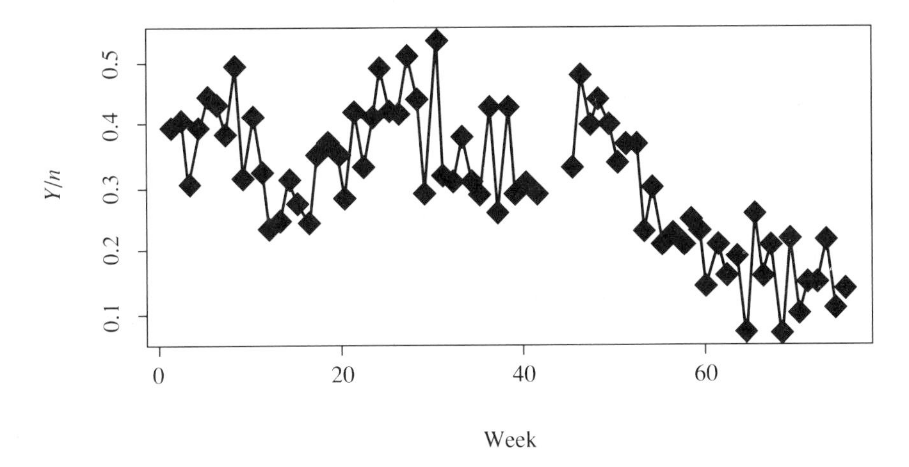

FIGURE 3. Awareness response as proportion: Y_t/n_t (weeks 42, 43, and 44 have missing observations).

The following parameters are used to model the total population awareness at time t:

α_t: lower threshold of awareness, i.e. the minimum level of awareness expected at time t.

β_t: upper threshold of awareness, i.e. the maximum level of awareness expected at time t.

E_t: effect of current and past weekly advertising, adding up with α_t to the total population awareness $\mu_t = E_t + \alpha_t$ at time t.

ρ_t: decay of awareness: if $X_t = 0$ then the effect of past advertising is expected to decay exponentially by: $E_t = \rho_t E_{t-1}$.

κ_t: penetration parameter: the meaning of this parameter becomes clear from the evolution equation.

The deterministic part of the evolution equation is given by:

$$E_t = \rho_t E_{t-1} + [1 - \exp(-\kappa_t X_t)][\beta_t - (\alpha_t + \rho_t E_{t-1})]. \tag{14}$$

The effect of past advertising is decaying by the factor ρ_t. The effect of current advertising is proportional to $[1 - \exp(-\kappa_t X_t)]$ and bounded above by $[\beta_t - (\alpha_t + \rho_t E_{t-1})]$.

Let $\theta_t = (\alpha_t, \beta_t, E_t, \rho_t, \kappa_t)'$ denote the parameter vector at time t. Assume a multivariate normal evolution noise with a covariance matrix corresponding to a discount factor of 0.97:

$$\omega_t \sim N(0, U_t) \quad \text{with } U_t = 0.03C_t,$$

where C_t is the variance of the current posterior $p(\theta_t|D_t)$. The concept of using a discount factor to specify W_t corresponds to specifying the relative increase of the current "uncertainty" C_t. See West and Harrison (1989), §2.4.2 for a more complete justification.

The complete model takes the form:

$$\begin{aligned} \text{Observation equation}: p(Y_t|\mu_t) &\propto \mu_t^{Y_t}(1-\mu_t)^{n-Y_i}, \\ \mu_t &= \alpha_t + E_t, \\ \text{Observation equation: } \theta_{t+1} &= g(\theta_t + \omega_t), \\ \omega_t &\sim N(0, U_t), \\ n = 66 \quad &\text{and} \quad t = 1, \dots, 75, \end{aligned}$$

where $g(\ )$ is the identity in α_t, β_t, ρ_t, and κ_t and determined by (14) for E_t. This is a nonlinear, nonnormal dynamic model with a system equation in the form of (10) and a sampling model slightly more general than (9).

Following West and Harrison the initial prior is chosen as

$$\theta_0|D_0 \sim N(m_0, C_0),$$
$$m_0 = (0.1, 0.85, 0.9, .02, 0.3)',$$
$$C_0 = 0.0001 \begin{bmatrix} 6.25 & 6.25 & 0 & 0 & 0 \\ 6.25 & 406.25 & 0 & 0 & 0 \\ 0 & 0 & 1 & 0 & 0 \\ 0 & 0 & 0 & 2.25 & 0 \\ 0 & 0 & 0 & 0 & 100 \end{bmatrix}.$$

The initial distribution is denoted here by $p(\theta_0|D_0)$ rather than $p(\theta_1|D_0)$ as in the description of Algorithm 1, because it models the beliefs on θ before the first evolution step.

4.2. Analysis as a dynamic generalized linear model. In their analysis of the advertising awareness model, West and Harrison first linearized the evolution equation:

$$\theta_t \approx g_t(m_{t-1}) + G_t * (\theta_{t-1} + \omega_t - m_{t-1}),$$

where

$$m_{t-1} = E(\theta_{t-1}|D_{t-1}),$$
$$G_t = \frac{\partial g_t(z)}{\partial z}|_{m_{t-1}}.$$

The linearized model was then analyzed as a dynamic generalized linear model, following the DGLM algorithm described in §2.3 The sampling model (7) of the DGLM specializes to (The analysis is easier in terms of the parameter μ_t rather than the natural parameter $\eta_t = log(\mu_t/(1-\mu_t)$.):

$$p(Y_t|\mu_t) \sim \text{ Binomial}(\mu_t, n),$$
$$\mu_t = (1, 0, 0, 0, 1)'\theta_t = \alpha_t + E_t.$$

Using the updating equations of the DGLM, as given e.g. in West and Harrison (1989) in §14.3.3., it is then straightforward to derive for each week $E(\theta_t|D_{t-1})$, $E(\theta_t|D_t)$, and other quantities of interest in the analysis of the model. In the next section the resulting data analysis is compared with the analysis obtained from the simulation algorithm.

4.3. Analysis by simulation. The same model was analyzed with the simulation algorithm suggested in §3. The simulation was started with a sample of size $n = 100$ from the initial distribution $p(\theta_0|D_0)$. (The prior specified by West and Harrison models the beliefs on θ before the first evolution step. Therefore we have to start with $p(\theta_0|D_0)$, rather than $p(\theta_1|D_0)$, as used in Algorithm 1.) By simulating the first evolution step, this sample was then transformed into a sample from the prior distribution $p(\theta_1|D_0)$. Then the information updating in the first observation step is simulated by Algorithm 2, leading to a sample from the posterior $p(\theta_1|D_1)$. From there the process iterates through all time steps, making at any time t a prior and a posterior sample available, which can be used to estimate prior and posterior integrals of the form (6) and (5).

The one step ahead forecast function gives the expected response in week $t+1$, based on the currently available information D_t:

$$f_t = E(\mu_{t+1}|D_t) = \int (\alpha_{t+1} + E_{t+1}) p(\theta_t|D_{t-1}) d\theta_{t+1}.$$

This is an integral of the form (6) and can be estimated by a Monte Carlo integral using the sample points from $p(\theta_t|D_{t-1})$. The forecasts are shown in Figure 4. The solid line plots the estimates of f_t; the dotted trajectory is the estimate of f_t using the dynamic generalized linear model. The actual responses are plotted as points. Figure 5 shows the estimated trajectory of the lower threshold parameter α_t. The solid line plots estimates of the posterior mean $E(\alpha_t|D_t)$, with the dashed lines representing one posterior standard deviation margins, i.e. $E(\alpha_t|D_t) + \sigma$ and $E(\alpha_t|D_t) - \sigma$, where $\sigma^2 = \text{var}(\alpha_t|D_t)$. (The posterior variance should be distinguished from the numerical variance, i.e. mean squared error, of the Monte Carlo estimator of $E(\alpha_t|D_t)$, which is σ^2/n, where $n = 100$ is the number of sample points.)

The dotted graph is the posterior estimate from the analysis as a DGLM. The DGLM estimates are well within the one posterior standard deviation margins of the simulation estimate. Around week 30 the posterior on α_t shifts to a slightly higher level. The DGLM analysis adjusts to this new level only towards the end of the campaign, when the exponentially decreasing response values give increasingly sharper information on the lower threshold.

Figures 6–8 (Figures 7 and 8 are on p. 160) give the same plots for the parameters β_t, ρ_t, and κ_t. Again, the estimates from the DGLM model and the simulation estimates stay always within one posterior standard deviation.

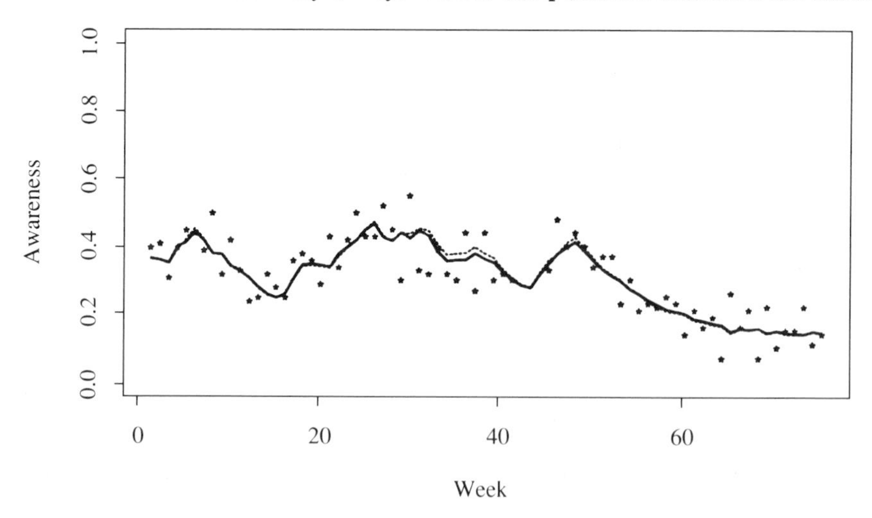

FIGURE 4. One step ahead forecasts.

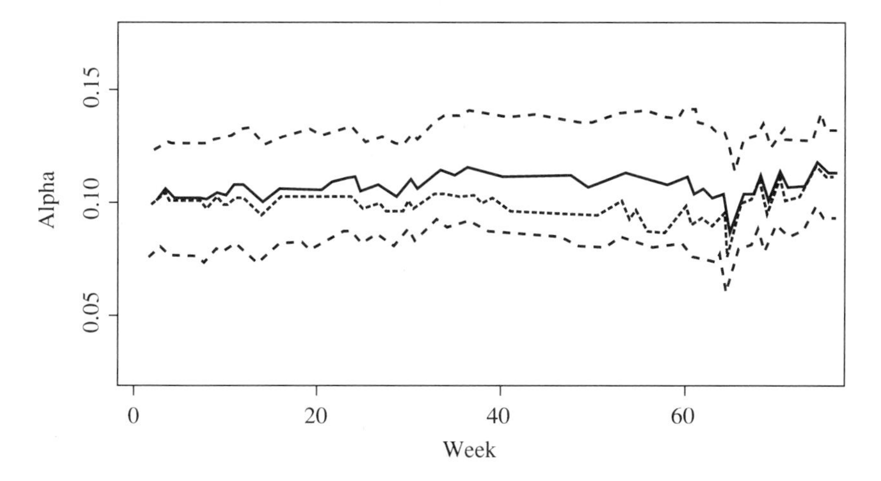

FIGURE 5. Lower threshold α_t.

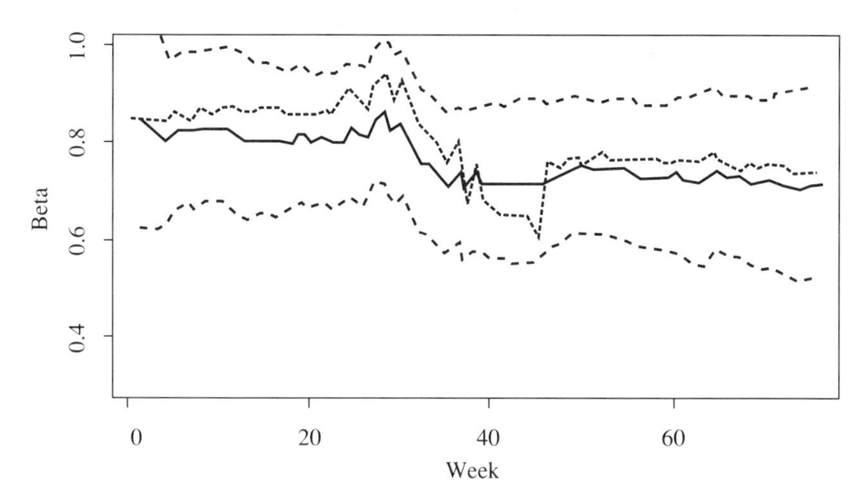

FIGURE 6. Upper threshold β_t.

The only noticeable difference occurs in the estimate for β_t around week 40. This downward jump in the trajectory of β_t corresponds to an overestimation of κ_t in the same time period. (The model takes a long time to readjust the posterior estimates, because the weeks 42, 43, and 44 have missing response values.)

The observed minor discrepancies between the DGLM analysis and the simulation analysis are due to approximations made in both algorithms.

In Figure 9, on p. 161, the beta density implicitly assumed for $p(\mu_{41}|D_{40})$ in the DGLM model is compared with a density estimate obtained from the sample points, which have been generated by the simulation algorithm.

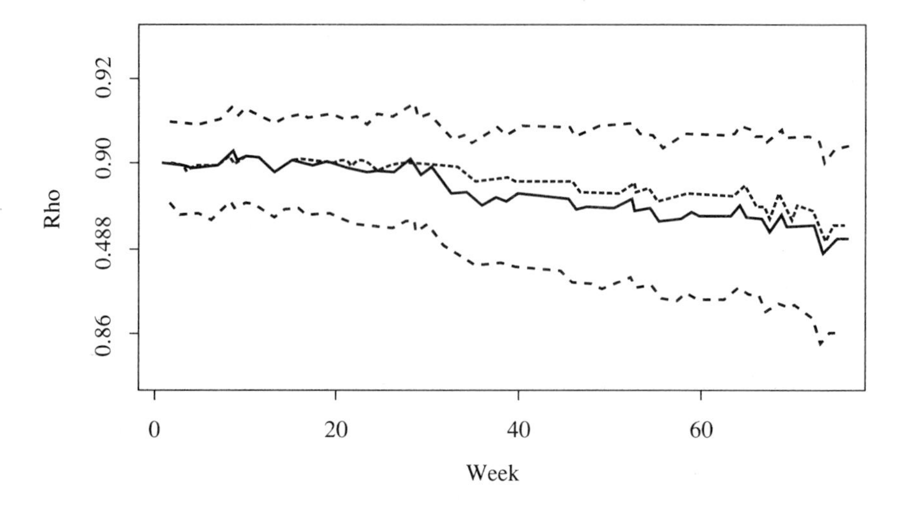

FIGURE 7. Decay parameter ρ_t.

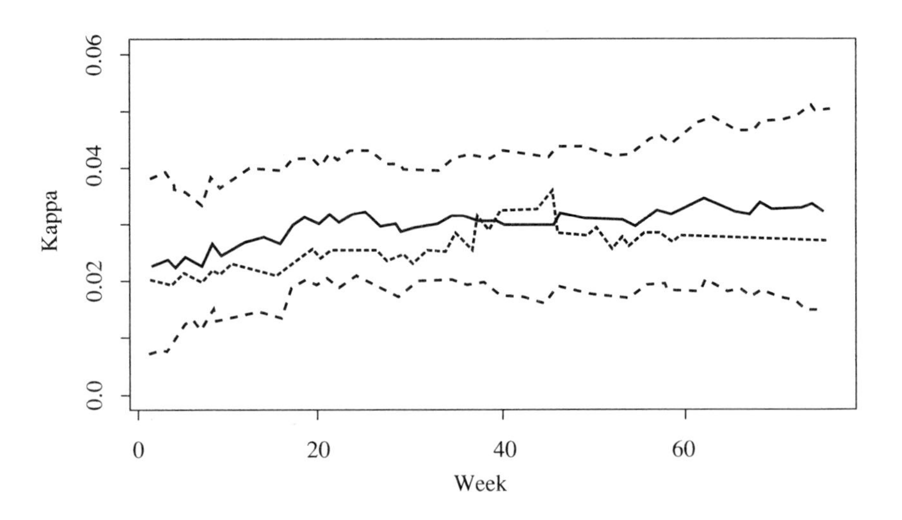

FIGURE 8. Penetration parameter ρ_t.

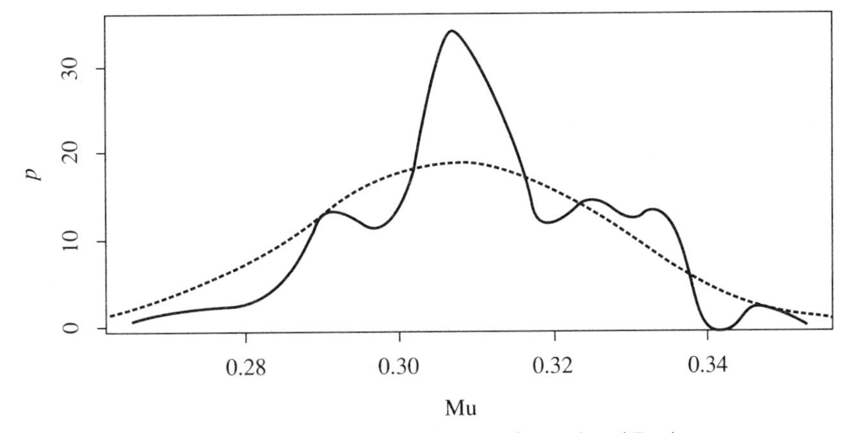

FIGURE 9. Density estimate for $p(\mu_{41}|D_{40})$.

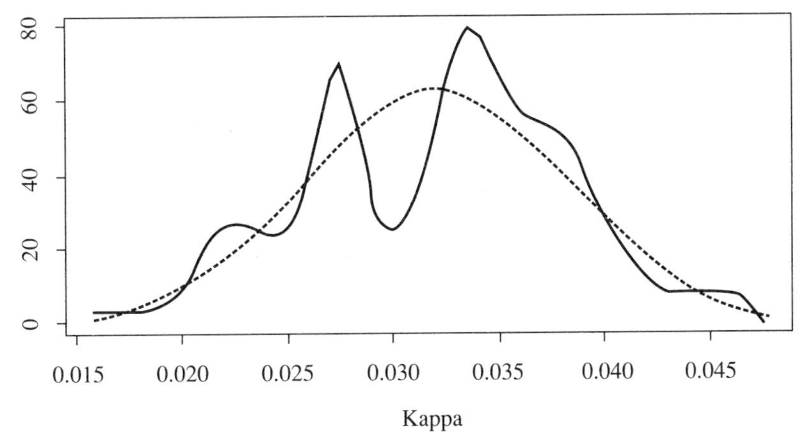

FIGURE 10. Density estimate for $p(\kappa_{40}|D_{40})$.

Figure 10 plots a density estimate for the posterior $p(\kappa_{40}|D_{40})$ together with the corresponding normal density. In both cases the unimodal beta, respectively normal distribution does not adequately reproduce the multimodal density. This does not contradict the DGLM, since this algorithm is designed for problems where the prior distributions and the evolution noise are not specified with densities, but only specified by their first and second moments.

5. Conclusion

While the simulation of the evolution step is a straightforward implementation of the system equation, the observation equation, i.e. Bayes' theorem, lacks such a natural interpretation as a mapping from points of a prior sample to points of a posterior sample. In this paper a two step procedure was used, first expanding the prior sample into a sample from an envelope density and then contracting by accept/reject to a sample from the posterior. This approach works well in the context of the dynamic model where the prior is essentially the posterior from the previous step and is typically not significantly changed by updating with the likelihood. For the simulation algorithm this means that only a moderate expansion of the sample size is required. In the application in §4 it was only rarely necessary to take more than one iteration step in the construction of the envelope density.

Appendix

Stopping rule for the iterative construction of the envelope density. The expected proportion of retained sample points in the accept/reject procedure is $\overline{w}/w^*$, where $\overline{w} = E_{I_M} w_M$ and $w^* = \sup(w_M)$. Let $\overline{X}|_A$ denote $E(X|A)$ and $\mu_m = I_m(L_m)$. Then $\overline{w}$ and w^* are given by:

$$\begin{aligned}\overline{w_m} &= (1-\mu_m)\overline{w_m}|_{L_m^c} + \mu_m \overline{w_m}|_{L_m} \\ &= (1-\mu_m)\overline{w_{m-1}}|_{L_m^c} + \mu_m q_m \overline{w_{m-1}}|_{L_m} \\ &= \alpha_m \overline{w_{m-1}}, \\ w_m^* &= \max(\overline{l}, q_m w_{m-1}^*).\end{aligned}$$

Then the expected proportion of retained sample points after shrinking takes the form:

$$\begin{aligned}\frac{\overline{w_m}}{w_m^*} &= \frac{\overline{l}}{w_m^*}\prod_{k=1}^{m}\alpha_k \\ &= \frac{1}{\max(\prod_1^m q_k l^*, \overline{l})}\overline{l}\prod_{k=1}^{m}\alpha_k.\end{aligned}$$

Since the sample size after the mth step is increased by a factor $1/\prod \alpha_k$ the following stopping rule will give an expected posterior sample size equal to the original prior sample size:

$$M = \min\left\{m : \prod_1^m q_k \le \frac{\overline{l}}{l^*}\right\}.$$

References

J. Geweke (1989), *Bayesian inference in econometric models using Monte Carlo integration,* Econometrica **57**, 1317–39.

P. S. Harrison and C. F. Stevens (1976), *Bayesian forecasting,* J. Roy. Statist. Soc., Ser. B **38**, 205–247.

Y. C. Ho and R. C. K. Lee (1964), *A Bayesian approach to problems in stochastic estimation and control,* IEEE Trans. Automat. Control **9**, 333–339.

A. Pole and M. West (1988), *Efficient numerical integration in dynamic models,* Warwick Research Report 136, Department of Statistics, University of Warwick.

A. Pole, M. West, and J. Harrison (1988), *Nonormal and nonlinear dynamic Bayesian modelling,* Bayesian Analysis of Time Series and Dynamic Models (J. C. Spall, ed.), Marcell Dekker, New York.

R. Y. Rubinstein (1981), *Simulation and the Monte Carlo Method,* J. Wiley, New York.

R. Serfling (1980), *Approximation Theorems of Mathematical Statistics,* J. Wiley, New York.

H. W. Sorenson and D. L. Alspach (1971), *Recursive Bayesian estimation using Gaussian sums,* Automatica 7, 465–479.

H. K. van Dijk (1984), *Posterior analysis of econometric models using Monte Carlo integration*, Reproduktie Wondestein, Erasmus University Rotterdam.

M. West and J. Harrison (1989), *Bayesian Forecasting and Dynamic Models*, Springer-Verlag, Berlin and New York.

M. West, P. J. Harrison, and H. S. Migon (1985), *Dynamic generalised linear models and Bayesian forecasting* (with discussion), J. Amer. Statist. Assoc. **80**, 73–97.

DEPARTMENT OF STATISTICS, PURDUE UNIVERSITY, WEST LAFAYETTE, INDIANA 47907
E-mail address: petr@pop.stat.purdue.edu

Contemporary Mathematics
Volume **115**, 1991

Monte Carlo Integration via Importance Sampling: Dimensionality Effect and an Adaptive Algorithm

MAN-SUK OH

ABSTRACT. The effect of dimension on the accuracy of Importance Sampling is studied for simple common cases by investigating the behavior of Importance Sampling estimates with a fixed number of iterations as a function of dimension p. In all the cases studied, the major dimensionality effect seems to be driven by the variance of the weight function, $\mathrm{var}(w)$. It is found that $\mathrm{var}(w)$ is exponentially increasing in p but with a good importance function $\mathrm{var}(w)$ is approximately linear function of p for moderate p. Also, reasonable choices, under various criteria, for location and scale parameters of the multivariate t importance function are discussed.

An adaptive Importance Sampling scheme is described as a flexible, yet mechanical way of selecting the parameters of the importance function. It runs ordinary Importance Sampling in an iterative fashion, continually updating the parameters of the importance function by simple linear operations. An example of a Bayesian hierarchical model is given to compare the adaptive Importance Sampling with ordinary Importance Sampling.

1. Introduction

We are interested in the numerical calculation of integrals of the form,

$$E\phi = \frac{\int \phi(\theta) f(\theta)\, d\theta}{\int f(\theta)\, d\theta}, \tag{1}$$

where $\theta \in R^p$, $\phi(\theta)$ is a measurable function, and $f(\theta)$ is proportional to a density function. In Bayesian analysis, the above integrals are crucial for posterior inferences, with $f(\theta)$ being the product of likelihood and prior for θ.

Monte Carlo integration via Importance Sampling approximates (1) by

$$\widehat{E}\phi = \frac{\sum_{i=1}^n \phi(\theta_i) w(\theta_i)}{\sum_{i=1}^n w(\theta_i)}, \tag{2}$$

1980 *Mathematics Subject Classification* (1985 *Revision*). Primary 62F15; Secondary 65D30.

Key words and phrases. Statistical multiple integration, numerical integration, importance function, Bayesian hierarchical model, posterior density function.

The final version of this paper will be submitted for publication elsewhere.

where $w(\theta)$, called the weight function, is $f(\theta)/g(\theta)$ for a density function $g(\theta)$, and θ_i, $i = 1, \ldots, n$, are independent and identically distributed random samples from $g(\theta)$.

Under easily attainable conditions, $\widehat{E}\phi$ converges to $E\phi$ with probability one and has approximate normal distribution with mean $E\phi$ and variance σ^2/n, where

$$(3)\ \sigma^2 = \frac{1}{(\int f(\theta)\,d\theta)^2}\int (\phi - E\phi)^2 \frac{f^2(\theta)}{g(\theta)}\,d\theta$$

$$(4)\ \quad = \frac{1}{(\int f(\theta)\,d\theta)^2}[\mathrm{var}_g(\phi w) - 2(E\phi)\,\mathrm{cov}_g(\phi w, w) + (E\phi)^2\,\mathrm{var}_g(w)],$$

the subscript g in $\mathrm{var}_g(\cdot)$, $\mathrm{cov}_g(\cdot)$ represents the variances and covariances as taken with respect to g so that

$$(5)\quad \mathrm{var}_g(\phi w) = \int \phi^2(\theta)\frac{f^2(\theta)}{g(\theta)}\,d\theta - \left(\int \phi(\theta) f(\theta)\,d\theta\right)^2,$$

$$(6)\ \mathrm{cov}_g(\phi w, w) = \int \phi(\theta)\frac{f^2(\theta)}{g(\theta)}\,d\theta - \left(\int \phi(\theta) f(\theta)\,d\theta\right)\left(\int f(\theta)\,d\theta\right),$$

$$(7)\quad \mathrm{var}_g(w) = \int \frac{f^2(\theta)}{g(\theta)}\,d\theta - \left(\int f(\theta)\,d\theta\right)^2,$$

In what follows, we will omit the subscript g unless it is necessary.

Discussions, error reduction techniques, and applications of Importance Sampling can be found in Davis and Rabinowitz (1975), Kloek and van Dijk (1978), Stewart (1979, 1983, 1984), van Dijk and Kloek (1980, 1983a, 1983b, 1985), van Dijk (1984, 1987), Rubinstein (1985), Geweke (1986, 1988), and references cited therein.

As shown in (3), the asymptotic variance σ^2 of an Importance Sampling estimate depends on the density function g, called the importance function. Indications of a good importance function are (i) simplicity in random variate generation, (ii) having thicker tails than f, (iii) being a good approximation to f (Kloek and van Dijk (1978)).

In this paper, we first investigate the effect of dimension on the accuracy of Importance Sampling estimates because dimensionality effect is a big issue in numerical integration. It is found that the major dimensionality effect comes from the variance of the weight function, $\mathrm{var}(w)$, given in (7). And $\mathrm{var}(w)$ is exponentially increasing in p. But if the importance function matches well with f, $\mathrm{var}(w)$ increases only linearly for moderate p. This stresses the importance of the selection of good importance function.

Next, an adaptive algorithm of Importance Sampling (Adaptive Importance Sampling) is described to improve the selection of importance function by cumulating informations about f during simulation. It runs ordinary Importance Sampling (Basic Importance Sampling, from here on) in an iterative way, repeatedly updates the parameters of g by pooling Importance

Sampling estimates about some features of f. The algorithm is simple and flexible, and can be run automatically with some plausible stopping rules. Additional calculations in Adaptive Importance Sampling seem to be cheap while the accuracy gained is superior.

2. Dimensionality effect

The accuracy of Importance Sampling can be measured by the variance of $\widehat{E}\phi$, $\mathrm{var}(\widehat{E}\phi) = \sigma^2/n$, where σ^2, given in (3), depends on ϕ, g, f, and the dimension p of θ. Thus, to determine the dimensionality effect on $\widehat{E}\phi$ with a fixed number of iterations, we must investigate how σ^2 varies with p, for various ϕ, g, and f.

In this section, we consider simple common models and choices of g and ϕ. In many Bayesian analyses, f is approximately normal. Thus, we will assume f the density function of $N(0, I)$ distribution, for simplicity.

The most common choices for g in a situation where f is approximately symmetric and unimodal are multivariate normal and multivariate t densities. The actual posterior mean and covariance matrix (with some adjustment, if necessary) would be natural choices for the location and scale parameters of such g, but they are, of course, unknown. Thus, the effect of dimension will be studied for normal and multivariate t importance functions, with various misspecified location and scale parameter values.

Natural ϕ's to look at are θ_i and θ_i^2, where θ_i is the ith element of θ, since these will yield the first two posterior moments in Bayesian analysis. Also, in many cases f is fairly sharp so that the Taylor expansion of ϕ up to quadratic terms gives a good approximation in the important region where f is concentrated. Then the behavior of $\mathrm{var}(\widehat{E}\theta_i)$ and $\mathrm{var}(\widehat{E}\theta_i^2)$ will determine the behavior of $\mathrm{var}(\widehat{E}\phi)$.

We also examine the behavior of $\mathrm{var}(w)$, because $\mathrm{var}(w)$ seems to reflect the behavior of $\mathrm{var}(\widehat{E}\phi)$ in considerable generality. Moreover, $\mathrm{var}(w)$ is itself necessary for the evaluation of $\mathrm{var}(\widehat{E}\phi)$ as shown in (4) and is a rough indicator of the accuracy of the Importance Sampling scheme. In fact, since f is assumed to be the $N(0, I)$ density function, from (3),

$$\mathrm{var}(\widehat{E}\phi) = \int (\phi(\theta) - E\phi)^2 \frac{f^2(\theta)}{g(\theta)}\, d\theta \tag{8}$$

$$= \frac{\int (\phi(\theta) - E\phi)^2 f^2(\theta)/g(\theta)\, d\theta}{\int f^2(\theta)/g(\theta)\, d\theta} \int \frac{f^2(\theta)\, d\theta}{g(\theta)\, d\theta} \tag{9}$$

$$= E_h(\phi(\theta) - E\phi)^2 (\mathrm{var}(w) + 1), \tag{10}$$

where $E_h(\cdot)$ represents expectation with respect to the density function:

$$h(\theta) = \frac{f^2(\theta)/g(\theta)}{\int f^2(\theta)/g(\theta)\, d\theta}. \tag{11}$$

In particular,

$$\text{var}(\hat{E}\theta_i) = E_h(\theta_i^2)(\text{var}(w)+1), \tag{12}$$

$$\text{var}(\hat{E}\theta_i^2) = E_h((\theta_i^2-1)^2)(\text{var}(w)+1). \tag{13}$$

We shall see that $E_h(\phi(\theta)-E\phi)^2$ is often fairly constant in p, so that $\text{var}(w)$ can be used to discern the dimensionality effect.

Finally, when f is symmetric and unimodal, the multivariate t density function is the most common choice for g because of its thick tails and convenience of random variate generation. Reasonable choices, under various criteria, for the location and scale parameters of the multivariate t importance function are discussed in §2.3.

2.1. Normal importance function. Consider first the case where g is itself a normal density function. The following lemma shows that unless the covariance matrix of g is too small compared to that of f, $h(\theta)$ is also a normal density function.

LEMMA 2.1. *If f and g are density functions of $N(0, I)$ and $N(\mu, \Sigma)$ distributions, respectively, and all characteristic roots of Σ are larger than $1/2$, then $h(\theta)$ is the density function of the $N(-(2\Sigma-I)^{-1}\mu, (2I-\Sigma^{-1})^{-1})$ distribution, where $h(\theta)$ is defined in* (11).

PROOF.

$$\begin{aligned} f^2(\theta)/g(\theta) &\propto \exp\{-\theta'\theta + \tfrac{1}{2}(\theta-\mu)'\Sigma^{-1}(\theta-\mu)\} \\ &= \exp\{-\tfrac{1}{2}(\theta'(2I-\Sigma^{-1})\theta + 2\mu'\Sigma^{-1}\theta) + \tfrac{1}{2}\mu'\Sigma^{-1}\mu\} \\ &\propto \exp\{-\tfrac{1}{2}(\theta + (2I-\Sigma^{-1})^{-1}\Sigma^{-1}\mu)'(2I-\Sigma^{-1})(\theta + (2I-\Sigma^{-1})^{-1}\Sigma^{-1}\mu)\} \\ &\qquad\qquad \text{if } (2I-\Sigma^{-1})^{-1} \text{ exists} \\ &= \exp\{-\tfrac{1}{2}(\theta + (2\Sigma-I)^{-1}\mu)'(2I-\Sigma^{-1})(\theta + (2\Sigma-I)^{-1}\mu)\} \\ &\qquad\qquad \text{since } (2I-\Sigma^{-1})^{-1}\Sigma^{-1} = (2\Sigma-I)^{-1}. \end{aligned}$$

Thus, if $(2I-\Sigma^{-1})$ is a positive definite matrix, then $h(\theta)$ is the $N(-(2\Sigma-I)^{-1}\mu, (2I-\Sigma^{-1})^{-1})$ density function. Clearly, $(2I-\Sigma^{-1})$ is positive definite if and only if all characteristic roots of Σ are larger than $1/2$. □

LEMMA 2.2. *Suppose the assumptions of Lemma 2.1 hold, and ϕ is a function of a fixed number of components, $\theta_{i_1}, \dots, \theta_{i_m}$, of θ. Also assume that*

(14) (a) *μ_i is independent of p, for $i = i_1, \dots, i_m$,*

(b) *σ_{ij} is independent of p, for all $i, j = i_1, \dots, i_m$,*

(15) (c) *$\sigma_{ij} = 0$, for all $i = i_1, \dots, i_m$, $j \neq i_1, \dots, i_m$,*

where μ_i *is the* ith *element of* μ *and* σ_{ij} *is the* (i, j) *th element of* Σ. *Then* $E_h(\phi(\theta_{i_1}, \dots, \theta_{i_m}) - E\phi)^2$ *is independent of* p.

PROOF. Without loss of generality, one may assume that $\phi(\theta)$ is a function of the first m elements, $\theta_1, \dots, \theta_m$, of θ. Then, by the assumptions, Σ becomes

$$\Sigma = \begin{pmatrix} \Sigma_1 & 0 \\ 0 & \Sigma_2 \end{pmatrix},$$

where Σ_1 is a $m \times m$ matrix with all elements independent of p and Σ_2 is a $(m-p) \times (m-p)$ matrix. By a simple calculation and from Lemma 2.1, $h(\theta)$ is the density function of the

$$N\left(\begin{pmatrix} -(2\Sigma_1 - I_1)^{-1}\mu_1 \\ -(2\Sigma_2 - I_2)^{-1}\mu_2 \end{pmatrix}, \begin{pmatrix} (2I_1 - \Sigma_1^{-1})^{-1} & 0 \\ 0 & (2I_2 - \Sigma_2^{-1})^{-1} \end{pmatrix}\right)$$

distribution, where μ_1 is the vector consisting of the first m elements of μ and μ_2 of the rest. From classical multivariate distribution theory, $(\theta_1, \dots, \theta_m)$ follows the $N(-(2\Sigma_1 - I_1)^{-1}\mu_1, (2I_1 - \Sigma_1^{-1})^{-1})$ distribution which is independent of p. □

Thus, when g and ϕ satisfy the conditions of Lemmas 2.1 and 2.2, the dimensionality effect is driven only by $\mathrm{var}(w)$ (see (10)). Even when ϕ doesn't satisfy the conditions of Lemma 2.2 there are many situations (for example, $\phi = \Sigma_{i=1}^p \theta_i/p$, $\phi = \sqrt[p]{\prod_{i=1}^p \theta_i}$), where the major dimensionality effect comes from $\mathrm{var}(w)$.

In many cases, the mean and covariance matrix of θ under f, which are desirable location and scale parameter values of g (with some modifications, if necessary), are unknown and one needs to use estimates of them. Thus, it is natural to consider cases where the location or scale of g is misspecified. For simplicity, assume that the scale parameter Σ of g is a diagonal matrix. To see the dimensionality effect clearly, assume that the location or scale of g is misspecified by the same amount in all directions. Thus, the case of g being the density function of the $N(\varepsilon\mathbf{1}, (1+\delta)I)$ distribution with $\delta > -1/2$, where $\mathbf{1} = (1, \dots, 1)$, will be considered in this section. We first record the expressions for (12) and (13) under these assumptions.

COROLLARY 2.1. *If* f *and* g *are the* $N(0, I)$ *and* $N(\varepsilon\mathbf{1}, (1+\delta)I)$, $\delta > -1/2$, *density functions, respectively, then*

$$\mathrm{var}^N_{\varepsilon,\delta}(\widehat{E}\theta_i) = \left(\frac{(1+\delta)(1+2\delta)+\varepsilon^2}{(1+2\delta)2}\right)(\mathrm{var}^N_{\varepsilon,\delta}(w)+1), \tag{16}$$

$$\mathrm{var}^N_{\varepsilon,\delta}(\widehat{E}\theta_i^2) = B(\varepsilon, \delta)(\mathrm{var}^N_{\varepsilon,\delta}(w)+1), \tag{17}$$

where

$$B(\varepsilon, \delta) = \frac{\varepsilon^4 + 2\varepsilon^2(2+\delta)(1+2\delta) + (1+2\delta)^2(3\delta^2+4\delta+2)}{(1+2\delta)^4},$$

and the scripts N, ε, δ *in* $\mathrm{var}^N_{\varepsilon,\delta}(\cdot)$ *are used to indicate the* $N(\varepsilon\mathbf{1}, (1+\delta)I)$ *density function* g.

PROOF. From Lemmas 2.1 and 2.2, $h(\theta)$ is the

$$N\left(-\frac{\varepsilon}{1+2\delta}\mathbf{1}, \frac{1+\delta}{1+2\delta}I\right)$$

density function. Thus $E_h(\theta_i^2)$ and $E_h((\theta_i^2-1)^2)$ can be easily calculated from (12), (13) and the results follow. □

Thus, in this situation the dimensionality effect is driven only by $\mathrm{var}^N_{\varepsilon,\delta}(w)$ for both $\widehat{E}\theta_i$ and $E\theta_i^2$ and we only need to investigate the behavior of $\mathrm{var}^N_{\varepsilon,\delta}(w)$.

- Variance of w. From (7),

$$\begin{aligned}
\mathrm{var}^N_{\varepsilon,\delta}(w) &= \left(\frac{1+\delta}{2\pi}\right)^{p/2}\int \exp\left\{\frac{1}{2(1+\delta)}(\theta-\varepsilon\mathbf{1})'(\theta-\varepsilon\mathbf{1})-\theta'\theta\right\}d\theta-1\\
&= \left(\frac{1+\delta}{2\pi}\right)^{p/2}\exp\left\{\frac{\varepsilon^2 p}{1+2\delta}\right\}\\
&\quad\times\int\exp\left\{\frac{1+2\delta}{2(1+\delta)}\left(\theta+\frac{\varepsilon}{1+2\delta}\mathbf{1}\right)\left(\theta+\frac{\varepsilon}{1+2\delta}\mathbf{1}\right)\right\}d\theta-1\\
&= \left(\frac{1+\delta}{\sqrt{1+2\delta}}\exp\left\{\frac{\varepsilon^2}{1+2\delta}\right\}\right)^p-1. \qquad (18)
\end{aligned}$$

As can be seen in (18), $\mathrm{var}^N_{\varepsilon,\delta}(w)$ is exponentially increasing in p. However, if $(\delta^2+2\varepsilon^2)/p(2(1+2\delta))$ is small, then

$$\begin{aligned}
\left(\frac{1+\delta}{\sqrt{1+2\delta}}\exp\left\{\frac{\varepsilon^2}{1+2\delta}\right\}\right)^p &= \left(1+\frac{\delta^2}{1+2\delta}\right)^{p/2}\exp\left\{\frac{\varepsilon^2 p}{1+2\delta}\right\}\\
&\approx\left(1+\frac{\delta^2 p}{2(1+2\delta)}\right)\left(1+\frac{\varepsilon^2 p}{1+2\delta}\right)\\
&\approx 1+\frac{\delta^2+2\varepsilon^2}{2(1+2\delta)}p.
\end{aligned}$$

Thus, $\mathrm{var}^N_{\varepsilon,\delta}(w)$ is approximately linear in p when the square of the misspecification error is small compared to p. However, one can expect exponential degradation for very large p. Note that if δ and ε were controlled to $O(p^{-1/2})$, there would be no degradation in performance as p increases.

The graph of $\mathrm{var}^N_{\varepsilon,\delta}(w)$ is given in Figure 1 for various values of ε, δ. (In each figure, $\delta=.2, -.1, .1, 0$, from the left to the right line.) Note that $\mathrm{var}^N_{\varepsilon,\delta}(w)$ is about linear for $\varepsilon\le .1$, $p\le 60$, but very rapidly increasing in p when $\varepsilon=.2$. Also note that, $\mathrm{var}^N_{\varepsilon,\delta}(w)$ is not affected much by different values of δ but it is very sensitive to different values of ε. This

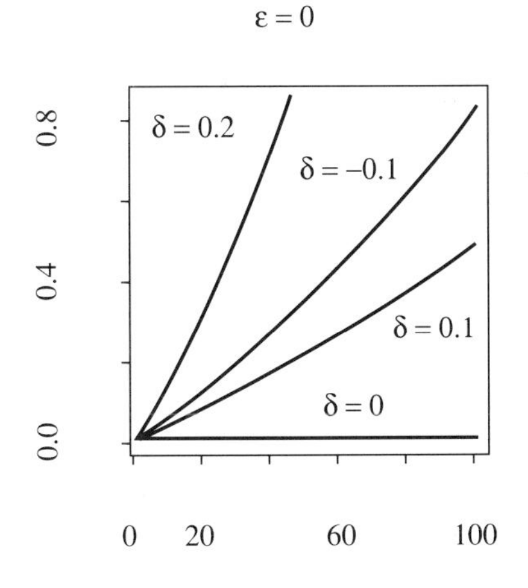

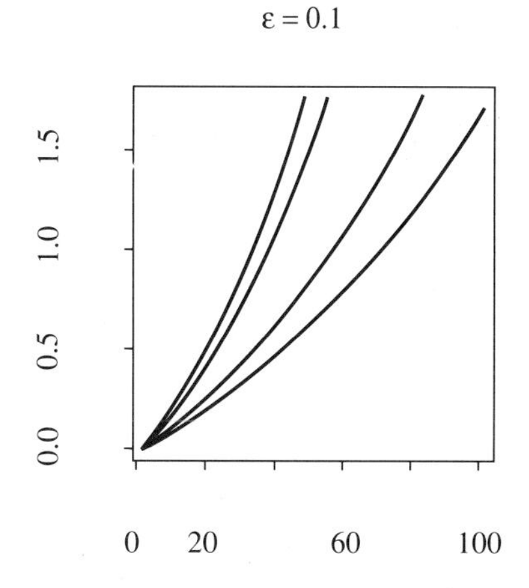

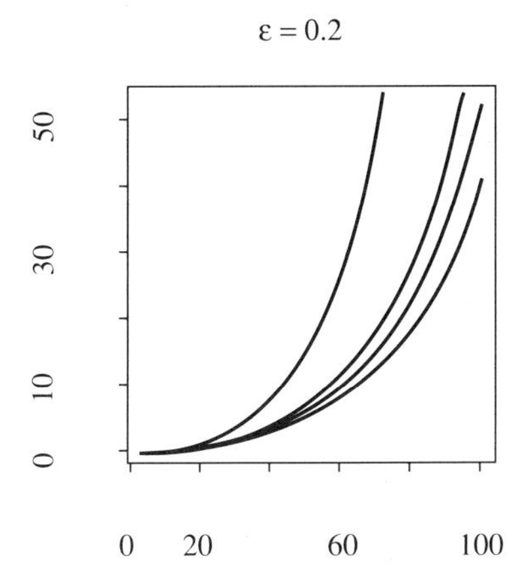

FIGURE 1. $\text{var}(w)$, $g = N(\varepsilon 1, (1+\delta)I)$.

implies that the misspecification of location is much more serious than the misspecification of scale. (Compare graphs with $(\varepsilon, \delta) = (0, 0)$, $(0, .1)$, $(.1, 0)$, $(.1, .1)$).

2.2. Multivariate t importance function. Consider next the more common case where g is the multivariate t density. As in the previous section, we consider misspecified location and scale so that g is the $\mathcal{T}_\alpha(\varepsilon\mathbf{1}, (1+\delta)I)$ density function. Simplifications such as (16) and (17) do not seem possible here, but evidence will be presented to the effect that the major dimensionality effect is still exhibited by $\text{var}(w)$.

LEMMA 2.3. *If f and g are the $N(0, I)$ and $\mathcal{T}_\alpha(\varepsilon\mathbf{1}, (1+\delta)I)$ density functions, respectively, and $\mathbf{z} = (z_1, \dots, z_p) \sim N(0, I)$, then*

(19)

$$\operatorname{var}^T_{\varepsilon,\delta,\alpha}(w) = K_{\alpha,p}(1+\delta)^{p/2} E\left[\left(1+\frac{1}{\alpha(1+\delta)}\sum_{j=1}^{p}\left(\frac{z_j}{\sqrt{2}}-\varepsilon\right)^2\right)^{\frac{\alpha+p}{2}}\right]-1,$$

(20)

$$\operatorname{var}^T_{\varepsilon,\delta,\alpha}(\widehat{E}\phi) = K_{\alpha,p}(1+\delta)^{p/2} E\left[(\phi(\mathbf{z}/2)-E\phi)^2\left(1+\frac{1}{\alpha(1+\delta)}\sum_{j=1}^{p}\left(\frac{z_j}{\sqrt{2}}-\varepsilon\right)^2\right)^{\frac{\alpha+p}{2}}\right]$$

where the constant $K_{\alpha,p}$ *is defined by*

$$K_{\alpha,p} = \left(\frac{\alpha}{4}\right)^{p/2}\frac{\Gamma(\alpha/2)}{\Gamma((\alpha+p)/2)}, \tag{21}$$

and the scripts T, ε, δ, α *in* $\operatorname{var}^T_{\varepsilon,\delta,\alpha}(\cdot)$ *are used to indicate the* $\mathscr{T}_\alpha(\varepsilon\mathbf{1}, (1+\delta)I)$ *density function* g.

PROOF. From (7),

$$\begin{aligned}\operatorname{var}^T_{\varepsilon,\delta,\alpha}(w) &= K_{\alpha,p}(1+\delta)^{p/2}(2\pi)^{-p/2}2^{p/2}\\ &\quad\times\int\left(1+\frac{1}{\alpha(1+\delta)}(\theta-\varepsilon\mathbf{1})'(\theta-\varepsilon\mathbf{1})\right)^{(\alpha+p)/2} e^{-\theta'\theta}\,d\theta-1\\ &= K_{\alpha,p}(1+\delta)^{p/2}(2\pi)^{-p/2}2^{p/2}\\ &\quad\times\int\left(1+\frac{1}{\alpha(1+\delta)}\sum_{j=1}^{p}(\theta_j-\varepsilon)^2\right)^{(\alpha+p)/2} e^{-\theta'\theta}\,d\theta-1\\ &= K_{\alpha,p}(1+\delta)^{p/2}E\left[\left(1+\frac{1}{\alpha(1+\delta)}\sum_{j=1}^{p}\left(\frac{z_j}{\sqrt{2}}-\varepsilon\right)^2\right)^{(\alpha+p)/2}\right]-1.\end{aligned}$$

Equation (20) can be derived in a similar way. □

• Variance of w. When $(\alpha+p)/2$ is an integer, the expectation in (19) can be represented as

$$E\left[\left(1+\frac{1}{\alpha(1+\delta)}\sum_{j=1}^{p}\left(\frac{z_j}{\sqrt{2}}-\varepsilon\right)^2\right)^{(\alpha+p)/2}\right] \tag{22}$$

$$= \sum_{k=0}^{(\alpha+p)/2}\binom{(\alpha+p)/2}{k}\left(\frac{1}{\alpha(1+\delta)}\right)^k E\left[\left(\sum_{j=1}^{p}\left(\frac{z_j}{\sqrt{2}}-\varepsilon\right)^2\right)^k\right]. \tag{23}$$

If we define

$$\mu_p^k = E\left[\left(\sum_{j=1}^{p}\left(\frac{z_j}{\sqrt{2}} - \varepsilon\right)^2\right)^k\right], \tag{24}$$

then

$$\mu_p^k = E\left[\left(\left(\frac{z_1}{\sqrt{2}} - \varepsilon\right)^2 + \sum_{j=2}^{p}\left(\frac{z_j}{\sqrt{2}} - \varepsilon\right)^2\right)^k\right] \tag{25}$$

$$= E\left[\sum_{l=0}^{k}\binom{k}{l}\left(\frac{z_1}{\sqrt{2}} - \varepsilon\right)^{2l}\left(\sum_{j=2}^{p}\left(\frac{z_j}{\sqrt{2}} - \varepsilon\right)^2\right)^{k-l}\right] \tag{26}$$

$$= \sum_{l=0}^{k}\binom{k}{l} E\left[\left(\frac{z1}{\sqrt{2}} - \varepsilon\right)^{2l}\right] E\left[\left(\sum_{j=2}^{p}\left(\frac{z_j}{\sqrt{2}} - \varepsilon\right)^2\right)^{k-l}\right] \tag{27}$$

$$= \sum_{l=0}^{k}\binom{k}{l}\mu_1^l \mu_{p-1}^{k-l}. \tag{28}$$

From (19) and (22)–(28), $\text{var}^T_{\varepsilon,\delta,\alpha}(w)$ can be calculated by recurrence and the graph is given in Figure 2, on p. 174, for various α, δ, and ε values. (In each figure, $\alpha = 4, 13, 19$, from the left to the right line.) When $(\alpha + p)/2$ is not an integer, we computed (22) by interpolation.

As can be seen from Figure 2, the location error ε has a great effect on $\text{var}^T_{\varepsilon,\delta,\alpha}(w)$ (see the maximum of $\text{var}^T_{\varepsilon,\delta,\alpha}(w)$ when $\varepsilon = 0, .1, .2$) compared to the effects of α and δ. If $|\varepsilon|$ is small ($|\varepsilon| \leq .1$), $\text{var}^T_{\varepsilon,\delta,\alpha}(w)$ is again approximately a linear function of p (slightly concave for $\varepsilon = 0$, $\alpha \leq 13$).

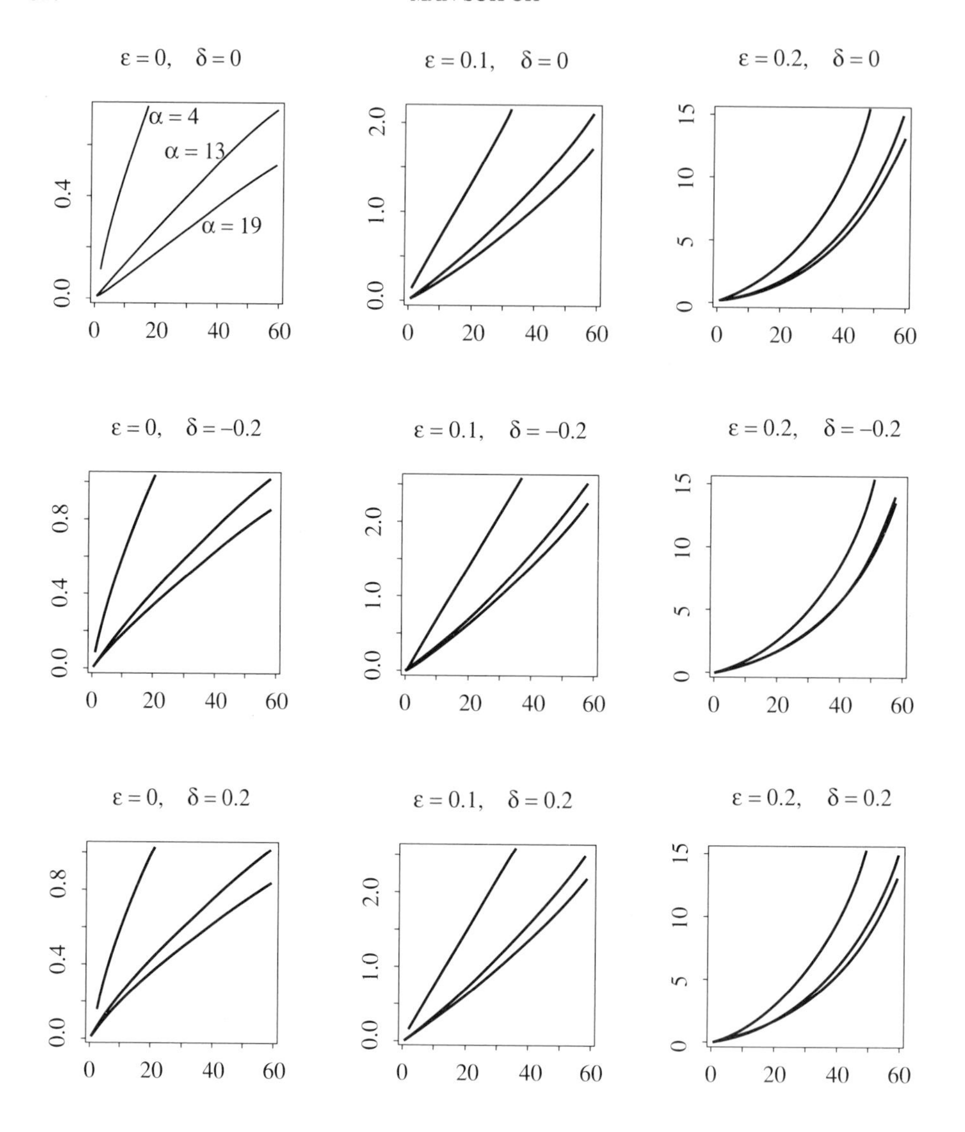

FIGURE 2. $\mathrm{var}(w)$, $g = \mathscr{T}_\alpha(\varepsilon 1, (1+g)I)$.

• <u>*Variance of* $\widehat{E}\theta_i$ *and* $\widehat{E}\theta_i^2$.</u> From Lemma 2.3,

(29)

$$\operatorname{var}^T_{\varepsilon,\delta,\alpha}(\widehat{E}\theta_i) = K_{\alpha,p}(1+\delta)^{p/2} E \times \left[\frac{z_i^2}{2} \left(1 + \frac{1}{\alpha(1+\delta)} \sum_{j=1}^{p} \left(\frac{z_j}{\sqrt{2}} - \varepsilon \right)^2 \right)^{(\alpha+p)/2} \right],$$

$$\operatorname{var}^T_{\varepsilon,\delta,\alpha}(\widehat{E}\theta_i^2) = K_{\alpha,p}(1+\delta)^{p/2} E \times \left[\left(\frac{z_i^2}{2} - 1 \right)^2 \left(1 + \frac{1}{\alpha(1+\delta)} \sum_{j=1}^{p} \left(\frac{z_j}{\sqrt{2}} - \varepsilon \right)^2 \right)^{(\alpha+p)/2} \right],$$

where $z_j \overset{\text{i.i.d.}}{\sim} N(0, 1)$, $j = 1, \ldots, p$, and $K_{\alpha,p}$ is a constant given in (21). We can compute $\operatorname{var}^T_{\varepsilon,\delta,\alpha}(\widehat{E}\theta_i)$ and $\operatorname{var}^T_{\varepsilon,\delta,\alpha}(\widehat{E}\theta_i^2)$ similarly to the case of $\operatorname{var}^T_{\varepsilon,\delta,\alpha}(w)$, and the graphs are given in Figures 3 and 4, on pp. 176 and 177. Again, in each figure, $\alpha = 4, 13, 19$, from the left to the right line. Both $\operatorname{var}^T_{\varepsilon,\delta,\alpha}(\widehat{E}\theta_i)$ and $\operatorname{var}^T_{\varepsilon,\delta,\alpha}(\widehat{E}\theta_i^2)$ are approximately linear for small $|\varepsilon|$, $|\delta|$ and exponentially increasing otherwise (slightly concave for $\varepsilon = 0, \alpha \le 13$).

Specifically, the shape of $\operatorname{var}^T_{\varepsilon,\delta,\alpha}(\widehat{E}\theta_i)$ is very similar to that of $\operatorname{var}^T_{\varepsilon,\delta,\alpha}(w)$ as can be seen in Figure 3. To see this more clearly, the graph of $E_h(\theta_i^2) = \operatorname{var}^T_{\varepsilon,\delta,\alpha}(\widehat{E}\theta_i)/(\operatorname{var}^T_{\varepsilon,\delta,\alpha}(w) + 1)$, from (12), is shown in Figure 5, on p. 178, for $\varepsilon = \delta = 0$, and for $\alpha = 4, 13, 19$, from the left to the right line. (Different values of ε, δ, ($|\varepsilon|, |\delta| \le 0.2$) gave similar results.) Note that $E_h(\theta_i^2)$ is approximately constant with respect to p, except for very small p. Thus, once again, the dimensionality effect seems to arise mainly through the effect on $\operatorname{var}(w)$.

In $\operatorname{var}^T_{\varepsilon,\delta,\alpha}(\widehat{E}\theta_i^2)$, the slope is much larger and the effect of scale misspecification δ is greater. Also, note that $E_h((\theta_i^2 - 1)^2)$, as shown in Figure 6, on p. 178, is a concave function of p. If we compare $E_h((\theta_i^2 - 1)^2)$ and $\operatorname{var}^T_{\varepsilon,\delta,\alpha}(w)$, we can see that $E_h((\theta_i^2 - 1)^2)$ has a moderate contribution to the dimensionality effect when p is not large or $|\varepsilon|$, $|\delta|$ are small, since both $E_h((\theta_i^2 - 1)^2)$ and $\operatorname{var}^T_{\varepsilon,\delta,\alpha}(w)$ are concave or linear functions. But when p is large or $|\varepsilon|$, $|\delta|$ are large, $\operatorname{var}^T_{\varepsilon,\delta,\alpha}(w)$ increases exponentially while $E_h((\theta_i^2 - 1)^2)$ is still a concave function of p; the major contribution to the dimensionality effect in such a case comes from $\operatorname{var}^T_{\varepsilon,\delta,\alpha}(w)$.

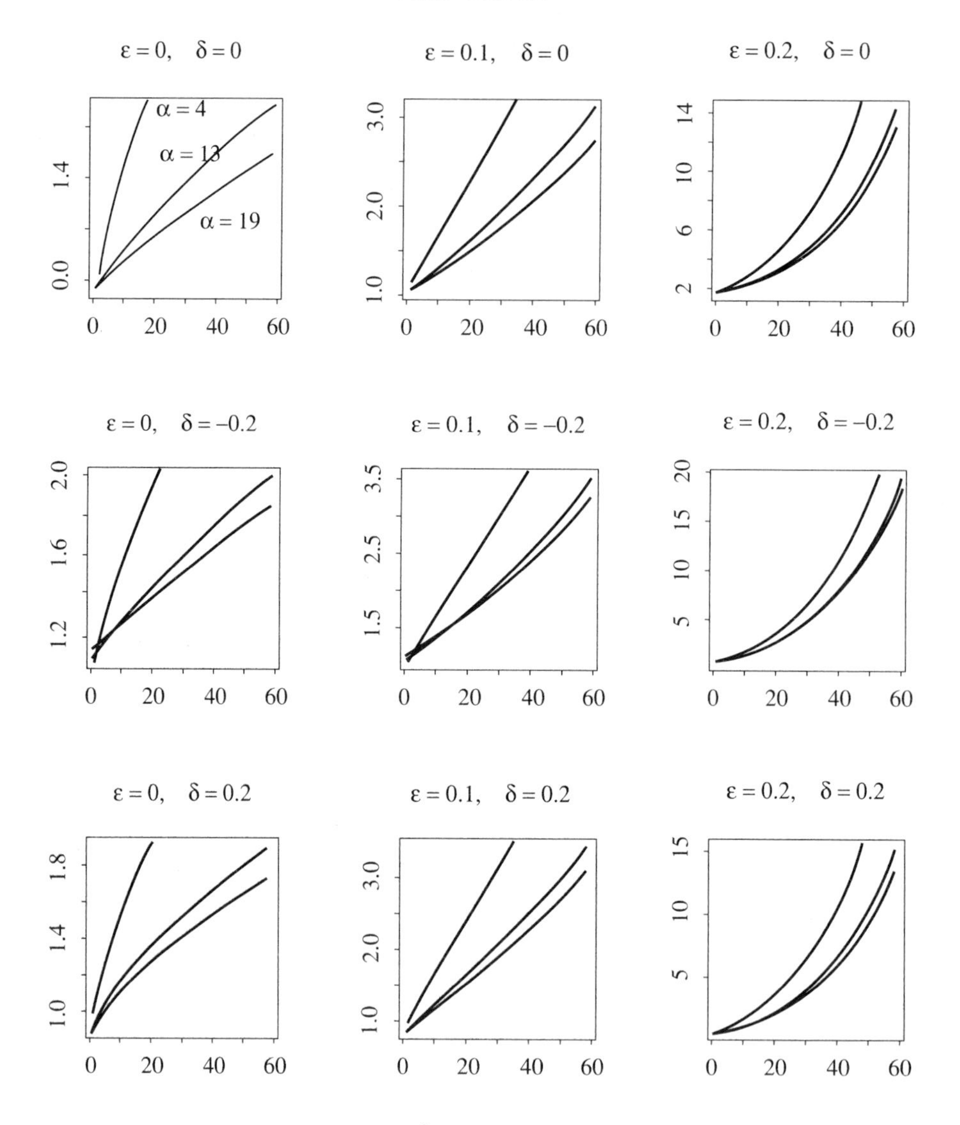

FIGURE 3. $\mathrm{var}(\widehat{E}\theta_i)$, $g\mathscr{T}_\alpha(\varepsilon\mathbf{1}, (1+\delta)I)$.

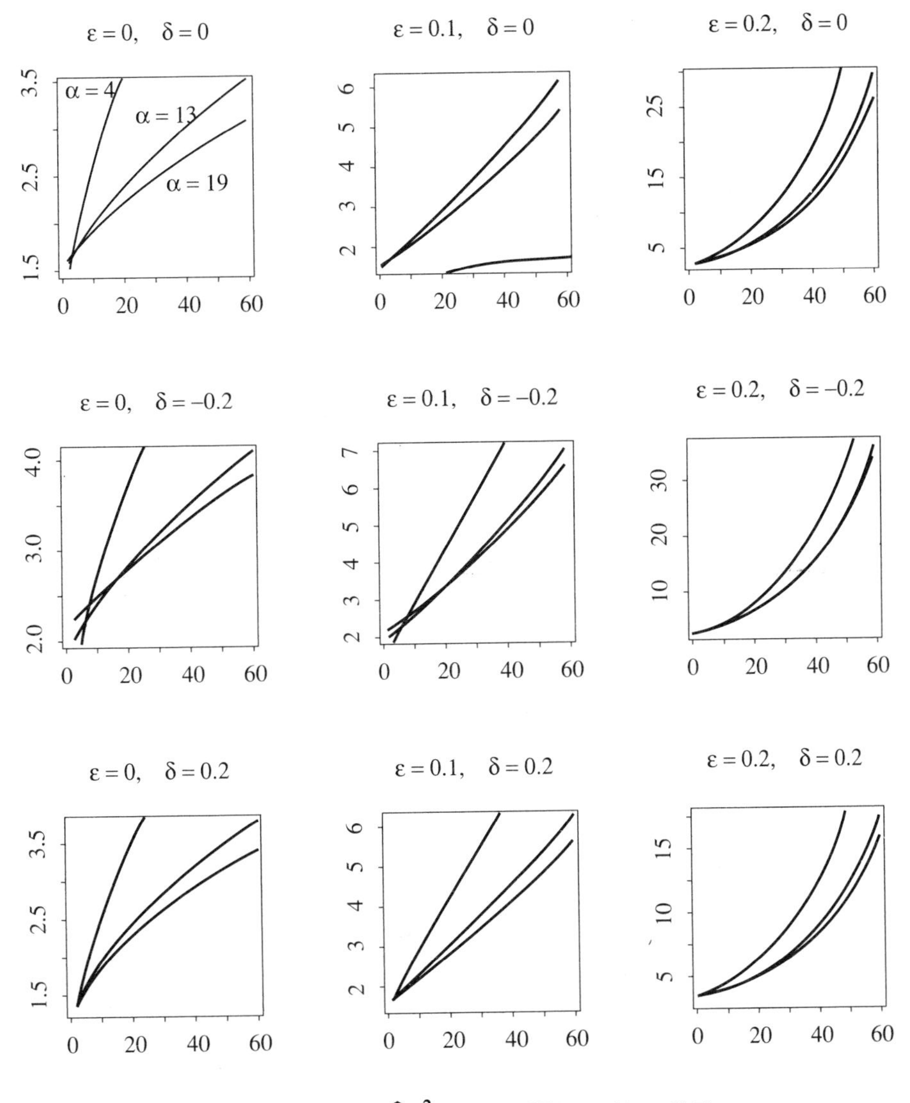

FIGURE 4. $\operatorname{var}(\widehat{E}\theta_i^2)$, $g = \mathscr{T}_\alpha(\varepsilon\mathbf{1}, (1+\delta)I)$.

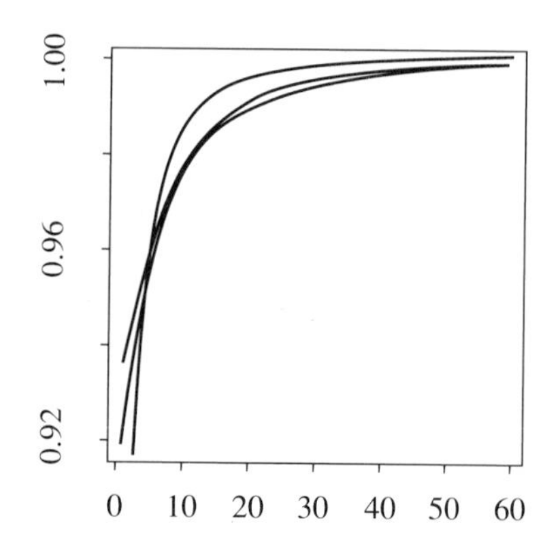

FIGURE 5. $E_h(\theta_i^2)$, $\varepsilon = 0$, $\delta = 0$.

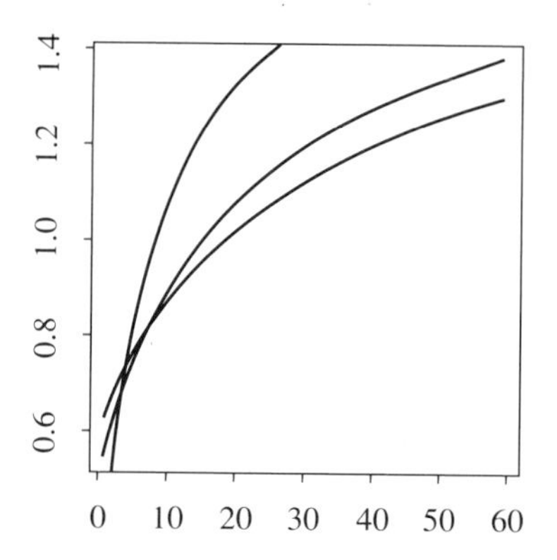

FIGURE 6. $E_h((\theta_i^2 - 1))^2$, $\varepsilon = 0$, $\delta = 0$.

2.3. Matching a t importance function with f. Assume that g is the density function of the $\mathscr{T}_\alpha(\mu, \Sigma)$ distribution. A simple and natural way of matching g and f would be matching the first and second moments, i.e., the mean and covariance matrix of θ under g with those under f. In such a case, recalling that the mean and covariance matrix of θ under $\mathscr{T}_\alpha(\mu, \Sigma)$ are μ and $\alpha/(\alpha-2)\Sigma$, respectively, the location and scale parameters of g will be $\mu = \mu_f$ and $\Sigma = ((\alpha-2)/\alpha)\Sigma_f$, respectively, where μ_f and Σ_f are the mean and covariance matrix of θ under f.

An alternative way of matching g and f is to match the mode and minus inverse Hessian matrix at the mode of $\log g$ with those of $\log f$.

PROPOSITION 2.1. *The minus inverse Hessian of log of the $\mathscr{T}_\alpha(\mu, \Sigma)$ density function at the mode is $\alpha/(\alpha+p)\Sigma$.*

PROOF. Clear from the definition and simple calculations. □

This would thus suggest choosing μ and Σ to be $\mu = \hat{\theta}_f$, the mode of f, and $\Sigma = ((\alpha+p)/\alpha)(-\widehat{I}_f^{-1})$, where $-\widehat{I}_f^{-1}$ is the minus inverse Hessian of $\log f$ at mode $\widehat{\theta}_f$. However, note that $(\alpha+p)/\alpha$ increases as p increases

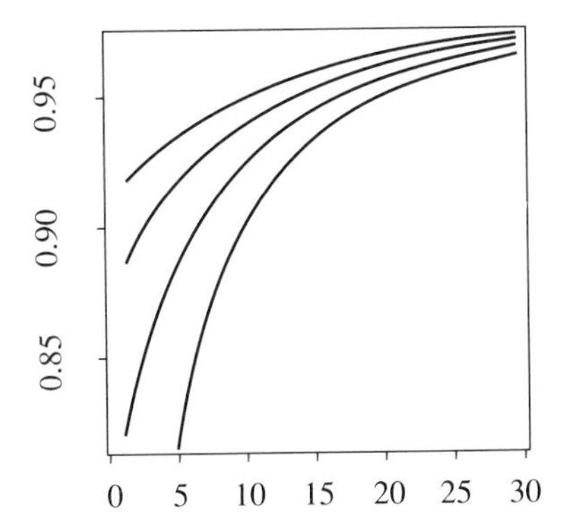

FIGURE 7. Optimal scaling constant ρ^*.

and $\mathcal{T}_\alpha(\widehat{\theta}_f, ((\alpha+p)/\alpha)(-\widehat{I}_f^{-1}))$ might yield a bad approximation to f unless f is itself approximately a multivariate t density function.

A third approach to matching g and f is based on the importance of $\mathrm{var}(w)$ as an indicator of the accuracy of Importance Sampling. Plausible choices of μ and Σ are $\mu = \mu_f$ and $\Sigma = \rho^*\Sigma_f$, where ρ^* is the constant which minimizes $\mathrm{var}(w)$ with respect to ρ when g is the $\mathcal{T}_\alpha(\mu_f, \rho\Sigma_f)$ density function.

Unfortunately, $\mathrm{var}(w)$ is typically analytically intractable. Suppose, however, that f is approximately normal, say $N(\mu_f, \Sigma_f)$. Then the optimal ρ^* can be determined numerically. In Figure 7, the graph of ρ^* for $\alpha = 19, 13, 10, 4$, from the left to the right line, as a function of p is given. It is clear that ρ^* converges to 1 as α increases, since $\mathcal{T}_\infty(\cdot, \cdot) = N(\cdot, \cdot)$. When α and p are very small, ρ^* is substantially less than 1. However, it is interesting to note that ρ^* seems to converge to 1 as p increases.

3. Adaptive importance sampling

Even though the selection of good importance function is the key issue in Importance Sampling as shown in §2, it is often hard to find a density function which satisfies all desirable properties of importance function presented in §1. A convenient and common way to choose an importance function is to first choose a parametric family of density function, $\mathcal{G} = \{g_\lambda; \lambda \in \Lambda\}$, which have simplicity in random variate generation and tails no sharper than f. Then select the parameter λ of g_λ to match some features of f. In this section, Adaptive Importance Sampling (AIS) is described to deal with the selection of parameters of importance function.

3.1. Algorithm. Let $\mathcal{G} = \{g_\lambda; \lambda \in \Lambda\}$ be a parametric family of density functions chosen for the importance function. Assume $\lambda = (\lambda_1, \dots, \lambda_m)'$ and that it would be desirable to choose λ equal to $E\xi = \int \xi(\theta)f(\theta)\,d\theta / \int f(\theta)\,d\theta$, where $\xi(\theta) = (\xi_1(\theta), \dots, \xi_m(\theta))$, in making g_λ a good approximation to f. For instance, choosing $\xi_1(\theta) = \theta$ would mean that it would be desirable to have λ_1 equal to the mean of f. Of course, $E\xi$ will itself typically be unknown (indeed, frequently some of the $\xi_i(\theta)$, $1 \le i \le m$, will be the target $\phi(\theta)$), in which case it is natural to choose λ to be an estimate of $E\xi$.

Assume that $\phi = (\phi_1, \dots, \phi_l)'$.

Stage 0. Choose a stopping rule (see §3.2) and an initial estimate $\lambda^{(0)}$ of $E\xi$. Let $g^{(0)}$ be $g_{\lambda(0)}$. (Often $\lambda^{(0)}$ is chosen by likelihood methods.)

Stage k $(k \ge 1)$. Draw n_k i.i.d. random drawings $\theta_1^{(k)}, \dots, \theta_{n_k}^{(k)}$ from $g^{(k-1)}$. Let

$$w^{(k)}(\theta) = f(\theta)/g^{(k-1)}(\theta), \tag{30}$$

and define a functional

$$N^{(k)}(h) = \sum_{i=1}^{n_k} h(\theta_i^{(k)}) w^{(k)}(\theta_i^{(k)}). \tag{31}$$

Compute $N^{(k)}(\xi_i)$, $i = 1, \dots, m$, $N^{(k)}(\phi_i)$, $i = 1, \dots, l$, and $N^{(k)}(1)$. Also, one might need to compute some statistics needed for a stopping rule; this will be discussed in the next section. Check a desired stopping rule to see if AIS should end. If so, go to "Conclusion". If not, set λ equal to

$$\lambda^{(k)} = \left(\frac{N^{(k)}(\xi_1)}{N^{(k)}(1)}, \dots, \frac{N^{(k)}(\xi_m)}{N^{(k)}(1)} \right)', \tag{32}$$

and go to the next stage with $g^{(k)} = g_{\lambda(k)}$.

Conclusion. When the stopping rule yields "stop AIS" (at the kth stage, for instance), estimate $E\phi$ by

$$\hat{E}^{(k)}\phi = \left(\frac{\sum_{j=1}^k N^{(j)}(\phi_1)}{\sum_{j=1}^k N^{(j)}(1)}, \dots, \frac{\sum_{j=1}^k N^{(j)}(\phi_l)}{\sum_{j=1}^k N^{(j)}(1)} \right). \tag{33}$$

Note that the AIS estimates of $E\xi$ and $E\phi$ are obtained by linear accumulation of the $N^{(j)}(\cdot)$. This linear accumulation of statistics from previous stages has several advantages. First, it is cheap: The major additional calculations in AIS, compared to the computations needed in Basic Importance Sampling (BIS), are summations and multiplications involved in (32) and the computations of $\xi_i(\theta)$, $i = 1, \dots, m$, in each stage. However, the additions and multiplications are typically cheap compared to the other computations, such as generation of random variates and computation of $f(\theta)$, $\phi(\theta)$ and $g^{(j)}(\theta)$. Also, note that the computations of the $\xi_i(\theta)$ are typically cheap because the $\xi_i(\theta)$ are often linear or polynomial functions. Finally, when (say) h of the ϕ_i are equal to ξ_i (as is often the case when the $\lambda_i s$ are moments), then the computations related to $\lambda_1, \dots, \lambda_h$ are not extra anymore.

Second, the sample sizes in the stages can be small or moderate, while still accumulating to give high accuracy overall. The advantage of small or moderate sample sizes in the stages is, of course, quicker adaptation of g_λ to f. Note that even a completely adaptive scheme, with each $\theta_i^{(j)}$ defining a new stage, is possible.

As mentioned in Section 1, BIS estimate $\widehat{E}\phi$ converges to $E\phi$ with probability one (as $n \to \infty$) and is approximately normally distributed with mean $E\phi$ and variance σ^2/n, where σ^2 is given by equations (3) and (4). The same thing seems to hold for AIS estimate $\widehat{E}^{(k)}\phi$ with n and g replaced by $\sum_{j=1}^k n_j$ and $g_{E\xi}$, respectively. Also, an estimate of σ^2 can be given by

(34)
$$\hat{\sigma}^{2(k)} = \frac{1}{(\overline{w}^{(k)})^2}(\hat{\text{var}}^{(k)}(\phi w) - 2(\widehat{E}^{(k)}\phi)\hat{\text{cov}}^{(k)}(\phi w, w) + (\widehat{E}^{(k)}\phi)^2\hat{\text{var}}^{(k)}(w)),$$

where

$$n^{(k)} = \sum_{j=1}^{k} n_j, \tag{35}$$

$$\overline{w}^{(k)} = \sum_{j=1}^{k} N^{(j)}(1)/n^{(k)}, \tag{36}$$

$$\hat{\text{var}}^{(k)}(\phi w) = \sum_{j=1}^{k} N^{(j)}(\phi^2 w^{(j)})/n^{(k)} - (\widehat{E}^{(k)}\phi)^2(\overline{w}^{(k)})^2, \tag{37}$$

$$\hat{\text{cov}}^{(k)}(\phi w, w) = \sum_{j=1}^{k} N^{(j)}(\phi w^{(j)})/n^{(k)} - \widehat{E}^{(k)}\phi(\overline{w}^{(k)})^2, \tag{38}$$

$$\hat{\text{var}}^{(k)}(w) = \sum_{j=1}^{k} N^{(j)}(w^{(j)})/n^{(k)} - (\overline{w}^{(k)})^2. \tag{39}$$

Note that the *new* sums in (37), (38), and (39) can be linearly accumulated between stages exactly as in the BIS algorithm.

3.2. Stopping rules for AIS. First, consider the case of scalar ϕ.

The simplest stopping rule would be to stop after a fixed number of stages with fixed sample sizes. Then the variance of $\widehat{E}\phi$ can be estimated by $\hat{\sigma}^{2(k)}/n^{(k)}$, where $\hat{\sigma}^{2(k)}$ and $n^{(k)}$ are given in (34)–(39).

To guarantee that

$$P\left(\frac{|\widehat{E}^{(k)}\phi - E\phi|}{|E\phi|} \leq \varepsilon\right) \approx 1 - \eta, \tag{40}$$

it follows from approximate normality that $n^{(k)} = \sum_{j=1}^k n_j$ should be chosen such that

$$\frac{\sigma^2}{n^{(k)}|E\phi|^2} \leq \left(\frac{\varepsilon}{c(\eta)}\right)^2, \tag{41}$$

where $c(\eta)$ is the $(1-\eta/2)$ th quantile of the standard normal distribution. A method for deciding when to stop AIS immediately suggests itself. Replace σ^2 and $E\phi$ in (41) by $\hat{\sigma}^{2(k)}$ and $\widehat{E}^{(k)}\phi$, respectively, and stop AIS when

$$\frac{\hat{\sigma}^{2(k)}}{n^{(k)}|\widehat{E}^{(k)}\phi|^2} \le \left(\frac{\varepsilon}{c(\eta)}\right)^2. \tag{42}$$

When $\phi = (\phi_1, \ldots, \phi_l)'$ is a vector, one could calculate $\hat{\sigma}^{2(k)}$ for each ϕ_i and stop when (42) is simultaneously satisfied for all components. This can be expensive, however, if l is large. A rough surrogate for the stopping rule in such a situation is to replace $\hat{\sigma}^{2(k)}/|\widehat{E}^{(k)}\phi|^2$ in (42) by $\widehat{\mathrm{var}}^{(k)}(w)/(\overline{w}^{(k)})^2$. This yields the rule: stop if

$$\frac{\widehat{\mathrm{var}}^{(k)}(w)}{n^{(k)}(\overline{w}^{(k0})^2} \le \left(\frac{\varepsilon}{c(\eta)}\right)^2. \tag{43}$$

Note that $\widehat{\mathrm{var}}^{(k)}(w)/(\overline{w}^{(k)})^2$ is the term of (34) which, when scaled by $(\widehat{E}^{(k)}\phi)^2$, does not involve ϕ.

3.3. Example and numerical comparison. We consider an example from Fong (1987), which analyzed a complete block design of Stenstrom (given in SAS (1985), p. 487) using a hierarchical Bayesian model. Data $\mathbf{y}$ was assumed to follow the model

$$y_{ijk} = \mu_{ij} + \varepsilon_{ijk}, \qquad i = 1, \ldots, I;\ j = 1, \ldots, J;\ k = 1, \ldots, K_j, \tag{44}$$

where $I = 3$, $J = 4$, $K_1 = K_2 = K_3 = 2$, $K_4 = 1$, and $\varepsilon_{ijk} \overset{\text{i.i.d.}}{\sim} N(0, \tau^2)$, for unknown τ^2. Suppose $\mu_{ij} = \mu + \alpha_i + \beta_j$, μ has $N(33.5, 9.0)$ for the prior density, and the prior distributions of α_i and β_j, given τ_α^2, τ_β^2, and τ^2, are

$$\alpha_i \overset{\text{i.i.d.}}{\sim} N(0, \tau_\alpha^2), \qquad \beta_j \overset{\text{i.i.d.}}{\sim} N(0, \tau_\beta^2); \tag{45}$$

the second stage prior density of $(\tau_\alpha^2, \tau_\beta^2, \tau^2)$ is

$$\pi(\tau_\alpha^2, \tau_\beta^2, \tau^2) = \pi(\tau_\alpha^2|\tau^2)\pi(\tau_\beta^2|\tau^2)\pi(\tau^2) \tag{46}$$

$$= 1/(\tau_\alpha^2 + \tau^2)(\tau_\beta^2 + \tau^2)\tau^2. \tag{47}$$

Determining the probabilities $P(\beta_j$ is the largest $|\mathbf{y})$ is of great interest in ranking and selection. In Fong (1987), it is shown that calculation of these probabilities reduces to determining the posterior expectations of

$$\psi_j(\tau^2, \tau_\alpha^2, \tau_\beta^2) = E^\beta E^{\phi_j} \prod_{s \neq j} \Phi\left(\frac{\phi_j - u_s}{\sqrt{V_s}}\right), \qquad j = 1, \ldots, J, \tag{48}$$

where

$$\beta \sim N(u^*, V^*), \quad \phi_j \sim N(u_j, V_j), \tag{49}$$

$$u^* = \tilde{y}\ldots - [1 + (\tau_\alpha^2 + 9.0I)\tau_b^{-2}]^{-1}(\tilde{y}\ldots - 33.5), \tag{50}$$

$$V^* = [I((\tau_\alpha^2 + 9.0I)^{-1} + \tau_b^{-2})]^{-1}, \tag{51}$$

$$\tilde{y}\ldots = \frac{\sum_{i=1}^{I}\sum_{j-1}^{J}\sum_{k=1}^{K_j} y_{ijk}/(\tau^2 + IK_j\tau_\beta^2)}{I\tau_b^{-2}}, \tag{52}$$

$$\tau_b^2 = \left(\sum_{j=1}^{J}\frac{1}{\tau^2/K_j + I\tau_\beta^2}\right)^{-1}, \tag{53}$$

$$u_j = \overline{y}_{.j.} - \frac{\tau^2/K_j}{\tau^2/K_j + I\tau_\beta^2}(\overline{y}_{.j.} - \beta), \tag{54}$$

$$V_j = \frac{(\tau^2/K_j)\tau_\beta^2}{\tau^2/K_j + I\tau_\beta^2}, \qquad j = 1, \ldots, J, \tag{55}$$

$\overline{y}_{.j.}$ is a constant obtained from data $\mathbf{y}$, and $\mathbf{\Phi}$ is the standard normal cumulative distribution function.

To compute the posterior expectations, first transform τ^2, τ_α^2, τ_β^2 to

$$\theta_1 = \log\tau^2, \qquad \theta_2 = \log\tau_\alpha^2, \qquad \theta_3 = \log\tau_\beta^2, \tag{56}$$

since this transformation often makes the posterior density function less skewed and changes the ranges of the variables to $(-\infty, \infty)$. Let $\theta = (\theta_1, \theta_2, \theta_3)'$ and $\phi_j(\theta) = \psi_j(e^{\theta_1}, e^{\theta_2}, e^{\theta_3})$, $j = 1, \ldots, J$. The transformed posterior density can be shown to be

$$\begin{aligned} P(\theta|\mathbf{y}) &\propto e^{-(IN_k+I+J-1)\theta_1/2}(e^{\theta_1} + N_k e^{\theta_2})^{-(I-1)/2'} \\ &\times \prod_{j=1}^{J}\left(\frac{e^{\theta_1}}{K_j} + I\varepsilon^{\theta_3}\right)^{-1/2}\left(A(e^{\theta_1}, e^{\theta_3}) + \frac{1}{\varepsilon^{\theta_2} + 9.0I}\right)^{-1/2} \\ &\times \exp\left[-\frac{1}{2}\left(\frac{S_3}{e^{\theta_1}} + \frac{N_k S_1}{e^{\theta_1} + N_k e^{\theta_2}} + I\sum_{j=1}^{J}\frac{(\overline{y}_{.j.} - \tilde{y}_{...}^2)^2}{e^{\theta_1}/K_j + Ie^{\theta_3}}\right)\right. \\ &\left. + \frac{I(\tilde{y}_{...} - 33.5)^2 A(e^{\theta_1}, e^{\theta_3})}{1 + (e^{\theta_2} + 9.0I)A(e^{\theta_1}, e^{\theta_3})}\right]\frac{e^{\theta_2+\theta_3}}{(e^{\theta_1} + e^{\theta_2})(e^{\theta_1} + e^{\theta_3})}, \end{aligned} \tag{57}$$

where

$$A(e^{\theta_1}, e^{\theta_3}) = \sum_{j=1}^{J}\frac{1}{e^{\theta_1}/K_j + Ie^{\theta_3}}, \tag{58}$$

S_1, S_3 are constants obtained from $\mathbf{y}$, $N_k = \sum_{j=1}^{J} K_j$, and $\tilde{y}_{...}$ is given in (52).

A multivariate t form with 7 degrees of freedom is chosen for the importance function because of its thick tails and convenience in generating random samples, so

$$(59)\quad \mathcal{G} = \{g_{\lambda^\dagger,\dagger\dagger}(\theta) \propto |\lambda^{\dagger\dagger}|^{-1/2}(1 + \tfrac{1}{7}(\theta - \lambda^\dagger)'\lambda^{\dagger\dagger-1}(\theta - \lambda^\dagger))^{-5}, \quad \lambda^\dagger \in R^3,\ \lambda^{\dagger\dagger} \text{ is pos. def.}\}.$$

Again, estimated moments, $\hat{\mu}$ and $\widehat{\Sigma}$, of the posterior yield values $\lambda^\dagger = \hat{\mu}$ and $\lambda^{\dagger\dagger} = (.08609)\widehat{\Sigma}$ for the parameters of the importance function (.8609 being the adjustment factor ρ^* introduced in §2.3, for multivariate t density function with 7 degrees of freedom in three dimensional problems). Thus, $\xi^\dagger(\theta) = \theta$ and $\xi^{\dagger\dagger}(\theta) = .8609(\theta\theta' - E\xi^\dagger E\xi^{\dagger'})$ will be used in the AIS algorithm so that $\lambda^\dagger = \widehat{E}\xi^\dagger$ and $\lambda^{\dagger\dagger} = \widehat{E}\xi^{\dagger\dagger}$.

A sample size if 200 was selected for each stage, and the stopping rule (42) with $\eta = .05$ and $\varepsilon = .2$ was chosen for each element of ϕ; thus sampling stopped when

$$(60)\qquad \frac{\hat{\mathrm{var}}(\widehat{E}^{(k)}\phi_j)}{(\widehat{E}^{(k)}\phi_j)^2} \le 1.04123 \times 10^{-2}, \quad \text{for } j = 1, \dots, 4.$$

Note that $\hat{\mathrm{var}}(\widehat{E}^{(k)}\phi_j)$ is of the form $\hat{\sigma}^{2(k)}/n^{(k)}$, where $\hat{\sigma}^{2(k)}$ is given in (34) with $\phi = \phi_j$. Finally, the importance function at stage 0 was chosen to have $\lambda^{\dagger(0)} = \hat{\theta}$ (the posterior mode) and $\lambda^{\dagger\dagger(0)} = (-.8609)\widehat{I}^{-1}(-\widehat{I}^{-1}$ being minus the inverse Hessian of $\log f$); these were determined using maximum likelihood methods.

TABLE 1. Comparison of BIS and AIS. *

n	32, 000	10, 600
time (sec)	1033.0	361.0
$\hat{\mu}$	(1.354, 1.619, 1.667)	(1.445, 2.021, 1.759)
$\widehat{\Sigma}$	.129	.121
	.195, 1.844	.021, 2.438
	.126, .144, .938	−0.15, .038, 1.375
$\widehat{E}\phi_1$	.0400	.0376
$\widehat{E}\phi_2$	.0411	.0419
$\widehat{E}\phi_3$	.9176	9.177
$\widehat{E}\phi_4$	.0037	.0028
$\hat{\mathrm{var}}(\widehat{E}\phi_1)/(\widehat{E}\phi_1)^2$	1.1640×10^{-3}	6.8301×10^{-4}
$\hat{\mathrm{var}}(\widehat{E}\phi_2)/(\widehat{E}\phi_2)^2$	4.7175×10^{-4}	7.0058×10^{-4}
$\hat{\mathrm{var}}(\widehat{E}\phi_3)/(\widehat{E}\phi_3)^2$	3.8291×10^{-6}	3.5244×10^{-6}
$\hat{\mathrm{var}}(\widehat{E}\phi_4)/(\widehat{E}\phi_4)^2$	4.5000×10^{-2}	1.0365×10^{-2}
$\hat{\mathrm{var}}^{(k)}(w)/\{n^{(k)}(\overline{w}^{(k)})^2\}$	2.2558×10^{-5}	1.4106×10^{-5}

**All computations were done on Unix 4.3 BSD at Purdue University.*

With the stopping rule mentioned above, AIS stopped after 53 stages ($n = 10,600$), 9 times earlier than BIS which stopped after 470 stages ($n = 94,000$). However, $\widehat{\text{var}}(\hat{E}\phi_j)/(\hat{E}\phi_j)^2$, $j = 1, 2, 3$, in BIS were about $1/3$ of those in AIS because large $\widehat{\text{var}}(\hat{E}\phi_4)/(\hat{E}\phi_4)^2$ controlled the stopping rule. Thus, to compare both schemes it would be more reasonable to run BIS with 160 stages ($n = 32,000$) and AIS with 53 stages so that most variances are roughly the same. The results are given in Table 1.

There are no regular relationships between $\widehat{\text{var}}^{(k)}(w)/\{n^{(k)}(\overline{w}^{(k)})^2\}$ and $\widehat{\text{var}}\,\hat{E}\phi_j/(\hat{E}\phi_j)^2$, $j = 1, \ldots, 4$, in this example. And because the $\widehat{\text{var}}^{(k)}(w)/\{n^{(k)}(\overline{w}^{(k)})^2\}$ are for the most part smaller than the $\widehat{\text{var}}\,\hat{E}\phi_j/(\hat{E}\phi_j)^2$, use of the stopping rule (43) would have caused the scheme to stop much earlier, resulting in less accuracy of the $\hat{E}\phi_j$, $j = 1, 2, 4$.

It is of interest to compare the mode and minus inverse Hessian (which are not only the parameter values used in BIS, but are also the estimates that would result from a likelihood approach to the problem) with AIS estimates of the mean and covariance matrix. Tables 2 and 3 show that there are substantial differences between the mode and mean, and between minus the inverse Hessian and the variance of θ_2.

TABLE 2. Mode and mean of posterior.

i	1	2	3
$\hat{\theta}_i$	1.355	1.619	1.667
$\hat{\mu}_i$	1.445	2.021	1.759

TABLE 3. Minus inverse Hessian and covariance matrix of posterior.

(i, j)	$(1, 1)$	$(1, 2)$	$(1, 3)$	$(2, 2)$	$(2, 3)$	$(3, 3)$
$-\hat{I}^{-1}$	.129	.195	.126	1.844	.144	.938
$\hat{\Sigma}$	.121	.021	−.015	2.438	.038	1.375

4. Conclusions and generalizations

One cannot, of course, draw firm conclusions about dimensionality effect from examination of just a few cases. However, the following observations seemed to hold across our examples, and hence may be indicative of general behavior.

First, most of the dimensionality effect seems to be driven by $\text{var}(w)$. Second, $\text{var}(\omega)$ increases exponentially as the dimension p increases. But if the location and scale of the importance function match well with those of f, $\text{var}(w)$ increases only approximately linearly for moderate p, and if the misspecification error is $O(p^{-1/2})$ then $\text{var}(w)$ is approximately constant in

p. Finally, misspecification of location was seen in the figures to be more serious than misspecification of scale, specially in high dimensions.

The advantage of Adaptive Importance Sampling, dealing with the selection of parameters of importance function, include its efficiency, its potential for being automated, and its flexibility. Its efficiency arises from the possibility of repeatedly improving the importance function during the simulation, from the utilization of *all* Monte Carlo observations in the computation, and from the utilization of cheap linear operations to update the importance function.

The automatic nature of AIS is attractive in that once the class of importance functions is chosen, and the sample sizes and desired accuracy specified, there is no need for further interaction with the statistician.

Finally, AIS is flexible in terms of its features, allowing any class of parameterized importance functions (with linearly estimable parameters) and arbitrary sample sizes in each stage. When the initial inputs are quite uncertain and/or the extra cost of AIS is relatively small, one can make the sample size in each stage very small (even one) so that parameters of the importance function are updated often. It is even possible to have the choice of the sample size for the next stage depend on the current estimates; for instance, if the importance function has stabilized and there is still a long way to go to achieve the desired accuracy, one might choose a larger sample size for the next stage.

REFERENCES

P. J. Davis and P. Rabinowitz (1975), *Methods of numerical integration*, Academic Press, New York.

D. K. H. Fong (1987), *Ranking and estimation of exchangeable means in balanced and unbalanced models: A Bayesian approach*, Ph. D. Thesis, Dept. of Stat., Purdue University, W. Lafayette, IN.

J. Geweke (1986), *Bayesian inference in econometric models using Monte Carlo integration*, Dept. of Econometrics, Duke Univ., Durham, NC.

——(1988), *Antithetic acceleration of Monte Carlo integration in Bayesian inference*, J. Econometrics **38**, 73–90.

K. Kloek and H. K. van Dijk (1978), *Bayesian estimates of equation system parameters: An application of integration by Monte Carlo*, Econometrica **46**, 1–20.

R. Y. Rubinstein (1981), *Simulation and the Monte Carlo method*, Wiley, New York.

SAS Institute Inc. (1985), *SAS User's Guide: Statistics*, 5th ed., SAS Institute Inc., Cary, NC.

L. Stewart (1979), *Multiparameter univariate Bayesian analysis*, J. Amer. Statist. Assoc. **76**, 684–693.

——(1983), *Bayesian analysis using Monte Carlo integration—a powerful methodology for handling some difficult problems*, The Statistician **32**, 195–200.

——(1984), *Bayesian analysis using Monte Carlo integration with an example of the analysis of survival data*, Lockheed Palo Alto Research Lab., Palo Alto, CA.

H. K. van Dijk (1984), *Posterior analysis of econometric models using Monte Carlo integration*, Reproduktie Woudestein, Erasmus Universiteit Rotterdam.

——(1987), *Some advances in Bayesian estimation methods using Monte Carlo integration*, Econometric Institute, Erasmus Univ., Rotterdam, and Center for Operations Research and Econometrics (CORE), Universite Catholique de Louvain.

H. J. van Dijk and T. Kloek (1980), *Further experience in Bayesian analysis using Monte Carlo integration*, J. Econometrics **14**, 307–328.
——(1983a), *Monte Carlo analysis of skew posterior distributions: an illustrative econometric example*, The Statistician **32**, 216–223.
——(1983b), *Experiments with some alternatives for simple importance sampling in Monte Carlo integration*, Bayesian Statistics 2, North-Holland, Amsterdam, pp. 511–530.
H. K. van Dijk, T. Kloek, and C. G. E. Boender (1985), *Posterior moments computed by mixed integration*, J. Econometrics **29**, 3–18.

Department of Statistics, University of California, Berkeley, Berkeley, California 94720

E-mail address: oh@stat.berkeley.edu

Contemporary Mathematics
Volume 115, 1991

Comparison of Simulation Methods in the Estimation of the Ordered Characteristic Roots of a Random Covariance Matrix

VESNA LUZAR AND INGRAM OLKIN

ABSTRACT. Several simulation methods are used to evaluate the expected values and correlations of the ordered characteristic roots of a sample covariance matrix arising from a normal distribution with zero mean and identity covariance matrix. The methods used are importance sampling coupled with regression and ratio estimates. The comparisons of the efficiency of these methods is determined, where efficiency is defined in terms of variability measures.

1. Introduction

It is well known that the ordered characteristic roots of a sample covariance matrix from a normal distribution are biased estimates of the ordered characteristic roots of the population covariance matrix. The purpose of this study is to examine the degree of bias and the correlation between the ordered characteristic roots. An exact computation of the moments of the ordered characteristic roots involves the evaluation of p-dimensional integrals. Such computations for small values of p are feasible, but become prohibitive (even when using a supercomputer) for larger values of p. An alternative is to use simulation methods, and it is this aspect on which we focus.

In carrying out a simulation it became apparent that different simulation methods yielded results with considerable variation in accuracy. Indeed, the computational problems then assumed an interest outside of the problem itself, and it is a comparison of alternative computational or simulation methods that is the focus of this paper.

1980 *Mathematics Subject Classification* (1985 *Revision*). Primary 65C05, 62H10.

Key words and phrases. Wishart distribution, importance sampling, Cholesky decomposition, regression simulations, ratio simulations, mixture sampling.

The first author was supported in part as a Fulbright Scholar, Stanford University.

The second author was supported in part by the National Science Foundation.

This paper is in final form and no version of it will be submited for publication elsewhere.

We first describe the underlying structure. Let $X = (x_{ij})$ be a $p \times n$ random matrix for which the p-dimensional columns are independently distributed with a common normal distribution with mean vector zero and positive definite covariance matrix Σ. The distribution of

$$V = XX' \tag{1.1}$$

is called the Wishart distribution and has a density

$$f(V) = c(p, n)(\det V)^{\frac{n-p-1}{2}} e^{-\frac{1}{2}\operatorname{tr}\Sigma^{-1}V}, \tag{1.2}$$

where $[c(p, n)]^{-1} = 2^{pn/2}\pi^{p(p-1)/4}|\Sigma|^{n/2}\prod_1^p \Gamma(\frac{1}{2}(n-i+1))$.

It is well known that the ordered characteristic roots $l_{(1)} \geq \cdots \geq l_{(p)}$ of the sample covariance matrix V/n are biased estimators of the ordered population characteristic roots $\lambda_{(1)} \geq \cdots \geq \lambda_{(p)}$ of Σ in the sense that

$$E(l_{(1)} + \cdots + l_{(k)}) \geq \lambda_{(1)} + \cdots + \lambda_{(k)},$$
$$E(l_{(p)} + \cdots + l_{(p-k+1)}) \leq \lambda_{(p)} + \cdots + \lambda_{(p-k+1)}, \qquad k = 1, \ldots, p,$$

but the degree of bias is not known. Some asymptotic expressions and other approximations do exist, but these generally deal with the case of large n or large p. There is almost no information about the covariances of the ordered sample characteristic roots.

We investigate the means and covariances of the ordered characteristic roots in the case that $\Sigma = I$. In this case the distribution of the characteristic roots $l_1, \ldots, l_p$ of V/n is

$$f(l_1, \ldots, l_p) = d(p, n)\prod_{i \neq j}|l_i - l_j|\prod_1^p l_i^{(n-p-1)/2} e^{-\Sigma\, l_i/2}, \tag{1.3}$$

where $d(p, n) = \pi^{p/2}2^{-pn/2}/\prod_1^p[\Gamma(\frac{1}{2}(n-i+1)\Gamma(\frac{i}{2})]$. (See Anderson (1984) or Muirhead (1982).)

There have been some attempts to develop methods for generating extreme characteristic roots of sample covariance matrices (see Marasinghe and Kennedy (1982)), but they are limited to some special cases of $p = 2, 3$.

Our concern is with the numerical determination of $El_{(i)}$ and with $\operatorname{Cov}(l_{(i)}, l_{(j)})$ when $\Sigma = I$. In order to compute these values we use several alternative simulation procedures, and provide comparisons between them. These methods raise some interesting issues concerning the accuracy of simulations in multivariate analyses. In order to fix ideas and to illustrate the issues we provide comparisons for the case $p = 3$, $n = 15$. Computations for other values of p and n indicate that this single case can well serve as a prototype to illustrate the pros and cons of various methods.

All numerical computations are based on 10^5 replications, and the results are plotted at each 1000 replications. This permits for a more visual comparison, and shows more clearly wherein the differences lie.

2. Generating random matrices

Perhaps the most natural direct procedure to obtain the expected values of the ordered sample characteristic roots is to generate random matrices $X = (x_{ij})$, then to form $V = XX'$, and finally to compute the characteristic roots. Although straightforward, this method is inefficient, especially for large n. However, this method does have a distinct advantage in that it permits comparisons of the results when some of the x_{ij} deviate from normality.

A modification of this is based on the Cholesky decomposition $V = TT'$, where $T = (t_{ij})$ is a $p \times p$ lower triangular matrix. Here the t_{ij}'s are independently distributed; each t_{ii}^2, $i = 1, \ldots, p$, has a chi-square distribution with $n - i$ degrees of freedom, and each $t_{ij} (i \neq j)$ has a standard normal distribution. (See Olkin (1985).)

The results of this simulation are given in Figure A.

It is interesting to observe the degree of variation. From 10,000 to 20,000 replications the results vary from 1.553 to 1.557. Even after 40,000 replications we reach a value of 1.559. Ultimately, the value settles down towards 1.558.

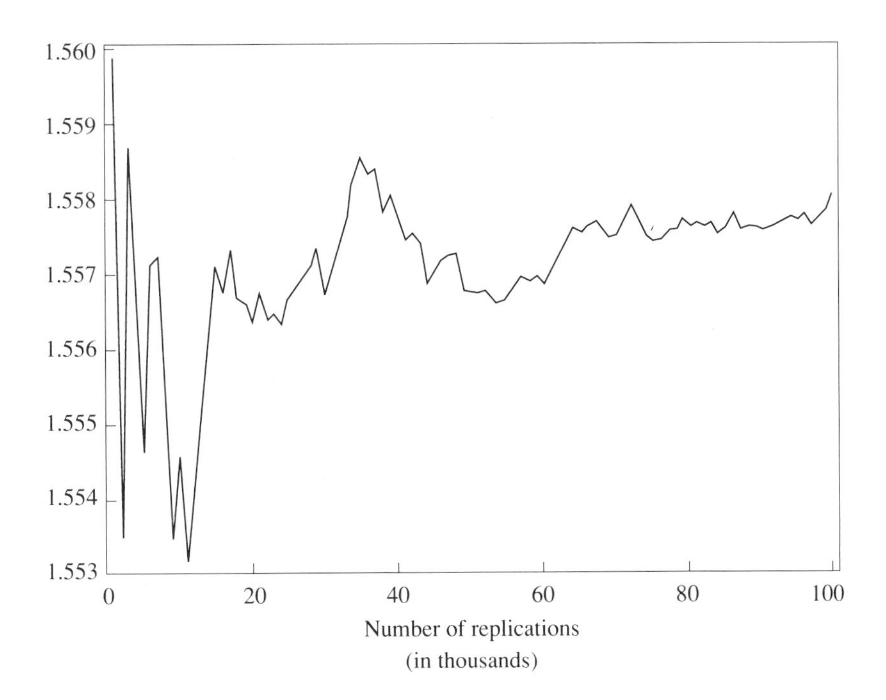

FIGURE A. Cumulative means of the expected value of the largest characteristic root ($p = 3$, $n = 15$), calculated by the Cholesky decomposition.

3. Importance sampling

From the joint distribution of the characteristic roots (1.3), we express an expected value as a multiple integral:

$$Eg(\underset{\sim}{l}) = \int g(\underset{\sim}{l})\, d(p, n) \prod_{i \neq j} |l_i - l_j| \prod_{1}^{p} l_i^{(n-p-1)/2} e^{\Sigma l_i/2} \Pi d l_i . \tag{3.1}$$

In order to determine $Eg(\underset{\sim}{l})$ we can generate random numbers according to the exponential density $\frac{1}{2}\exp\left(-\frac{1}{2}x\right)$ and average over

$$g_1(\underset{\sim}{l}) \equiv 2^p g(\underset{\sim}{l}) d(p, n) \prod_{i \neq j} |l_i - l_j| \prod_{1}^{p} l_i^{(n-p-1)/2} . \tag{3.2}$$

This method appeared at first to be the most expeditious, since exponential random variables can be obtained very readily from uniform random variables. However, this method led to considerable variation as seen in Figure B, where the curve with the largest variation (curve C_1) corresponds to expected values obtained by generating exponentially distributed random numbers.

What we see here is that exponential random variables yield somewhat unstable results even up to 20,000 replications. This variability might be

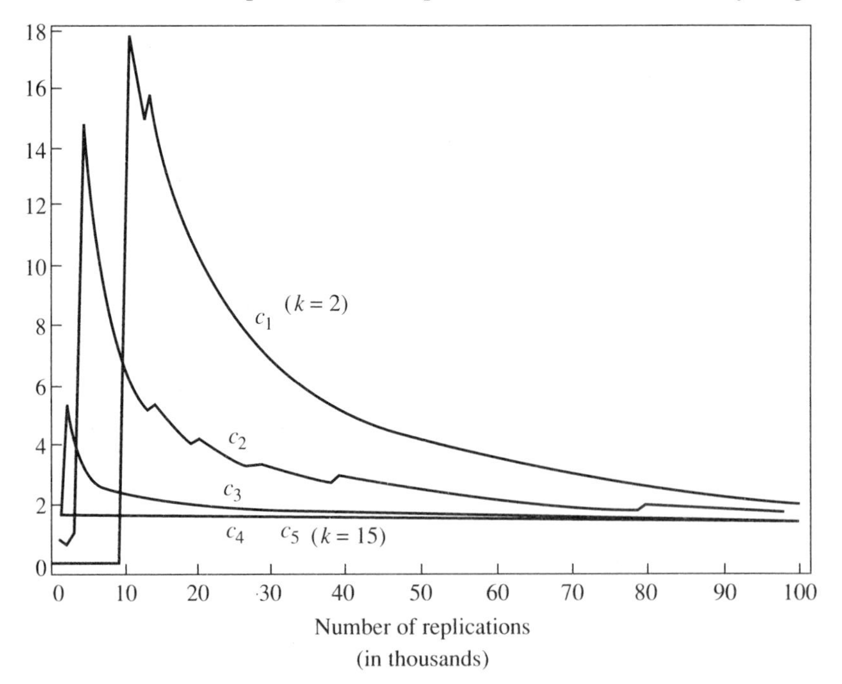

FIGURE B. Cumulative means of the expected value of the largest characteristic root $(p = 3, n = 15)$, using importance sampling with a chi-square distribution with $k = 2, 6, 10, 13, 15$ d.f.

alleviated by using a truncated exponential distribution. Because of this early instability it takes many additional replications before convergence occurs.

A review of (3.1) suggests that instead of an exponential distribution, we can incorporate the product term and generate random numbers according to a gamma, or equivalently a chi-square density

$$c_1 x^{(n-p-1)/2} e^{-x/2},$$

where c_1 is a normalizing constant. With this method we average over

$$g_2(\underset{\sim}{l}) = c_1^{-1} g(\underset{\sim}{l})\, d(p, n) \prod_{i \neq j} |l_i - l_j|. \tag{3.3}$$

Indeed, the principle underlying importance sampling is to express $Eg(\underset{\sim}{l})$ as

$$Eg(\underset{\sim}{l}) = \int \frac{g(\underset{\sim}{l}) f(\underset{\sim}{l})}{\varphi(\underset{\sim}{l})} \varphi(\underset{\sim}{l}) \Pi dl_i, \tag{3.4}$$

and to choose $\varphi(\underset{\sim}{l})$ in some appropriate way. (See Hammersley and Hanscomb (1964) for a discussion of importance sampling.)

Here we let $\varphi(\underset{\sim}{l})$ be a chi-square density with k degrees of freedom and let k vary over 2, 6, 10, 13, 15. These results are provided in Figure B.

As seen from Figure B this method gives wide variation before 20,000 replications, and only begins to settle down after a considerable number of additional replications. The variation diminishes as the number of degrees of freedom increases.

To compare the variances of the estimates corresponding to different choices for the degrees of freedom in the χ^2 distribution, ratios of the variances for $k = 2, 6, 10, 13$ to the variance for $k = 15$ (based on 1000 replications) are presented in Table 1. The increase in efficiency with the degrees of freedom is readily visible from this table.

TABLE 1. Efficiency as measured by the ratio: $\text{Var}(\chi_k^2 \text{ method})/\text{Var}(\chi_{15}^2 \text{ method})$ for $k = 2, 6, 10, 13$.

d:	2	6	10	13
E:	42, 578	10, 100	325	1.3

In Figure C, on page 194, the result for $k = 15$ is provided on an enlarged scale. This shows convergence to a value of 1.56. It is clear from the figures is that both methods yield values correct to two decimals. What is at issue is which methods yields accuracy to three decimals, and at what rate does the convergence take place.

4. Regression techniques

In this method we make use of known auxiliary information. Recall that $\Sigma = I$, in which case a straightforward computation yields

$$E \operatorname{tr} V = E(l_1 + \cdots + l_p) = pn, \tag{4.1a}$$

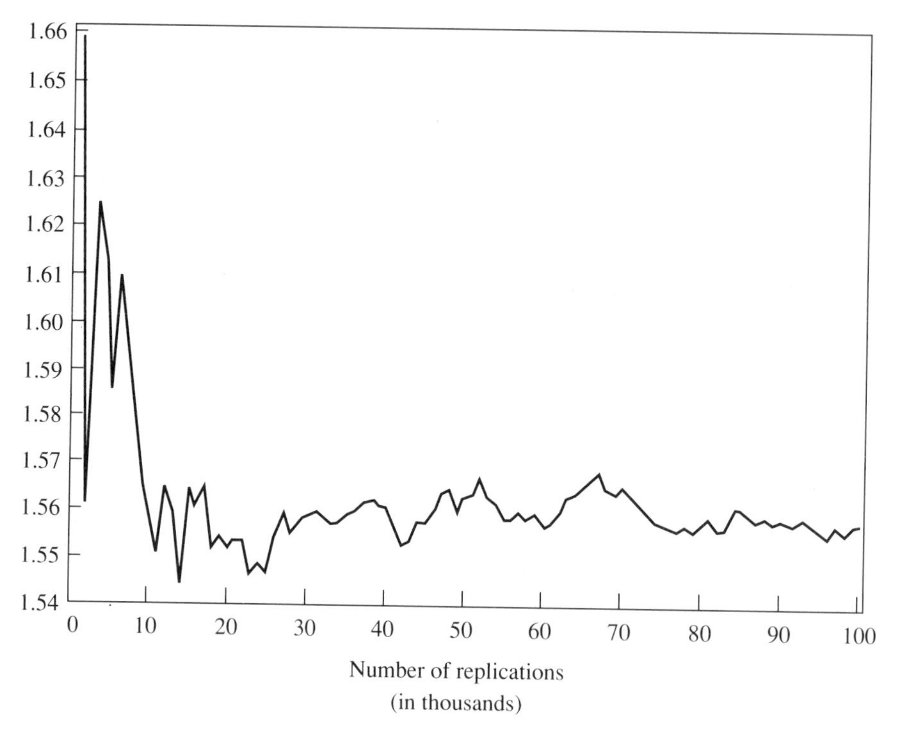

FIGURE C. Cumulative means of the expected value of the largest characteristic root ($p = 3, n = 15$), using importance sampling with a chi-square distribution with $k = 15$ d.f.

$$E \det V = E(l_1 \cdots l_p) = c(p, n)/c(p, n+2). \tag{4.1b}$$

We also can use a constant, i.e., a weight variable W such that $P\{W = 1\} = 1$, in which case

$$EW = 1. \tag{4.1c}$$

These values are used in a regression format as follows.

Let $\hat{l}_j$ be the estimate of the jth root using importance sampling. Then the regression estimate $\hat{l}_j^{reg}$ is defined as:

$$\hat{l}_j^{reg} = \hat{l}_j - \hat{\beta}_j(\hat{x} - E(x)), \tag{4.2}$$

where for x we can take any of the control variables $\mathrm{tr} V$, $\det V$, W, each of which has a known expected value as given in (4.1a,b,c). (See Rubinstein (1981).) Coefficients $\hat{\beta}_j$ are estimated from the regressions of x on l_j. This regression adjustment should have less variability than $\hat{l}_j$ by itself. To see this we compare and exhibit the return of three methods in Figure D:

(i) importance sampling with a chi-square distribution with $k = 15$ degrees of freedom,
(ii) regression methods using (4.1a),
(iii) regression methods using (4.1c).

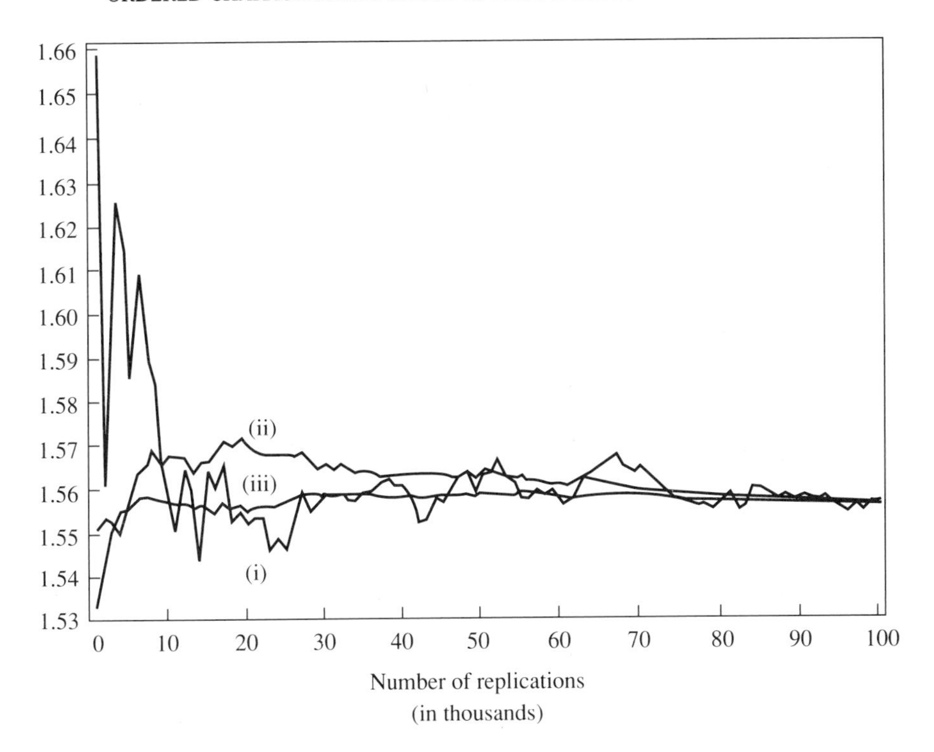

FIGURE D. Cumulative means of the expected value of the largest characteristic root ($p = 3, n = 15$), generated by (i) importance sampling with a chi-square distribution with $k = 15$ d.f., (ii) regression using (i) and a weight variable as regressor, (iii) regression using (i) and trace as regressor.

The most stable method is based on (4.1a), which yields the value of 1.56 almost at the outset. There is very little variation. Use of (4.1c) has more variation, but converges fairly rapidly to 1.56. In either case, the result 1.56 is achieved by 30,000 replications Use of regression has the effect of curtailing the high variation, especially in the early stages of replication, in the importance sampling method.

5. Ratio techniques

The same control variables as in (4.1a,b,c) can be used to yield ratio estimates. One advantage of this method is that the computation of the normalizing constant can be avoided. The ratio estimate $\hat{l}_j^{rat}$ is defined as:

$$\hat{l}_j^{rat} = \frac{\hat{l}_j}{\hat{x}} E(x),$$

with x being one of the control variables. The asymptotic variance of the ratio estimate is given by

$$\mathrm{var}(\tilde{l}_j^{rat}) = \mathrm{var}(\tilde{l}_j) - 2\frac{\tilde{l}_j}{\hat{x}}\mathrm{cov}(\tilde{x}, \tilde{l}_j) + \frac{\hat{l}_j^2}{\hat{x}^2}\mathrm{var}(\tilde{x}).$$

(See Cochran (1977).) The results using a ratio estimate with the weight variable and with the trace serving as the control variable are shown in Figure E.

As in the case of regression estimates, use of ratio estimates diminishes the variation, and use of the trace variable yields very good results with convergence to 1.56 at 20,000 replications.

Both the regression and ratio estimates are biased, with the bias of order $O(1/n)$. Asymptotic variances are smaller than in simple importance sampling, depending on the correlations between $\hat{l}_j$ and the control variables in (4.1a, b, c).

In Figure F we provide a 95% confidence interval for the mean of the largest characteristic root. This is based on importance sampling together with a ratio estimate using the trace as a control variable. The confidence interval is approximately (1.548, 1.566) with a point estimate of 1.557.

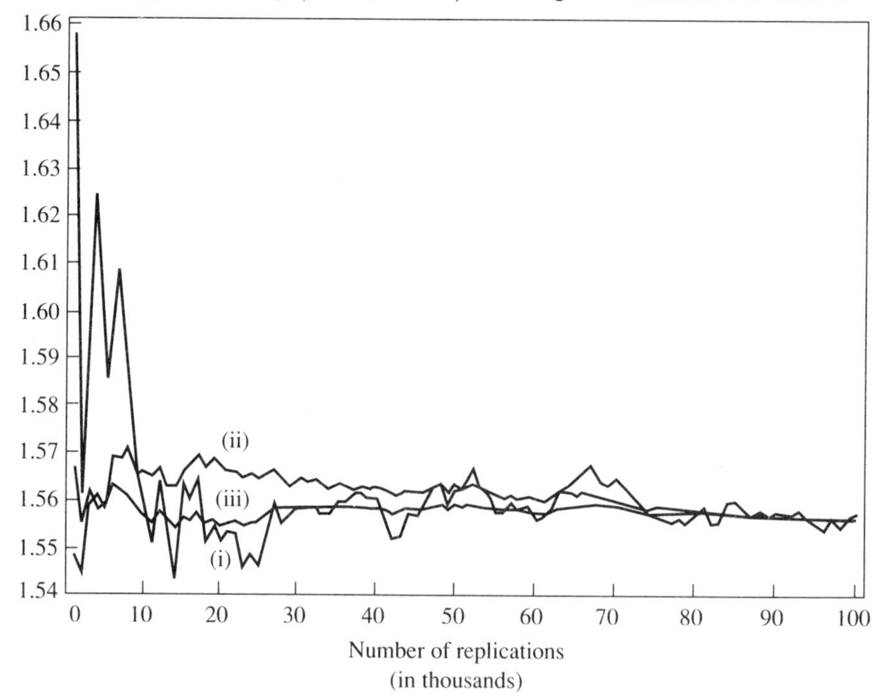

FIGURE E. Cumulative means of the expected value of the largest characteristic root ($p = 3$, $n = 15$), generated by (i) importance sampling with a chi-square distribution with $k = 15$ d.f., (ii) ratio estimates using (i) and a weight variable, (iii) ratio estimates using (i) and trace.

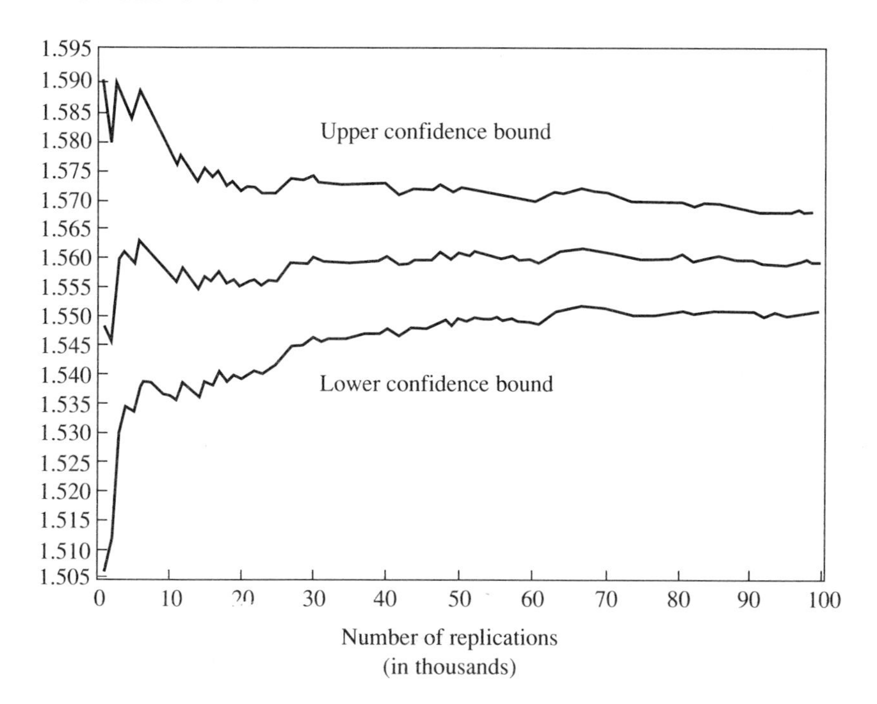

FIGURE F. Upper and lower 95% confidence bounds for the cumulative means of the expected value of the largest characteristic root ($p = 3$, $n = 15$) using importance sampling with chi-square distribution with $k = 15$ d.f. and ratio estimate based on the trace.

6. Multivariate regression and ratio estimates

Instead of using a single control variable, we can use several control variables to obtain multiple regression or ratio estimates. For example, the bivariate ratio estimate $\hat{l}_j^{brat}$ is:

$$\hat{l}_j^{brat} = \pi \hat{l}_j^{rat_1} + (1 - \pi)\hat{l}_j^{rat_2}$$

where $\hat{l}_j^{rat_1}$ and $\hat{l}_j^{rat_2}$ are the univariate ratio estimates of l_j based on each of two different x variables, and where the mixing constant π is estimated by minimizing the variance of $\hat{l}_j^{brat}$. (See Olkin (1958).)

The bivariate regression estimate $\hat{l}_j^{breg}$ is

$$\hat{l}_j^{breg} = \hat{l}_j - \hat{\beta}_{j1}(\hat{x}_1 - E\hat{x}_1) - \hat{\beta}_{j2}(\hat{x}_2 - E\hat{x}_2).$$

Because of high correlations among the control variables $\operatorname{tr} V$ and $\det V$, say, we cannot expect much improvement in the use of bivariate over univariate regression. Ratio estimates depend on correlations in a more circuitous

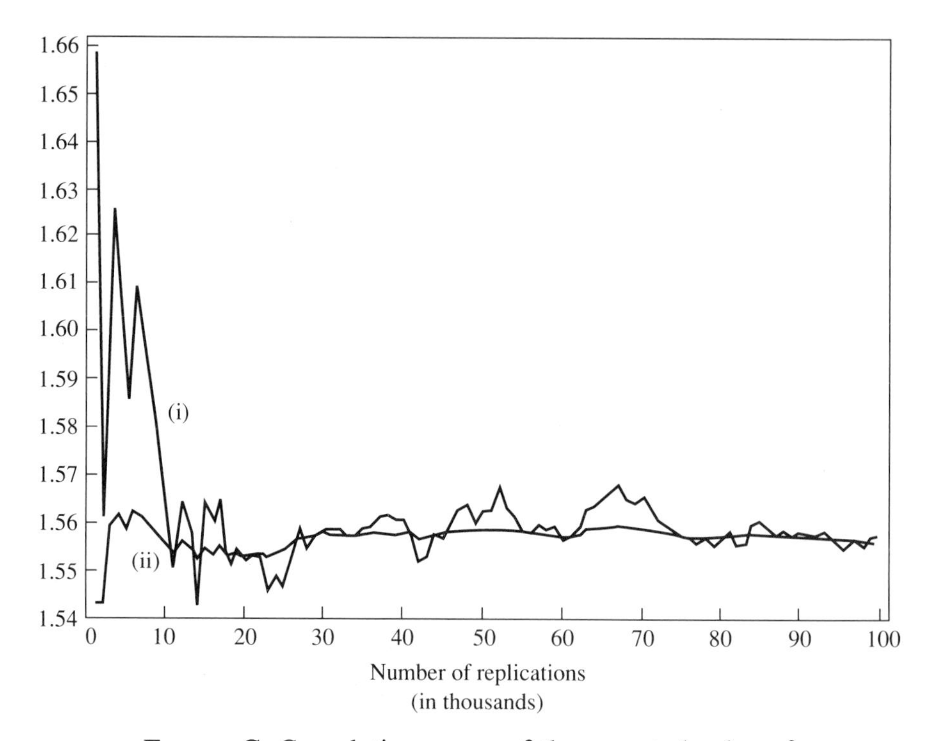

FIGURE G. Cumulative means of the expected value of the largest characteristic root ($p = 3$, $n = 15$) based on (i) importance sampling with a chi-square distribution with $k = 15$ d.f. and (ii) multivariate ratio estimate based on a weight variable and the trace.

manner, and here we present results for bivariate ratio estimates in Figure G. The bivariate ratio estimates are exceedingly stable with convergence to 1.56 at 5,000 replications. The efficiency of each variance reduction technique described in sections 4–6 are summarized in Table 2.

TABLE 2. Efficiency as measured by the ratio: $\mathrm{Var}(\chi^2_{15}$ method$)/\mathrm{Var}$ (reduction methods).

Method	
ratio estimate using W	11.7
regression estimate using W	14.4
ratio using trace V	38.5
bivariate ratio estimate	40.3
regression estimate using trace V	82.9

7. Mixture sampling

Yet another type of importance sampling is based on generating random numbers according to some mixture of distributions. An obvious choice is a mixture of chi-square distributions for which the density in (3.4) is

$$\varphi(l_1, \dots, l_p) = \prod_{j=1}^{p} \left[\pi_1 c_1 l_j^{\frac{1}{2}-1} e^{-\frac{1}{2}l_j} + \dots + \pi_n c_n l_j^{\frac{n}{2}-1} e^{-\frac{1}{2}l_j}\right],$$

where $\pi_i > 0$, $\sum_{i=1}^{n} \pi_i = 1$ are mixing coefficients, and c_i is a normalizing constant for the χ_i^2 distribution. In Table 3 we compare the variances for different choices of $\underset{\sim}{\pi} = (\pi_1, \pi_2, \dots, \pi_n)$.

TABLE 3. Efficiencies as measured by the ratio: Var($\underset{\sim}{\pi}$ mixture sampling)/ Var(χ_{15}^2 method).

Mixing Weights	
$(1/n, 1/n, \dots, 1/n)$	4.998
$(0, 0, \dots, 0, .3, .7)$	1.696
$(0, 0, \dots, 0, .2, .8)$	1.685
$(0, .1, 0, \dots, 0, .2, .7)$	1.608
$(0, 0, \dots, 0, .1, .2, .7)$	1.496
$(0, 0, \dots, .1, .9)$	1.307
$(.1, 0, \dots, 0, .9)$	1.288

Unlike the efficiencies in Tables 1 and 2, efficiencies of different mixture sampling are based on only 3,000 replications and are therefore not directly comparable to results in the first two tables which are based on 10^5 replications. Nevertheless, the results clearly indicate that "contaminating" chi-square distributions with n degrees of freedom with one or more chi-square distributions with $k < n$ degrees of freedom yields an increase in the variance. Efficiencies in Table 3 would be even more favorable for χ_n^2 distributions had the significant increase in the amount of computing time involved in sampling from the mixture been taken into account.

8. First differences

Instead of using

$$g(\underset{\sim}{l}) = l_{(i)}, \qquad i = 1, \dots, p,$$

the ith characteristic root, we can use the gap statistic

$$g(\underset{\sim}{l}) = l_{(i)} - l_{(i+1)}, \qquad i = 1, \dots, p-1,$$

and apply any of the methods previously mentioned to the differences of the ordered roots. Because the roots are positively correlated, the difference of two roots should have less variability than the sum of the variances of the individual roots. The marginal distributions of the differences tend to be less

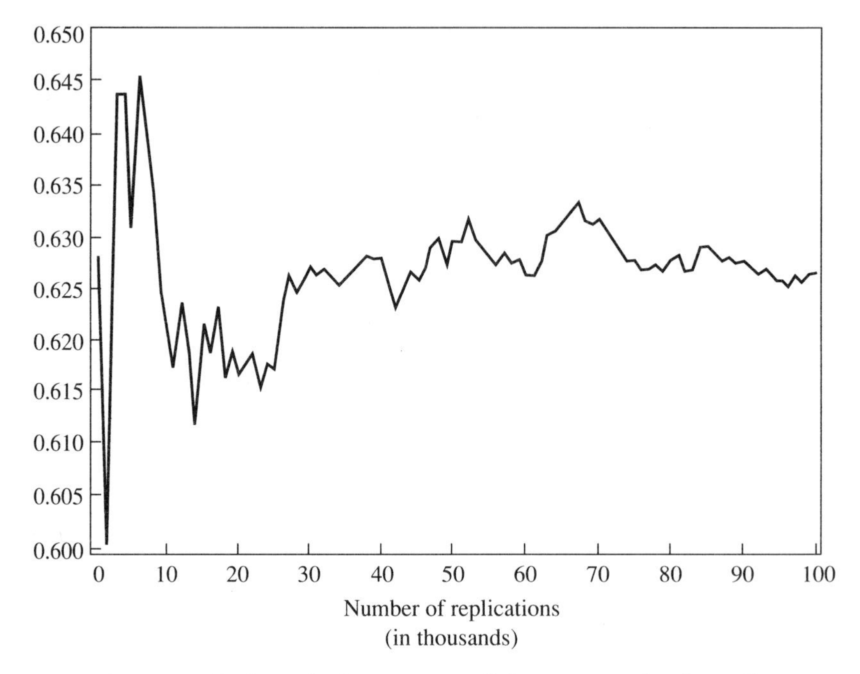

FIGURE H. Cumulative means of the expected value of the difference between the largest and the second largest characteristic root ($p = 3, n = 15$) generated by importance sampling with a chi-square distribution with $k = 15$ d.f.

skewed and therefore the mean value curve, presented in Figure H is more smooth.

Commentary

Variances of different variance reduction estimates are compared in Table 3. The largest reduction in variability is achieved with a regression using the trace variable (83 times smaller variance than the variance of simple importance sampling with χ_n^2). The ratio estimate using the trace is approximately half as efficient as the regression using the trace, but is still much better than the regression estimate with a weight variable. Using a bivariate ratio estimate instead of a univariate ratio estimate reduces the variance, but is still less efficient than using a regression estimate with trace as the only regressor. The amount of precision achieved with 10^5 replications can be seen from Figure I where the cumulative means stabilized at 20,000 replications. It seems that after 30,000 replications the expected value is still fluctuating, but only between 1.558 and 1.560.

From what has been said and from the experience of some other authors (see for example, Bratley, Fox, and Schrage (1983) and Hesterberg (1987)) it appears that the simulation in general, and importance sampling in partic-

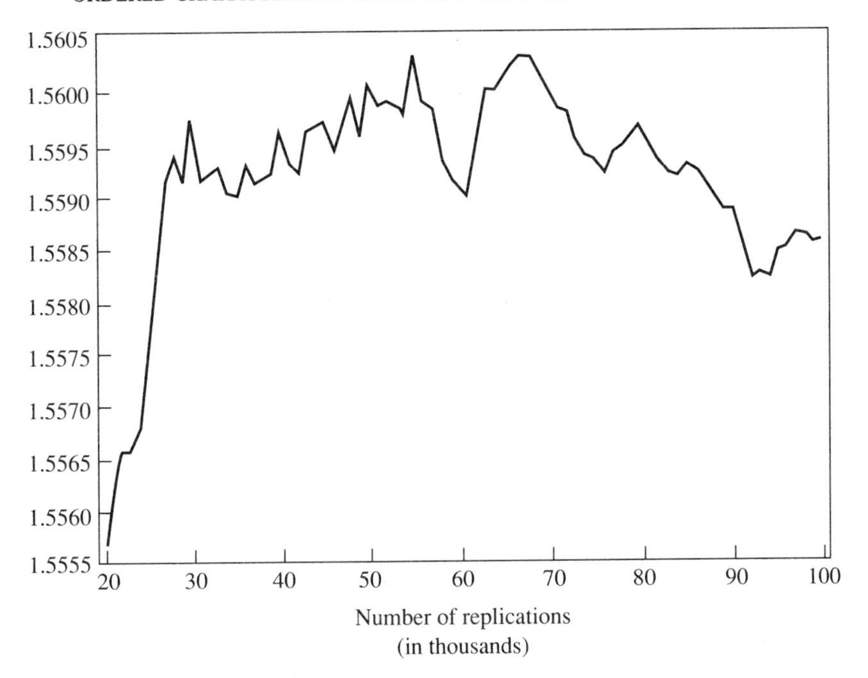

FIGURE I. Cumulative means of the expected value of the largest and the second largest characteristic root ($p = 3$, $n = 15$) generated by importance sampling with a chi-square distribution with $k = 15$ d.f.

ular, can be applied in multivariate problems only with extreme care. Even in the tridimensional case it is not advisable to use importance sampling without some additional variance reduction technique. It is hardly possible to make any generalization on the best choice of the variance reduction method. Whereas in some problems mixture sampling may achieve a significant variance reduction, in other problems multiple regression estimation may give better results than either simple regression or ratio estimation. Because the distributions of the examined and control variables, as well as the correlations among control variables, usually determine the optimal variance reduction method, examining the sampling distribution of each variable, correlations and all possible regressions and ratios may give more insight in the problem.

When the number of variables in the integrand increases, the number of replications required to yield the same degree of precision must be increased considerably. Nevertheless, even with the increased number of replications and computations required, the use of importance sampling, together with regression estimation, yields a more efficient procedure for estimating the mean of the ordered characteristic roots than does the method of generating random matrices.

References

T. W. Anderson (1984), *Introduction to multivariate statistical analysis*, (second ed.), John Wiley & Sons, Inc., New York.

P. Bratley, B. L. Fox and L. E. Schrage (1983), *A guide to simulation*, Springer-Verlag, New York.

W. G. Cochran (1977), *Sampling techniques*, John Wiley & Sons, Inc., New York.

J. M. Hammersley and D. C. Hanscomb (1964), *Monte Carlo methods*, Methuen, London.

T. Hesterberg (1987), *Importance sampling in multivariate problems*, Proc. Stat. Comp. Section ASA, pp. 412–417.

M. G. Marsinghe and W. J. Kennedy, Jr. (1982), *Direct methods for generating extreme characteristic roots of certain random matrices*, Commun. Statist.—Simula. Computa. **11**, 527–542.

R. J. Muirhead (1982), *Aspects of multivariate statistical theory*, John Wiley & Sons, Inc., New York.

I. Olkin (1958), *Multivariate ratio estimation for finite populations*, Biometrika **45**, 154–165.

——(1985), *Estimating a Cholesky decomposition*, Linear Algebra and Its Applications **67**, 201–205.

R. Y. Rubinstein (1981), *Simulation and Monte Carlo method*, John Wiley & Sons, Inc., New York.

University Computing Centre, 41000 Zagreb, Engelsova bb, Yugoslavia
E-mail address: I73LUZO@YUZGSC21.BITNET

Department of Statistics, Stanford University, Stanford, California 94305
E-mail address: iolkin@playfair.stanford.edu

Contemporary Mathematics
Volume **115**, 1991

A Stationary Stochastic Approximation Method

JOHN F. MONAHAN AND ROGER F. LIDDLE

ABSTRACT. A common problem in Bayesian statistical decision analysis is the optimization of an expected loss or gain function, where the expectation is taken with respect to the posterior distribution. Viewed mathematically, it is the optimization of a function defined by an integral. In high dimensions, the integration step must be done using Monte Carlo methods, and stochastic approximation combines both integration and optimization for the solution of such a problem. However, stochastic approximation is difficult to implement, especially the practicalities of stopping and verifying conditions. In response to these difficulties, we propose a stationary version, which is less sophisticated mathematically, but much easier to implement.

1. Introduction

Consider the general problem of finding θ, the minimum of maximum of a function of p variables. This function of θ is defined by an integral over another set of r variables x, that cannot be evaluated analytically. The integrand includes a function of both sets of variables $g(x;\theta)$ and a weight function $w(x)$ does not depend on θ. An alternative to optimizing the integral:

$$\min_\theta \textit{ or } \max_\theta \int g(x;\theta)w(x)\,dx \tag{1.1}$$

is to take the derivative with respect of θ, leading to solving a system of nonlinear equations

$$0 = \int h(x;\theta)w(x)\,dx. \tag{1.2}$$

The latter is usually preferred because gains in faster convergence rates analogous to bisection versus golden section are expected. As a result, this paper will focus on solving the system of nonlinear equations (1.2). In some

1980 *Mathematics Subject Classification* (1985 *Revision*). Primary 62L20, 62M10, 65C05.

This paper is a synopsis of a paper previously published in 1988 under the same title in Journal of Econometrics, Volume 38, p. 91–102.

situations, the integration could be imbedded within the optimization/root-finding, but in high dimensions where $r \geq 3$, the curse of dimensionality would force the use of Monte Carlo methods. Moreover, most optimization methods require accurate evaluation of the function, in this case, accurate integration. While this could certainly be done with hard work, accurate evaluations early in the search would be a wasted effort.

This problem arises in Bayesian decision applications where the goal is to take an action θ to minimize expected loss. Here θ represents action or decision variables, and x corresponds to the unknown parameters of the statistical model $f(y \mid x)$. The integrand $g(x;\theta)$ encompasses both a loss function $L(\theta, x)$ and the posterior density $\pi(x|y) \propto \pi(x)f(y|x)$ for data y, and prior distribution $\pi(x)$. In nonstandard problems that require a numerical solution, importance sampling methods would be employed, yielding $g(x;\theta) = L(\theta, x)\pi(x \mid y)/w(x)$. While solving the general Bayesian decision problem served as the main goal of this work, the immediate motivation was the computation of Minimum Hellinger Distance (MHD) location and dispersion parameters (Tamura and Boos (1986)), where the distance between the kernel density estimate $\hat{f}(x)$ and the parameterized density $f_\theta(x)$ is measured by

$$d(\hat{f}, f_\theta) = \int [\hat{f}(x)^{1/2} - f_\theta(x)^{1/2}]^2\, dx.$$

The integration involves the same dimension r as the observations. The dimension for optimization is $r + r(r+1)/2$, so that for $r = 2$ or 3, then the optimization is over 5 or 9 dimensions. No method other than the one proposed shows any promise computing the MHD estimates, even for small r.

2. Stochastic approximation and the stationary alternative

When the integration is to be done by Monte Carlo, where observations X_i are obtained independently and identically distributed (iid) with density $w(x)$, the natural approach is to use stochastic approximation (SA). Stochastic approximation is designed to find the root t^* of a regression function $E\{Y|t\} = m(t)$, where a response Y could be obtained from any value of t. Robbins and Monro (1951) invented this root-finding method; Kiefer and Wolfowitz (1952) extended the technique for optimization. Fitting the problem at hand into this scheme leads to the general strategy of taking $Y = h(X;t)$ and generating random variables X iid with density $w(x)$. SA has rarely been applied, with Ruppert, Reish, Deriso, and Carroll (1984) being a notable exception which follows this general strategy.

The motivation for SA can be seen by considering the search for the root of a function using bisection with averages $\overline{Y}(t)$ at particular points t in the place of evaluations of the unknown function $m(t)$. Starting in an interval known to include the root of an increasing function, suppose we have $\overline{Y}(t_1)$, when do we decide that $m(t_1)$ is greater or less than zero? A natural approach is to perform a t-test, but at what level? Since bisection, and many other search routines, cannot handle any evaluation errors, this problem becomes very complicated quickly. Stochastic approximation says to move constantly, $t_{k+1} = t_k - a_k Y_k$, where $a_k = a/k$, and the optimal $a = 1/b = 1/m'(t)$. In the multivariate case, this update becomes slightly more complicated,

$$t_{k+1} = t_k - a_k A Y_k \tag{2.1}$$

where both t_k and Y_k are p-dimensional vectors, a_k is a scalar $O(k^{-1})$, and the p-by-p matrix A may be diagonal to provide rescaling, or reorient Newton-like to improve convergence.

The asymptotic properties of the SA process t_k provides the key to its difficulties, especially with assessing accuracy. Under the right conditions, Robbins and Monro showed

$$\sqrt{n}(t_n - t^*) \to N(0, a^2\sigma^2/(2ab - 1)) \tag{2.2}$$

for the univariate case, and constructing a confidence interval based on (2.2) only requires estimation of σ^2 and the slope b. But the estimation of b also serves to check the condition $2ab > 1$ that is required for both normality and the $1/\sqrt{n}$ rate of convergence. Since $\sum(x_i - \overline{x})^2$ is $O(\log n)$, the convergence of the natural regression estimate $\hat{b}$ is slow, $O_p((\log n)^{-1/2})$. Moreover, the initial $\hat{b}$ overestimates the slopes, and that bias dissipates slowly. Because the slope is overestimated, the problem appears easier than it really is, and the condition that $2ab > 1$ appears to be verified when it does not hold (Liddle (1989)). The convergence of t_n is a two-edged sword, with the small variation in the design leading to slow convergence of the slope estimate.

Stationary stochastic approximation (SSA) gives up the convergence on t_k to improve the slope estimate, by following the iteration

$$t_{k+1} = t_k - aAY_k \tag{2.3}$$

with both the scalar a and matrix A fixed. If the response surface is linear (affine), then the multivariate regression model applies

$$Y_k = b + Bt_k + e_k, \tag{2.4}$$

where b is a p-dimensional intercept vector and B is a p-by-p slope matrix. The probability calculus is much simpler than SA, since combining (2.3) and (2.4) leads to a multivariate autoregressive time series process:

$$\begin{pmatrix} Y_{k+1} \\ t_{k+1} \end{pmatrix} = \begin{pmatrix} b \\ 0 \end{pmatrix} + \begin{bmatrix} 0 & B \\ -aA & I_p \end{bmatrix} \begin{pmatrix} Y_k \\ t_k \end{pmatrix} + \begin{pmatrix} e_k \\ 0 \end{pmatrix}, \tag{2.5}$$

where I_p is the $p \times p$ identity matrix. For stationarity, the eigenvalues of $(I - aAB)$ must be less than one in absolute value. Since taking $A = B^{-1}$ leads to eigenvalues of $(1 - a)$, the strategy is to estimate the slope matrix B using the multivariate regression, use $A = (\widehat{B})^{-1}$ in (2.3), and to take the scalar a small enough to insure the affine model (2.4) is acceptable. SSA and SA differ in many ways. SSA has more accurate estimates of b and B, but t_n does not converge, so that a new solution must be found. While SSA depends heavily on the assumption of linearity (2.4), the scalar a can be chosen to govern the domain of linearity. Taking multiple observations Y at each t provides a model-free estimate of the variance (pure error) and permits the simple regression lack of fit test for model adequacy. In place of t_n, that does not converge, some other estimate must be found for the solution t^*. Another candidate is the root of the estimated regression function $-(\widehat{B})^{-1}\hat{b}$, but constructing confidence regions for this solution becomes very complicated. Additionally, testing whether a root has been found $H\colon M(t^*) = 0$ using $-(\widehat{B})^{-1}\hat{b}$ turns out to be vacuous.

We have chosen the process average $\bar{t}_n$ as the estimate of the solution t^*. Confidence intervals can be constructed easily using the asymptotic result:

$$\operatorname{Cov}(\bar{t}_n) = n^{-1}B^{-1}\Sigma B^{-T} + O(n^{-2}). \tag{2.6}$$

Testing $H\colon M(t) = 0$ using $\bar{t}_n$ leads to a very powerful test, because

$$\operatorname{Cov}(\overline{Y}) = \operatorname{Cov}(\hat{b} + \widehat{B}\bar{t}_n) = O(n^{-2}). \tag{2.7}$$

At any alternative this test is likely to reject, so that acceptance is strong evidence that the root has been reached. The algorithm for Stationary Stochastic Approximation can be summarized as follows:

(1) Begin with moderately small a and $A = I$.
(2) Iterate the process (2.3) for a while.
(3) Stop for estimation and testing:
  (i) Estimate $\hat{b}$ and $\widehat{B}$.
  (ii) Test whether a root has been reached using (2.7). If not, return to (2).
  (iii) Test whether surface is linear. If not, return to (2) with reduced a.
(4) Continue until desired accuracy is achieved.
(5) Compute covariance matrix of solution estimate $\bar{t}_n$ using (2.6).

3. Performance in applications

As previously mentioned, the problem that immediately motivated this work was computing minimum Hellinger distance estimates. Two three-dimensional problems were attempted with SSA. In the first, 100 observations in a real application were used; the convergence was slow owing to a bumpy surface perhaps with many local minima. In the second problem, 800 observations were generated from a Normal $(0, I_3)$ distribution. Despite a sample size 8 times larger than the real application, slowing each evaluation, the smoothness of the problem lead to convergence in less time than the first.

To evaluate the performance of SSA in a Bayesian decision problem, an example by Cox (1970, Table 5.3) was modified to obtain a significant effect. Observed were y_i ingots out of m_i not ready for rolling after heating time θ_{1i} and soaking time θ_{2i}. The probability function $p(\theta, x)$ of the binomial response y_i is modelled as logit $p(\theta, x) = x_0 + \theta_1 x_1 + \theta_2 x_2$, so that x_0, x_1, x_2 are parameters of the model, with design/decision variables θ_1, θ_2. The loss function was chosen to take the form $L(\theta, x) = c_1/\theta_1 + c_2\theta_2 = c_3 p(\theta, x)$. The integration in the three model variables was improved by sampling from the normal approximation to the posterior. SSA swiftly found the solution.

4. Conclusions

While stationary stochastic approximation shows great promise in solving nonlinear integral equations, the ultimate goal of a black box problem solver has not yet been achieved. Further work is needed on determining useful options in the SSA algorithm and extensions for the optimization problem. A mystery is that SSA seems unaffected by changes in sign and direction, while multivariate SA methods usually require that the matrix A be positive definite, or could be made so with appropriate permutations.

References

D. R. Cox (1970), *The analysis of binary data*, Methuen, London.

J. Kiefer and J. Wolfowitz (1952), *Stochastic estimation of the maximum of a regression function*, Ann. of Math. Stat. **23**, 462–466.

R. F. Liddle and J. F. Monahan (1988), *A stationary stochastic approximation method*, J. Econometrics **38**, 91–102.

R. F. Liddle (1989), *Stochastic approximation for the optimization of integrals*, Ph. D. Dissertation, North Carolina State University.

H. Robbins and S. Monro (1951), *A Stochastic approximation method*, Ann. of Math. Stat. **22**, 400–407.

D. Ruppert, R. L. Reish, R. B. Deriso, and R. J. Carroll (1984), *Optimization using stochastic approximation and Monte Carlo simulation* (*with application to harvesting of Atlantic Menhaden*), Biometrics **40**, 535–545.

R. N. Tamura and D. D. Boos (1986), *Minimum Hellinger distance estimation for multivariate location and covariance*, J. Amer. Statist. Assoc. **67**, 223–229.

Department of Statistics, North Carolina State University, Raleigh, North Carolina 27695-8203

Glaxo Incorporated, 5 Moore Drive, Research Triangle Park, North Carolina 27709

Contemporary Mathematics
Volume **115**, 1991

Inequalities and Bounds for a Class of Multiple Probability Integrals, with Applications

Y. L. TONG

ABSTRACT. In this paper we present a survey of certain useful bounds and inequalities for a class of multiple probability integrals over one-sided and two-sided n-dimensional rectangles and ellipsoids. The mathematical tool used involves majorization and Schur-concavity, and the theorems apply to a large class of distributions. Special applications to permutation-symmetric multivariate normal and t distributions are discussed.

1. Introduction and motivation

Let $\mathbf{X} = (X_1, \dots, X_n)$ denote an n-dimensional random variable with density function $f(\mathbf{x})$ that is absolutely continuous with respect to Lebesgue measure, and let $A \subset \mathfrak{R}^n$ be a measureable subset. In many statistical applications, the probability content of the form

$$P[\mathbf{X} \in A] = \int_A f(\mathbf{x})\, d\mathbf{x} \tag{1.1}$$

is of interest for given A and $f(\mathbf{x})$. In some special cases, such as when $f(\mathbf{x})$ is the density function of a multivariate normal distribution and A is an n-dimensional rectangle, various numerical methods already exist for approximating such a probability integral on the computer, and statistical tables are also available.

In almost all of the statistical tables published, however, the numerical values are given only for some special cases, and the number of parameters in those tables are kept to a minimum. For example, if $\mathbf{X}$ has a multivariate normal distribution with a common mean zero, a common variance one, and a common correlation coefficient $\rho \geq 0$, then the numerical values of

1980 *Mathematics Subject Classification* (1985 *Revision*). Primary 60E15, 62H99, 65D20.

Key words and phrases. Multiple probability integrals, probability inequalities, majorization and Schur-concavity, multivariate normal and t distributions, statistical computing.

This research was supported in part by NSF Grants DMS-8801327, A01 and DMS-9001721.

This paper is in final form and no version of it will be submitted for publication elsewhere.

$P[\bigcap_{i=1}^n \{X_i \leq a\}]$ have been tabulated for selected values of n and a; such a table contains only two parameters (n and a), and can be printed on one page for fixed ρ (see, e.g., the Appendix of Tong (1990)). On the other hand, the tabulation of $P[\bigcap_{i=1}^n \{X_i \leq a_i\}]$ seems difficult when the a_i's are different because it will involve $n+1$ parameters (n and $a_1, \ldots, a_n$); and even for $n = 3$ or 4 the table will have to be 4- or 5-dimensional. A question of interest is then this: Suppose that in a given application the values of the a_i's are not the same. If we use the table value of $P[\bigcap_{i=1}^n \{X_i \leq \bar{a}\}]$ as an approximation to the true value of $P[\bigcap_{i=1}^n \{X_i \leq a_i\}]$, where $\bar{a} = \frac{1}{n}\sum_{i=1}^n a_i$ is the arithmetic mean, can we claim that the approximated value is actually a lower or upper bound? If the answer is in the affirmative, then we can proceed to obtain a conservative or liberal solution for a given statistical decision problem *without* evaluating the true value of this probability integral.

The above discussion illustrates one of the many applications of multivariate probability inequalities in multivariate analysis and probability integrals. In this paper we present a survey of some existing inequalities and bounds for a large class of multiple probability integrals, with special emphasis on the multivariate normal probability integrals. Most of the results stated in this paper have become available only recently; the mathematical tool used is majorization and Schur-concavity, and the geometric regions (defined by the subset A) under consideration include one-sided and two-sided n-dimensional rectangles and ellipsoids.

The class of probability density functions under consideration mainly concerns Schur-concave functions. Thus the notion of majorization and Schur-concavity are briefly reviewed in §2. In §3 we state the main results without proofs, and a convenient reference for the proofs of the results is Tong (1988) or Tong (1990, Chapters 4 and 7). Section 4 contains some special applications to the multivariate normal and t distributions, and the results also involve the mean vector and the covariance matrix of a multivariate normal distribution.

2. Majorization and Schur-concave functions

Majorization deals with the diversity of the components of a vector. For given $n \geq 2$ let

$$\mathbf{a} = (a_1, \ldots, a_n), \qquad \mathbf{b} = (b_1, \ldots, b_n) \tag{2.1}$$

be two real vectors. Let

$$a_{[1]} \geq a_{[2]} \geq \cdots \geq a_{[n]}, \qquad b_{[1]} \geq b_{[2]} \geq \cdots \geq b_{[n]} \tag{2.2}$$

denote their ordered components.

DEFINITION 2.1. $\mathbf{a}$ is said to majorize $\mathbf{b}$, in symbols $\mathbf{a} \succ \mathbf{b}$, if

$$\sum_{i=1}^m a_{[i]} \geq \sum_{i=1}^m b_{[i]} \quad \text{holds for } m = 1, 2, \ldots, n-1 \tag{2.3}$$

and $\sum_{i=1}^{n} a_i = \sum_{i=1}^{n} b_i$.

This definition provides a partial ordering; namely, $\mathbf{a} \succ \mathbf{b}$ implies that (for a fixed sum) the a_i's are more diverse than the b_i's. For example, it is obvious that the following two facts are true:

FACT 2.2. $\mathbf{a} \succ (\bar{a}, \ldots, \bar{a})$ *holds for all* $\mathbf{a}$ *where* $\bar{a} = \frac{1}{n}\sum_{i=1}^{n} a_i$.

FACT 2.3. *If* $a_i \geq 0$ $(i = 1, \ldots, n)$, *then* $(\sum_{i=1}^{n} a_i, 0, \ldots, 0) \succ \mathbf{a}$ *holds.*

Furthermore, it is known that

FACT 2.4. *If* $\mathbf{a} \succ \mathbf{b}$ *holds, then* $\frac{1}{n-1}\sum_{i=1}^{n}(a_i - \bar{a})^2 \geq \frac{1}{n-1}\sum_{i=1}^{n}(b_i - \bar{b})^2$ *holds. Thus majorization is stronger than the variance concept in measuring the diversity of the components of a vector.*

And, by applying Birkhoff's theorem, we have

FACT 2.5. $\mathbf{a} \succ \mathbf{b}$ *holds if and only if there exists a doubly stochastic matrix* $\mathbf{Q}$ *such that* $\mathbf{b} = \mathbf{aQ}$.

In view of Fact 2.5, we may regard the linear transformation $\mathbf{b} = \mathbf{aQ}$ as an averaging process when the sum of the vector components is kept fixed. It is in this sense we say that if $\mathbf{a} \succ \mathbf{b}$, then $\mathbf{b}$ is an "average" of $\mathbf{a}$ (Hardy, Littlewood, and Pólya (1934, 1952, p. 49).

An important application of majorization in the theory of inequalities involves Schur functions.

DEFINITION 2.6. A real-valued function $f(\mathbf{x}) : \mathfrak{R}^n \to \mathfrak{R}$ is said to be Schur-concave if $\mathbf{x} \succ \mathbf{y}$ implies $f(\mathbf{x}) \leq f(\mathbf{y})$ for all $\mathbf{x}, \mathbf{y} \in \mathfrak{R}^n$.

That is, $f(\mathbf{x})$ is a Schur-concave function if the functional value of f becomes larger when the components of $\mathbf{x}$ are less diverse in the sense of majorization.

The notion of majorization and Schur functions appears to be systemmatically studied for the first time by Hardy, Littlewood, and Pólya (1934, 1952). In their 1979 monograph Marshall and Olkin presented a most comprehensive treatment of this subject, and gave many useful applications in the theory of inequalities. The reader is referred to their monograph for details.

Since most of the results given in §3 depend on the assumption that the density function $f(\mathbf{x})$ is a Schur-concave function of $\mathbf{x} \in \mathfrak{R}^n$, in the following we discuss how Schur-concave density functions and other classes of density functions are related. For the proofs of these results see, e.g., Tong (1990, Chapter 4).

First we restate a definition of unimodal density functions due to Anderson (1955):

DEFINITION 2.7. A density function $f(\mathbf{x}) : \mathfrak{R}^n \to [0, \infty)$ is said to be unimodal if $\{\mathbf{x} : f(\mathbf{x}) \geq \lambda\}$ is a convex set in $\mathfrak{R}^n$ for all $\lambda > 0$.

The following fact shows how permutation-symmetric and unimodal density functions and Schur-concave density functions are related.

FACT 2.8. *If the density function* $f(\mathbf{x}) : \mathfrak{R}^n \to [0, \infty)$ *of* $\mathbf{X}$ *is permutation symmetric and unimodal, then it is a Schur-concave function of* $\mathbf{x}$.

A density function $f(\mathbf{x})$ is said to be log-concave if $\log f(\mathbf{x})$ is a concave function of $\mathbf{x} \in \mathfrak{R}^n$. Log-concave density functions play an important role in statistics, and the following result shows how log-concave density functions and unimodal density functions are related.

FACT 2.9. *All log-concave density functions are unimodal.*

As a consequence of Facts 2.8 and 2.9, we have

FACT 2.10. *If the density function* $f(\mathbf{x})$ *of* $\mathbf{X}$ *is permutation symmetric and log-concave, then it is a Schur-concave function of* $\mathbf{x}$.

FACT 2.11. *Assume that* $X_1, \ldots, X_n$ *are i.i.d. random variables with a continuous common marginal density* $h(x) : \mathfrak{R} \to [0, \infty)$ *such that the support of* h *is an interval* I. *If* $h(x)$ *is a log-concave function of* x *for* $x \in I$, *then the joint density function* $f(\mathbf{x}) = \prod_{i=1}^n h(x_i)$ *is a Schur-concave function of* $\mathbf{x}$ *for* $\mathbf{x} \in I \times \cdots \times I$.

Since the density function of a multivariate normal distribution is both unimodal and log-concave, as a special consequence of either Fact 2.8 or 2.10 we have

FACT 2.12. *Let* $f(\mathbf{x}; \boldsymbol{\mu}\boldsymbol{\Sigma})$ *denote the density function of an* $N_n(\boldsymbol{\mu}, \boldsymbol{\Sigma})$ *distribution. If* f *is permutation symmetric, i.e.,*

$$\mu_1 = \cdots = \mu_n, \qquad \sigma_1^2 = \cdots = \sigma_n^2,$$
$$\rho_{ij} = \rho \in \left(-\frac{1}{n-1}, 1\right) \quad \textit{for all } i \neq j,$$

then $f(\mathbf{x}; \boldsymbol{\mu}, \boldsymbol{\Sigma})$ *is a Schur-concave function of* $\mathbf{x}$.

It is known that the class of Schur-concave density functions includes most of the important densities in statistical applications. In particular, it is known that, in addition to the multivariate normal density, the density functions of permutation-symmetric (i) multivariate t distribution, (ii) multivariate χ^2 and gamma distributions, (iii) multivariate F distribution, (iv) multinomial, and Dirichlet distributions, and many other distributions, are all Schur-concave. Thus Theorems 3.1, 3.2, and 3.4 stated in §3 apply to all of those distributions.

3. Statements of the main results

In this section we state the main results concerning bounds for the probability integral given in (1.1). The results are given under the assumption that $f(\mathbf{x})$, the density function of $\mathbf{X}$, is either a Schur-concave or a log-concave function of $\mathbf{x} \in \mathfrak{R}^n$. The geometric regions of the probability integral, specified by the subset $A \subset \mathfrak{R}^n$, is assumed to be an n-dimensional rectangle or ellipsoid. There exist useful results when A is neither a rectangle nor an ellipsoid but, in order not to overload this paper, such results are not given here.

In one of the earlier papers on majorization in multivariate distributions, Marshall and Olkin (1974) proved the following result:

THEOREM 3.1. *If* $f(\mathbf{x})$ *(the density function of* $\mathbf{X} = (X_1, \ldots, X_n)$*) is a Schur-concave function of* $\mathbf{x}$*, then*

$$F(\mathbf{a}) = P\left[\bigcap_{i=1}^{n}\{X_i \le a_i\}\right] = \int_{A_1(\mathbf{a})} f(\mathbf{x})\, d\mathbf{x} \tag{3.1}$$

is a Schur-concave function of $\mathbf{a}$*, where* F *is the distribution function of* $\mathbf{X}$ *and*

$$A_1(\mathbf{a}) = \{\mathbf{x} : \mathbf{x} \in \mathfrak{R}^n,\ x_i \le a_i \text{ for } i = 1, \ldots, n\} \tag{3.2}$$

is a one-sided n*-dimensional rectangle.*

Consequently, for $\mathbf{a} \succ \mathbf{b}$ we have

$$F(\mathbf{a}) \le F(\mathbf{b}) \le F(\bar{\mathbf{a}}), \tag{3.3}$$

where

$$\bar{\mathbf{a}} = (\bar{a}, \ldots, \bar{a}), \qquad \bar{a} = \frac{1}{n}\sum_{i=1}^{n} a_i. \tag{3.4}$$

Motivated by this result, Tong (1982) obtained a two-sided analog of Theorem 3.1:

THEOREM 3.2. *If* $f(\mathbf{x})$ *is a Schur-concave function of* $\mathbf{x}$*, then*

$$G(\mathbf{a}) = P\left[\bigcap_{i=1}^{n}\{|X_i| \le a_i\}\right] = \int_{A_2(\mathbf{a})} f(\mathbf{x})\, d\mathbf{x} \tag{3.5}$$

is a Schur-concave function of $\mathbf{a}$ *for* $a_i > 0$ $(i = 1, \ldots, n)$*, where* G *is the distribution function of* $|\mathbf{X}| = (|X_1|, \ldots, |X_n|)$ *and*

$$A_2(\mathbf{a}) = \{\mathbf{x} : \mathbf{x} \in \mathfrak{R}^n,\ |x_i| \le a_i \text{ for } i = 1, \ldots, n\} \tag{3.6}$$

is a two-sided n*-dimensional rectangle.*

Consequently, for $\mathbf{a} \succ \mathbf{b}$ we have

$$G(\mathbf{a}) \le G(\mathbf{b}) \le G(\bar{\mathbf{a}}). \tag{3.7}$$

The rectangle $A_2(\mathbf{a})$ defined in (3.6) is centered at the origin, and can be treated as a special case of the class of n-dimensional rectangles of the form

$$A_3(\mathbf{a}_1, \mathbf{a}_2) = \{\mathbf{x} : \mathbf{x} \in \mathfrak{R}^n,\ a_{1i} \le x_i \le a_{2i} \text{ for } i = 1, \ldots, n\}$$

with $a_{1i} = -a_{2i}$, where $\mathbf{a}_1 = (a_{11}, \ldots, a_{1n})$, $\mathbf{a}_2 = (a_{21}, \ldots, a_{2n})$ are such that $a_{1i} < a_{2i}$ $(i = 1, \ldots, n)$. An interesting question is whether the probability content of $A_3(\mathbf{a}_1, \mathbf{a}_2)$ becomes larger when $A_3(\mathbf{a}_1, \mathbf{a}_2)$ is closer to being an n-dimensional cube when the perimeter is kept fixed. This problem was studied independently by Karlin and Rinott (1983) and Tong (1983),

(1989), and the following theorem was obtained:

THEOREM 3.3. *If* $f(\mathbf{x})$ *is a log-concave function of* $\mathbf{x}$, *and if there exists a doubly stochastic matrix* $\mathbf{Q}$ *such that*

$$\mathbf{b}_1 = \mathbf{a}_1\mathbf{Q} \quad \text{and} \quad \mathbf{b}_2 = \mathbf{a}_2\mathbf{Q}, \tag{3.8}$$

then

$$\int_{A_3(\mathbf{a}_1, \mathbf{a}_2)} f(\mathbf{x})\,d\mathbf{x} \le \int_{A_3(\mathbf{b}_1, \mathbf{b}_2)} f(\mathbf{x})\,d\mathbf{x} \tag{3.9}$$

or equivalently,

$$P[\mathbf{X} \in A_3(\mathbf{a}_1, \mathbf{a}_2)] \le P[\mathbf{X} \in A_3(\mathbf{b}_1, \mathbf{b}_2)]. \tag{3.10}$$

As a special case, we have

$$P\left[\bigcap_{i=1}^{n}\{a_{1i} \le X_i \le a_{2i}\}\right] \le P\left[\bigcap_{i=1}^{n}\{\overline{a}_1 \le X_i \le \overline{a}_2\}\right], \tag{3.11}$$

where $\overline{a}_j = \frac{1}{n}\sum_{i=1}^{n} a_{ji}$ $(j = 1, 2)$.

Note that the right-hand side of (3.11) is the probability content of an n-dimensional cube when the perimeter is kept fixed. Furthermore, note that (by Fact 2.10) the condition on $f(\mathbf{x})$ in Theorem 3.3 is stronger than Schur-concavity. It is not yet known to the author whether the conclusion of Theorem 3.3 holds true for all Schur-concave density functions.

The results in Theorems 3.1–3.3 concern inequalities for the probability contents of (one-sided and two-sided) n-dimensional rectangles. In the following we state a result for ellipsoids. The proof of this result can be found in Tong (1982).

THEOREM 3.4. *Let* $\mathbf{a}^2 = (a_1^2, \ldots, a_n^2)$ *be a real vector such that* $a_i > 0$ $(i = 1, \ldots, n)$, *and define*

$$A_4 = \left\{\mathbf{x} : \mathbf{x} \in \mathfrak{R}^n,\ \sum_{i=1}^{n}(x_i/a_i)^2 \le \lambda\right\}, \tag{3.12}$$

where λ *is arbitrary but fixed. If* $f(\mathbf{x})$ *is a Schur-concave function of* $\mathbf{x}$, *then*

$$P[\mathbf{X} \in A_4(\mathbf{a}^2)] = P\left[\sum_{i=1}^{n}(X_i/a_i)^2 \le \lambda\right] \tag{3.13}$$

is a Schur-concave function of $\mathbf{a}^2$. *Consequently we have*

$$P\left[\sum_{i=1}^{n}(X_i/a_i)^2 \le \lambda\right] \le P\left[\sum_{i=1}^{n}X_i^2 \le \lambda\overline{a^2}\right], \tag{3.14}$$

where $\overline{a^2} = \frac{1}{n}\sum_{i=1}^{n} a_i^2$.

Note that the left-hand side of (3.14) is the probability content of an ellipsoid, and the right-hand side is that of a sphere in $\mathfrak{R}^n$.

4. Applications to the multivariate normal distribution

The multivariate normal distribution has played a predominant role in statistics, and has made its appearance in various areas of applications. Thus the multivariate normal probability integrals have received special attention in the literature, and the evaluation of and approximations to such an integral constitutes an integral part of statistical computing. In particular, the numerical values of the one-sided and two-sided probability integrals have been tabulated. For a historical development, methods for evaluating and approximating such probability integrals, and the one-sided and two-sided tables, see the recent book by Tong (1990, Chapter 8 and Appendix).

In this section we observe some useful results concerning bounds and inequalities for the multivariate normal probability integrals. We first observe that, by Fact 2.12, we have

FACT 4.1. *If* $\mathbf{X} = (X_1, \ldots, X_n)$ *has an* $\mathcal{N}_n(\boldsymbol{\mu}, \boldsymbol{\Sigma})$ *distribution such that*

$$\mu_1 = \cdots = \mu_n, \qquad \sigma_1^2 = \cdots = \sigma_n^2,$$
$$\rho_{ij} = \rho \in (-1/(n-1), 1) \quad \textit{for all } i \neq j,$$

then Theorems 3.1, 3.2, *and* 3.4 *apply. Consequently, the inequalities in* (3.3), (3.7), (3.11), *and* (3.14) *hold.*

As a special application of Theorem 3.1, we immediately obtain the following result.

THEOREM 4.2. *If* $\mathbf{X}$ *has an* $\mathcal{N}_n(\boldsymbol{\mu}, \boldsymbol{\Sigma})$ *distribution such that*

$$\sigma_1^2 = \cdots = \sigma_n^2 \quad \textit{and} \quad \rho_{ij} = \rho \in (-1/(n-1), 1) \quad \textit{for all } i \neq j,$$

then $P_{\boldsymbol{\mu}}[\bigcap_{i=1}^n \{X_i \leq a\}]$ *is a Schur-concave function of* $\boldsymbol{\mu}$ *for every fixed* $a \in \mathfrak{R}$.

Consequently, it is maximized when $\mu_1 = \cdots = \mu_n = \frac{1}{n}\sum_{i=1}^n \mu_i \equiv \overline{\mu}$ for fixed $\overline{\mu}$.

Theorem 4.1 illustrates how a probability content depends on the mean vector. In the following theorem we apply Theorem 3.4 to obtain a bound via the variances of the normal variables.

THEOREM 4.3. *If* $\mathbf{X}$ *has an* $\mathcal{N}_n(\boldsymbol{\mu}, \boldsymbol{\Sigma})$ *distribution such that*

$$\mu_1 = \cdots = \mu_n \quad \textit{and} \quad \rho_{ij} = \rho \in (-1/(n-1), 1),$$

then $P[\sum_{i=1}^n X_i^2 \leq \lambda]$ *is a Schur-concave function of* $(\sigma_1^{-2}, \ldots, \sigma_n^{-2})$ *for every fixed* $\lambda > 0$. *In particular, we have*

$$P\left[\sum_{i=1}^n X_i^2 \leq \lambda\right] \leq P\left[\sum_{i=1}^n Z_i^2 \leq \frac{\lambda}{n}\sum_{i=1}^n \sigma_i^{-2}\right], \tag{4.1}$$

where $(Z_1, \ldots, Z_n)$ *has a multivariate normal distribution with a common mean* μ, *a common variance one, and a common correlation coefficient* $\rho \in (-1/(n-1), 1)$.

Note that if $\mu = 0$ and $\rho = 0$, then (4.1) becomes

$$P\left[\sum_{i=1}^{n} X_i^2 \leq \lambda\right] \leq P\left[\chi^2(n) \leq \frac{\lambda}{n}\sum_{i=1}^{n} \sigma_i^{-2}\right],$$

where $\chi^2(n)$ is a chi-squared random variable with n degrees of freedom. In this special case, it is known that a better bound is possible. That result is known as the Okamato–Marshall–Olkin Inequality (see Marshall and Olkin (1979, p. 303)), and is stated below:

THEOREM 4.4. *If* $\mathbf{X}$ *has an* $\mathcal{N}_n(\mathbf{0}, \mathbf{\Sigma})$ *distribution when* $\mathbf{\Sigma}$ *is a diagonal matrix with diagonal elements* $\sigma_1^2, \ldots, \sigma_n^2$, *then, for every fixed* $\lambda > 0$, $P[\sum_{i=1}^{n} X_i^2 \leq \lambda]$ *is a Schur-concave function of* $(\log \sigma_1^2, \ldots, \log \sigma_n^2)$.

Consequently we have

$$P\left[\sum_{i=1}^{n} X_i^2 \leq \lambda\right] \leq P\left[\chi^2(n) \leq \lambda\left(\prod_{i=1}^{n} \sigma_i^2\right)^{-1/n}\right]. \tag{4.2}$$

The right-hand side of (4.2) yields a better bound because by the arithmetic mean-geometric mean inequality we have

$$\frac{1}{n}\sum_{i=1}^{n} \sigma_i^{-2} \geq \left(\prod_{i=1}^{n} \sigma_i^{-2}\right)^{1/n} = \left(\prod_{i=1}^{n} \sigma_i^2\right)^{-1/n}.$$

Another result that also involves the geometric mean, given by Das Gupta and Rattihalli (1984), yields a better bound than that in (3.7) for the special case when (i) $\boldsymbol{\mu} = \mathbf{0}$ and (ii) the random variables are independent:

THEOREM 4.5. *If* $\mathbf{X}$ *has an* $\mathcal{N}_n(\mathbf{0}, \mathbf{\Sigma})$ *such that*

$$\sigma_1^2 = \cdots = \sigma_n^2 \quad \textit{and} \quad \rho_{ij} = 0 \quad \textit{for all } i \neq j,$$

then for $a_i > 0$ $(i = 1, \ldots, n)$ *we have*

$$P\left[\bigcap_{i=1}^{n}\{|X_i| \leq a_i\}\right] \leq P\left[\bigcap_{i=1}^{n}\{|X_i| \leq \overline{a}^*\}\right], \tag{4.3}$$

where $\overline{a}^* = (\prod_{i=1}^{n} a_i)^{1/n}$.

Since $\overline{a} > \overline{a}^*$, the inequality in (4.3) yields a better result than that in (3.7) for this special case; but, of course, Theorem 4.5 does not apply when the common mean μ is not zero or the random variables $X_1, \ldots, X_n$ are not independent.

REMARK 4.6. Let $\mathbf{X} = (X_1, \ldots, X_n)$ have an $\mathcal{N}_n(\mathbf{0}, \mathbf{R})$ distribution with $\sigma_1^2 = \cdots = \sigma_n^2 = 1$ and $\mathbf{R} = (\rho_{ij})$ a correlation matrix. Let V have a chi-square distribution with $r > 2$ degrees of freedom such that $\mathbf{X}$ and V are independent. Then the n-dimensional random variable

$$\mathbf{t} = (t_1, \ldots, t_n), \qquad t_i = X_i/\sqrt{V/r} \qquad (i = 1, \ldots, n)$$

is called a multivariate t variable in the literature. It is known that the components of $\mathbf{t}$ have a common mean 0, a common variance $r/(r-2)$, and correlation matrix $\mathbf{R}$, and that the distribution of $\mathbf{t}$ has important applications in statistics (see e.g., Tong (1990, Chapter 9)). Furthermore, when $\rho_{ij} = \rho \in (-1/(n-1), 1)$ for all $i \neq j$, the equi-coordinate one-sided and two-sided probability integrals for this distribution have also been tabulated. It is easy to see that the joint density function of $\mathbf{t}$ is Schur-concave when the random variables are equally correlated. Thus the results in Theorems 3.1, 3.2, and 3.4 also apply to this distribution.

We conclude this section by considering two numerical examples:

EXAMPLE 4.7. Suppose that $\mathbf{X} = (X_1, \ldots, X_4)$ has an $\mathcal{N}_4(\boldsymbol{\mu}, \boldsymbol{\Sigma})$ distribution such that

$$\mu_1 = \cdots = \mu_4 = 0, \qquad \sigma_1^2 = \cdots = \sigma_4^2 = 1.$$

(a) If $\rho_{ij} = \frac{1}{2}$ for all $i \neq j$, then from Theorem 3.1 we have

$$\gamma \equiv P[X_1 \leq 1.90,\ X_2 \leq 2.00,\ X_3 \leq 2.08,\ X_4 \leq 2.02] \leq P\left[\bigcap_{i=1}^{4}\{X_i \leq 2.00\}\right].$$

Combining this result with the trivial lower bound, we have (from Tong (1990, p. 237, Table C-5))

$$P\left[\bigcap_{i=1}^{4}\{X_i \leq 1.90\}\right] = 0.91172 \leq \gamma \leq 0.92845 = P\left[\bigcap_{i=1}^{4}\{X_i \leq 2.00\}\right].$$

(b) If $\rho_{ij} \leq \frac{1}{2}$ for all $i \neq j$, then combining with Slepian's inequality we have

$$\begin{aligned} &P_{\Sigma}[X_1 \leq 1.90,\ X_2 \leq 2.00,\ X_3 \leq 2.08,\ X_4 \leq 2.02] \\ &\quad \leq P_{\rho_{ij}=1/2}[X_1 \leq 1.90,\ X_2 \leq 2.00,\ X_3 \leq 2.08,\ X_4 \leq 2.02] \\ &\quad \leq P_{\rho_{ij}=1/2}\left[\bigcap_{i=1}^{4}\{X_i \leq 2.00\}\right] = 0.92845. \quad \square \end{aligned}$$

EXAMPLE 4.8. Let $\mathbf{X} = (X_1, X_2, X_3)$ have an $\mathcal{N}_3(\boldsymbol{\mu}, \boldsymbol{\Sigma})$ distribution such that

$$\mu_1 = \mu_2 = \mu_3 = 2, \qquad \sigma_1^2 = \sigma_2^2 = \sigma_3^2 = 1,$$
$$\rho_{ij} = 0.6 \quad \text{for all } i \neq j.$$

Then from Theorem 3.3 we have

$$\begin{aligned}
P[0.85 \le X_1 \le 2.92,\ \ 1.10 \le X_2 \le 3.09,\ \ 1.05 \le X_3 \le 2.99] \\
\le P\left[\bigcap_{i=1}^{3}\{1.00 \le X_i \le 3.00\}\right] \\
= P\left[\bigcap_{i=1}^{3}\{|Y_i| \le 1.00\}\right] = 0.40365,
\end{aligned}$$

where (Y_1, Y_2, Y_3) has an $\mathcal{N}_3(\mathbf{0}, \mathbf{\Sigma})$ distribution and the numerical value of the upper bound is taken from Tong (1990, p. 252, Table D-6). □

References

T. W. Anderson (1955), *The integral of a symmetric unimodal function over a symmetric convex set and some probability inequalities*, Proc. Amer. Math. Soc. **6**, 170–176.

S. Das Gupta and R. N. Rattihalli (1984), *Probability content of a rectangle under normal distribution: Some inequalities*, Sankhyā, Ser. A, **46**, 454–457.

G. H. Hardy, J. E. Littlewood, and G. Pólya (1934, 1952), *Inequalities*, 1st ed., 2nd ed., Cambridge University Press, Cambridge, England.

S. Karlin and Y. Rinott (1983), *Comparison of measures, multivariate majorization, and applications to statistics*, Studies in econometrics, time series and multivariate analysis (S. Karlin, T. Amemiya and L. A. Goodman, eds.), Academic Press, New York, pp. 465–489.

A. W. Marshall and I. Olkin (1974), *Majorization in multivariate distributions*, Ann. Statist. **2**, 1189–1200.

——(1979), *Inequalities: Theory of Majorization and its Applications*, Academic Press, New York.

Y. L. Tong (1982), *Rectangular and elliptical probability inequalities for Schur-concave random variables*, Ann. Statist. **10**, 637–642.

——(1983), *Probability inequalities for n-dimensional rectangles via multivariate majorization*, Technical Report No. 189, Department of Statistics, Stanford University, Stanford, CA.

——(1988), *Some majorization inequalities in multivariate statistical analysis*, SIAM Review **30**, 602–622.

——(1989), *Probability inequalities for n-dimensional rectangles via multivariate majorization*. Contributions to Probability and Statistics: Essays in Honor of Ingram Olkin (L. J. Gleser, M. D. Perlman, S. J. Press, and A. R. Sampson, eds.), Springer-Verlag, New York and Berlin, 146–159.

——(1990), *The multivariate normal distribution*, Springer-Verlag, New York and Berlin.

School of Mathematics, Georgia Institute of Technology, Atlanta, Georgia 30332-0160

E-mail address: ma201yt@gitvmi.bitnet

Contemporary Mathematics
Volume **115**, 1991

A Gaussian Cubature Formula for the Computation of Generalized B-splines and its Application to Serial Correlation

V. K. KAISHEV

ABSTRACT. The generalized B-spline is defined as the probability density function of a linear combination of Dirichlet distributed random variables with arbitrary parameters.

In this paper, Gaussian cubature formulae with error bounds are given for the representation of generalized B-splines and integrals of them. Monte Carlo methods for the numerical evaluation of the latter are derived as well. Special cases of these formulae are shown to supply means for the computation of the distributions of some serial correlation coefficients.

1. Introduction

It has recently been established (see Ignatov and Kaishev (1985, 1989), Karlin, Micchelli, and Rinott (1986)) that the classical B-spline coincides with the probability density function of a linear combination of random variables having a joint Dirichlet distribution with all parameters integers. Such linear combinations have been studied by Bloch and Watson (1967) and Margolin (1977) in the case when one of the parameters of the Dirichlet distribution is assumed real, the rest integers. As noted in both works, the corresponding formulae for the distribution and density functions are "too cumbersome for practical use". These formulae also assume specific conditions for the coefficients of the linear combination. Different expressions involving divided differences of certain truncated power functions were recently obtained by Ignatov and Kaishev (1987c). They are free of any restrictions on the coefficients of the linear combination and seem to be numerically more appealing. A general divided difference expression for the density and

1980 *Mathematics Subject Classification* (1985 *Revision*). Primary 62G30; Secondary 60E05.

Key words and phrases. B-spline, Gaussian cubature, Dirichlet distribution, serial correlation coefficient.

Research supported by the Bulgarian Academy of Sciences Grant No. 100 1006, and in part by the Committee of Science and Education Grant No 935.

This paper is in final form and no version of it will be submitted for publication elsewhere.

the distribution function of a linear combination of Dirichlet distributed random variables with arbitrary parameters was given in Ignatov and Kaishev (1987b). It represents a divided difference of a certain multivariate integral function which has a complex form and is difficult to compute.

In §3 of this paper, Gauss–Jacobi cubature formulae for the density and the distribution function of a linear combination of Dirichlet distributed random variables with arbitrary parameters are derived. Such a density is interpreted as a generalized B-spline, i.e., one allowing noninteger multiplicities of the knots. As known, (see e.g., Missovskih (1985), Stroud and Secrest (1965)) quadrature rules of the Gaussian type possess highest degree of algebraic accuracy and in most cases are numerically quite accurate. Monte Carlo methods are derived in §3 as alternative means for the computation of generalized B-splines and integrals of them. In §4, it is shown how the results of §3 can be applied to express and compute unknown distributions of certain noncircular serial correlation coefficients. In the next section we give some necessary background material on divided differences, B-splines, Dirichlet distribution, and Gauss–Jacobi quadrature rules.

2. Preliminaries

Given the bi-infinite sequence of points $\{t_i\}$, $-\infty < i < \infty$ on the real line, i.e., $t_i \in R^1$ and a sufficiently smooth function $\varphi(u)$. Its nth order divided difference over the points $t_i, \dots, t_{i+n}$ is defined as

$$(2.1) \quad [t_i, \dots, t_{i+n}]_u \varphi(u) = \frac{[t_{i+1}, \dots, t_n]_u \varphi(u) - [t_i, \dots, t_{i+n-1}]_u \varphi(u)}{t_{i+n} - t_i},$$

provided $t_i \neq t_{i+n}$. If $t_i = t_{i+1} = \cdots = t_{i+n}$, then

$$(2.2) \qquad [t_i, \dots, t_{i+n}]_u \varphi(u) = D^n \varphi(t_i)/n!,$$

where $D^n \varphi(t_i)$ is the nth derivative of $\varphi(u)$ at $u = t_i$, $n \geq 0$, $(D^0 \varphi(t_i) := \varphi(t_i))$.

Curry and Schoenberg (1966) used divided differences to define polynomial B-splines. The B-spline $M(t; t_i, \dots, t_{i+n})$ of degree $n-1$ with knots $t_i, \dots, t_{i+n} \in R^1$ is the nth order divided difference of the truncated power function $\Phi(u) = n(u-t)_+^{n-1}$, i.e.,

$$(2.3) \qquad M(t; t_i, \dots, t_{i+n}) := [t_i, \dots, t_{i+n}]_u \Phi(u),$$

where $(z)_+ = \max\{0, z\}$.

Widely used for computations is the normalized B-spline:

$$(2.4) \qquad N_{i,n}(t) = (t_{i+n} - t_i) M(t; t_i, \dots, t_{i+n})/n,$$

with jth derivative $(j \geq 0)$ of which can be efficiently computed by means of the stable de Boor–Cox recurrence relation (see e.g. de Boor (1976)).

The integral of a B-spline can also be recurrently computed, using the formula (c.f. de Boor (1976)):

$$\int_{-\infty}^{t} M(x; t_i, \dots, t_{i+n})\, dx = \sum_{j=i}^{i+\tau} N_{j,n+1}(t), \qquad t \le t_{i+\tau+1}. \tag{2.5}$$

If some of the knots of the B-spline coincide, the notation $M\begin{pmatrix} t; t_0, \dots, t_l \\ g_0, \dots, g_l \end{pmatrix}$ will be adopted. The quantity g_i equals the number of repetitions of the knot t_i, $i = 0, \dots, l$ and is called multiplicity of the latter.

Let us note that the polynomial B-spline $M\begin{pmatrix} t; t_0, \dots, t_l \\ g_0, \dots, g_l \end{pmatrix}$ with multiple knots coincides with the density function of the linear combination

$$S = \sum_{i=0}^{l} \theta_i t_i, \tag{2.6}$$

where the joint distribution of $\theta_0, \dots, \theta_l$ is Dirichlet with parameters $g_0, \dots, g_l$ (cf Ignatov and Kaishev (1985), Karlin, Micchelli and Rinott (1986)). Recall that the random variables $\theta_0, \dots, \theta_l$ have the joint Dirichlet distribution $\mathscr{D}(g_0, \dots, g_l)$ with parameters

$$g_0 > 0,\ g_1 > 0, \dots, g_l > 0, \tag{2.7}$$

i.e., $((\theta_0, \dots, \theta_l) \in \mathscr{D}(g_0, \dots, g_l))$ if $\theta_0 = 1 - \theta_0 - \dots - \theta_l$ and the joint probability density of $\theta_1, \dots, \theta_l$ with respect to the Lebesgue measure on the simplex $S^l = \{(u_1, \dots, u_l) \colon u_i \ge 0,\ \sum_{i=1}^{l} u_i \le 1\}$, $u_0 = 1 - u_1 - \dots - u_l$ is

$$\frac{\Gamma(g_0 + \dots + g_l)}{\Gamma(g_0) \cdots \Gamma(g_l)} (1 - u_1 - u_2 - \dots - u_l)^{g_0 - 1} u_1^{g_1 - 1} \cdots u_l^{g_l - 1}.$$

($\Gamma(\cdot)$ is the well known Gamma function).

If the parameters $g_0, \dots, g_l$ are real as in (2.7) then the density of S in (2.6) can be viewed as a generalized B-spline, i.e., one allowing noninteger multiplicities of the knots as well. Generalized B-splines were independently introduced in Karlin, Micchelli, and Rinott (1986), and Ignatov and Kaishev (1987a). A representation of the generalized B-spline as a divided difference of a certain multivariate integral function was given in Ignatov and Kaishev (1987b). However, the corresponding expression is too heavy to allow for a direct numerical implementation. In the next section we derive Gauss–Jacobi cubature formulae which seem to provide better means for numerical dealing with generalized B-splines. Let us recall shortly some basic formulae related to quadrature rules of the Gauss–Jacobi type that we need in the next section.

The Gauss–Jacobi elementary quadrature formula relative to the interval $[-1, 1]$ and to the weight $(1-x)^\alpha (1+x)^\beta$ with $\alpha > -1$, $\beta > -1$ is given by

$$\int_{-1}^{1} (1-x)^\alpha (1+x)^\beta G(x)\, dx = \sum_{i=1}^{q} H_{q,i}^{(\alpha,\beta)} G(x_{q,i}^{(\alpha,\beta)}) + R(G), \tag{2.8}$$

where $R(G) = 0$ when $G(x)$ is a polynomial of degree $\leq 2q-1$; $x_{q,i}^{(\alpha,\beta)}$, $i = 1, \ldots, q$ are the zeros of the Jacobi polynomial $P_q^{(\alpha,\beta)}(x)$; $x_{q,1}^{(\alpha,\beta)} < \cdots < x_{q,q}^{(\alpha,\beta)}$; and

$$(2.9)\quad H_{q,i}^{(\alpha,\beta)} = \frac{2^{\alpha+\beta+1}\Gamma(\alpha+q+1)\Gamma(\beta+q+1)}{q!\Gamma(\alpha+\beta+q+1)[1-(x_{q,i}^{(\alpha,\beta)})^2]}\left[\frac{d}{dx}P_q^{(\alpha,\beta)}(x)\right]_{x=x_{q,i}^{(\alpha,\beta)}}^{-2},$$

$i = 1, \ldots, q$, are the Christoffel numbers. As known, numerically convenient recurrence formulae for the computation of Jacobi polynomials and their derivatives exist.

If the $2q$th derivative $G^{(2q)}(x)$ is bounded in $[-1, 1]$, then for the remainder term in (2.8) we have

$$(2.10)\qquad |R(G)| \leq C \sup_{-1\leq x\leq 1} |G^{(2q)}(x)|,$$

where

$$C = \frac{2^{\alpha+\beta+2q+1} q!\Gamma(\alpha+q+1)\Gamma(\beta+q+1)\Gamma(\alpha+\beta+q+1)}{(2q)!(\alpha+\beta+2q+1)[\Gamma(\alpha+\beta+2q+1)]^2}.$$

As shown by DeVore and Scott (1984), if $G^{(s)}(x)(1-x^2)^{s/2}$ is integrable on $[-1, 1]$ for all integers $s \leq 2q$ then

$$(2.11)\qquad |R(G)| \leq C_s q^{-s} \int_{-1}^{1} |G^{(s)}(x)|(1-x^2)^{s/2}\,dx,$$

where C_s is a constant independent of q and G.

Let us note that this bound is appropriate in cases when $G(x)$ has singularities at the end points -1, 1.

For a more detailed account on Gaussian quadrature, Jacobi polynomials, and related subjects we refer to the book by Ghizzetti and Ossicini (1970).

3. A Gauss–Jacobi cubature formula for generalized B-splines

Denote by $\hat{g}_i$ the integer part of g_i, $\overline{g}_i = g_i - \hat{g}_i$. Suppose, $g_0, \ldots, g_n$, $g_{n+1}, \ldots, g_{n+m}$ are such that $\overline{g}_i > 0$, $i = 0, \ldots, n$; $\overline{g}_i = 0$, $i = n+1, \ldots, n+m$, and let $l = \sum_{i=0}^{n+m} \hat{g}_i$, $(l \geq 2)$.

Consider the random variable $S = \sum_{i=0}^{n+m} t_i\theta_i$, where $(\theta_0, \ldots, \theta_{n+m}) \in \mathscr{D}(g_0, \ldots, g_{n+m})$ and $t_i \in R^1$, $i = 0, \ldots, n+m$. As shown in Ignatov and Kaishev (1987b) the probability density function of S $f_{\mathscr{S}}(x)$, i.e., the generalized B-spline $M_G(x)$ with knots $t_0, \ldots, t_{n+m}$ of multiplicities, correspondingly, $g_0, \ldots, g_{n+m}$ coincides with the divided difference

$$(3.1)\qquad \begin{bmatrix} t_0, \ldots, t_n, t_{n+1}, \ldots, t_{n+m} \\ \hat{g}_0, \ldots, \hat{g}_n, g_{n+1}, \ldots, g_{n+m} \end{bmatrix}_u H(u),$$

if $x \in [B]$ and is zero otherwise,

$$H(u) = \frac{\Gamma(g_0 + \cdots + g_{n+m})}{\Gamma(l-1)\Gamma(\overline{g}_0)\cdots\Gamma(\overline{g}_n)} \times \int_{\Delta_n} \left(u - x + \sum_{i=0}^{n}(t_i - u)y_i\right)_+^{l-2} y_0^{\overline{g}_0 - 1}\cdots y_n^{\overline{g}_n - 1}\, dy_0 \cdots dy_n,$$

$\Delta_n = \{(y_0, \dots, y_n) \colon 0 \le y_i,\ i = 0, \dots, n,\ y_0 + \cdots + y_n \le 1\}$ and B is the set of all $t_i's$ for which $\hat{g}_i \ge 1$, $[B]$ denotes the convex hull of B.

It can be easily verified that if $\overline{g}_i = 0$, $i = 0, \dots, n+m$ then the generalized B-spline in (3.1) coincides with the classical one, given by (2.3). It can also be seen from (3.1) that the generalized B-spline admits a representation as a divided difference of an elementary function only if one of the knots is of noninteger multiplicity, i.e., when $n = 0$. In this case it has been shown (cf Ignatov and Kaishev (1987b)) that if $t_0 < t_i$, $i = 1, \dots, m$, then

$$f_{\mathscr{S}}(x) = \begin{bmatrix} t_0, t_1, \dots, t_m \\ \hat{g}_0, g_1, \dots, g_m \end{bmatrix}_u^{H'(u)},$$

where $H'(u) = (l + \overline{g}_0 - 1)(u - t_0)^{-\overline{g}_0}(u - x)_+^{l + \overline{g}_0 - 2}$.

It is obvious however, that (3.1) cannot directly serve for computations. A cubature formula which suits better these purposes is given by the following

THEOREM 3.1. *The generalized B-spline $M_G(x)$ with knots $t_0, \dots, t_n$, $t_{n+1}, \dots, t_{n+m}$ of multiplicities, g_i, $i = 0, \dots, n+m$, i.e.,*

$$(3.2)\qquad M_G\begin{pmatrix} x; t_0, \dots, t_n, t_{n+1}, \dots, t_{n+m} \\ g_0, \dots, g_n, g_{n+1}, \dots, g_{n+m} \end{pmatrix} = D \sum_{i_0, \dots, i_n = 1}^{q} H_{q,i_0}^{(\alpha_0, \beta_0)} H_{q,i_1}^{(\alpha_1, \beta_1)} \cdots H_{q,i_n}^{(\alpha_n, \beta_n)} M\begin{pmatrix} \gamma; t_0, \dots, t_n, t_{n+1}, \dots, t_{n+m} \\ \hat{g}_0, \dots, \hat{g}_n, g_{n+1}, \dots, g_{n+m} \end{pmatrix} + R,$$

if $x \in [B]$ and is zero otherwise. Here B is the set of all $t_i's$ for which $\hat{g}_i \ge 1$, $[B]$ denotes the convex hull of B,

$$\gamma = \frac{2^{n+1}x - \sum_{r=0}^{n} 2^{n-r} t_r (1 + x_{q,i_r}^{(\alpha_r, \beta_r)}) \prod_{k=0}^{r-1}(1 - x_{q,i_k}^{(\alpha_k, \beta_k)})}{\prod_{j=0}^{n}(1 - x_{q,i_j}^{(\alpha_j, \beta_j)})},$$

$$\alpha_i = l - 2 + \sum_{j=i+1}^{n} \overline{g}_j, \quad \beta_i = \overline{g}_i - 1, \qquad i = 0, \dots, n;$$

$H_{q,j_i}^{(\alpha_i, \beta_i)}$, $j_i = 1, \dots, q$, $i = 0, \dots, n$ are the Christoffel numbers, $x_{q,j_i}^{(\alpha_i, \beta_i)}$ are the zeros of the Jacobi polynomial: $P_q^{(\alpha_i, \beta_i)}(\cdot)$, $x_{q,j_i}^{(\alpha_i, \beta_i)} < \cdots < x_{q,j_i}^{(\alpha_i, \beta_i)}$,

$j_i = 1, \dots, q$, $i = 0, \dots, n$;

$$M\begin{pmatrix} \gamma; t_0, \dots, t_n, t_{n+1}, \dots, t_{n+m} \\ \hat{g}_0, \dots, \hat{g}_n, g_{n+1}, \dots, g_{n+m} \end{pmatrix}$$

is the polynomial B-spline of order $l-1$;

$$D = \frac{\Gamma\left(\sum_{i=0}^{n+m} g_i\right)}{\Gamma(l)\prod_{i=0}^{n}\Gamma(\overline{g}_i)2^{(n+1)(l-2)+g}}, \qquad g = \sum_{i=0}^{n}(i+1)\overline{g}_i,$$

R *is a remainder term and if* $t_0 < t_i$, $i = 1, \dots, n+m$ *and* $q = \text{integer}\{(l-g-1)/2\}$, $g = \max\{\hat{g}_0, \dots, \hat{g}_n, \dots, g_{n+1}, \dots, g_{n+m}\}$ *it has a bound*

(3.3)

$$|R| \le D' \sum_{p=0}^{n} C_p \sum_{i_0, \dots, i_{p-1}=1}^{q} H_{q,i_0}^{(\alpha_0,\beta_0)} \cdots H_{q,i_{p-1}}^{(\alpha_{p-1},\beta_{p-1})} L(x, i_0, \dots, i_{p-1})$$

$$\times \sup_{-1 \le z_p \le 1} \left| \frac{1}{(1-z_p)^{2q+1}} \int_{\Delta_{n-(p+1)}} M^{(2q)}(\gamma_p) z_{p+1}^{\overline{g}_{p+1}-1} \cdots \right.$$

$$\left. \times z_n^{\overline{g}_n} \left(1 - \sum_{i=p+1}^{n} z_i\right)^{l-3} dz_{p+1} \cdots dz_n \right|,$$

where

$$L(x, i_0, \dots, i_{p-1}) = |\mathscr{W}/\mathscr{T} - 2t_p|, \qquad \mathscr{T} = \prod_{k=0}^{p-1}(1 - x_{q,i_k}^{(\alpha_i,\beta_k)}),$$

$$\mathscr{W} = 2^{p+1}x - \sum_{r=0}^{p-1} 2^{p-r} t_r (1 + x_{q,i_r}^{(\alpha_r,\beta_r)}) \prod_{j=0}^{r-1}(1 - x_{q,i_j}^{(\alpha_j,\beta_j)}),$$

$$D' = D2^{(n+1)(l-2)+g},$$

$$C_p = \frac{2^{s(p)} q! \Gamma(\alpha_p + q + 1)\Gamma(\beta_p + q + 1)\Gamma(\alpha_p + \beta_p + q + 1)}{(\alpha_p + \beta_p + 2q + 1)[\Gamma(\alpha_p + \beta_p + 2q + 1)]^2},$$

$$s(p) = \alpha_p + \beta_p + 2q + 1 - (p+1)(l-2) - \sum_{i=0}^{p}(i+1)q_i - (p+1)\sum_{i=p+1}^{n} g_i,$$

$$\Delta_{n-r} = \{(z_r, \dots, z_n) : 0 \le z_i, i = r, \dots, n; z_r + \cdots + z_n \le 1\},$$

$$r = 1, \dots, n+1$$

$M^{(2q)}(\gamma_p)$ *is the* $(2q)$*th derivative of the polynomial B-spline*

$$M\begin{pmatrix} \gamma_p; t_0, \dots, t_n, t_{n+1}, \dots, t_{n+m} \\ \hat{g}_0, \dots, \hat{g}_n, g_{n+1}, \dots, g_{n+m} \end{pmatrix}$$

$$\gamma_p = \frac{\mathscr{W} - [t_p(1+z_p) + (1-z_p)\sum_{j=p+1}^{n} t_j z_j]\mathscr{T}}{(1-z_p)(1-z_{p+1} - \cdots - z_n)\mathscr{T}}.$$

REMARK. All sums in (3.2) and (3.3) should be considered vanishing when the corresponding lower limit of summation exceeds the upper one. Respectively, products in a similar case are assumed equal to unity. When $p = n$, integration in (3.3) is not carried out and the integral should be replaced by $M^{(2q)}(\gamma_p)$.

PROOF. Consider representation (3.1) of the generalized B-spline. Let us multiply and divide the integrand of $H(u)$ by $(l-1)(1-\sum_{i=0}^{n} y_i)^{l-2}$. After moving the divided difference operator under the integral, slightly rewriting the integrand function and having in mind that

$$\begin{aligned}(3.4)\qquad &\begin{bmatrix} t_0, \dots, t_n, t_{n+1}, \dots, t_{n+m} \\ \hat{g}_0, \dots, \hat{g}_n, g_{n+1}, \dots, g_{n+m} \end{bmatrix}_u (l-1)(u-\tilde{\tilde{\gamma}}_0)_+^{l-2} \\ &= M\begin{pmatrix} \tilde{\tilde{\gamma}}_0; t_0, \dots, t_n, t_{n+1}, \dots, t_{n+m} \\ \hat{g}_0, \dots, \hat{g}_n, g_{n+1}, \dots, g_{n+m} \end{pmatrix},\end{aligned}$$

where $\tilde{\tilde{\gamma}}_0 = (x - \sum_{i=0}^{n} t_i y_i)/(1-\sum_{i=0}^{n} y_i)$ we get

$$\begin{aligned}(3.5)\qquad M_G(x) = D' \int_{\Delta_n} & M(\tilde{\tilde{\gamma}}_0; t_0, \dots, t_{n+m}) y_0^{\overline{g}_0 - 1} \cdots y_n^{\overline{g}_n - 1} \\ &\times \left(1 - \sum_{i=0}^{n} y_i\right)^{l-2} dy_0 \cdots dy_n,\end{aligned}$$

where for short, we have denoted the polynomial B-spline on the right-hand side of (3.4) by $M(\tilde{\tilde{\gamma}}_0; t_0, \dots, t_{n+m})$ and $D' = \Gamma(\sum_{i=0}^{n+m} g_i)/\Gamma(l)\prod_{i=0}^{n}\Gamma(\overline{g}_i)$.

Denote by $\Delta_{n-1}(a)$ the cross-section of Δ_n with the hyperplane $y_0 = a$. For $0 \le a < 1$ this would be the $(n-1)$-dimensional simplex: $\{(y_1, \dots, y_n):$ $y_1 \ge 0, \dots, y_n \ge 0,\ y_1 + \dots + y_n \le 1 - a\}$. From (3.5) we have

(3.6)

$$\begin{aligned}M_G(x) = D' \int_0^1 y_0^{\overline{g}_0 - 1}\, dy_0 \int_{\Delta_{n-1}(y_0)} & M(\tilde{\tilde{\gamma}}_0; t_0, \dots, t_{n+m}) y_1^{\overline{g}_1 - 1} \cdots y_n^{\overline{g}_n - 1} \\ &\times \left(1 - \sum_{i=0}^{n} y_i\right)^{l-2} dy_1 \cdots dy_n.\end{aligned}$$

Let us change variables of integration $y_i = (1-y_0)z_i$, $i = 1, \dots, n$ on the right-hand side of (3.6). Then, the simplex $\Delta_{n-1}(y_0)$ is transformed into the standard simplex $\Delta_{n-1} = \{(z_1, \dots, z_n): z_i \ge 0,\ i = 1, \dots, n,\ z_1 + \dots + z_n \le 1\}$ and (3.6) can be rewritten as

(3.7)

$$M_G(x) = D' \int_0^1 y_0^{\bar{g}_0 - 1}(1 - y_0)^{l-2+\bar{g}_1+\dots+\bar{g}_n}\, dy_0$$

$$\times \int_{\Delta_{n-1}} M(\tilde{\gamma}_0; t_0, \dots, t_{n+m}) z_1^{\bar{g}_1 - 1} \cdots z_n^{\bar{g}_n - 1} \left(1 - \sum_{i=1}^n z_i\right)^{l-2} dz_1 \cdots dz_n,$$

$$\text{where } \tilde{\gamma}_0 = \frac{x - t_0 y_0 - (1 - y_0)\sum_{i=1}^n t_i z_i}{(1 - y_0)\left(1 - \sum_{i=1}^n z_i\right)}.$$

Let $y_0 = (1 + z_0)/2$. Then, rewriting (3.7) we get

(3.8)

$$M_G(x) = D' D_0 \int_{-1}^1 (1 + z_0)^{\bar{g}_0 - 1}(1 - z_0)^{l-2+\bar{g}_1+\dots+\bar{g}_n}\, dz_0$$

$$\times \int_{\Delta_{n-1}} M(\gamma_0; t_0, \dots, t_{n+m}) z_1^{\bar{g}_1 - 1} \cdots z_n^{\bar{g}_n - 1} \left(1 - \sum_{i=1}^n z_i\right)^{l-2} dz_1 \cdots dz_n,$$

$$\text{where } \gamma_0 = \frac{2x - t_0(1 + z_0) - (1 - z_0)\sum_{i=1}^n t_i z_i}{(1 - z_0)\left(1 - \sum_{i=1}^n z_i\right)},\quad D_0 = 1/2^{l-2+\bar{g}_0+\dots+\bar{g}_n}.$$

Denote by $G_0(z_0)$ the last integral on the right-hand side of (3.8). It can be easily checked that $G_0(z_0) \in C[-1, 1]$ and we can apply the Gauss–Jacobi quadrature formula (2.8) to (3.8) and obtain

$$M_G(x) = D' D_0 \left\{ \sum_{i_0=1}^q H_{q, i_0}^{(\alpha_0, \beta_0)} G_0(x_{q, i_0}^{(\alpha_0, \beta_0)}) + R_0(G_0(z_0)) \right\}, \tag{3.9}$$

where $x_{q,1}^{(\alpha_0, \beta_0)} < \dots < x_{q,q}^{(\alpha_0, \beta_0)}$ are the zeros of the Jacobi polynomial $P_q^{(\alpha_0, \beta_0)}$ of degree q with $\alpha_0 = l-2+\bar{g}_1+\dots+\bar{g}_n$, $\beta_0 = \bar{g}_0 - 1$, $H_{q,1}^{(\alpha_0, \beta_0)}, \dots, H_{q,q}^{(\alpha_0, \beta_0)}$ are the corresponding Christoffel numbers.

The sth derivative of $G_0(z_0)$ is

(3.10)

$$G_0(z_0)^{(s)} = \frac{2s!(x - t_0)}{(1 - z_0)^{s+1}} \int_{\Delta_{n-1}} M^{(s)}(\gamma_0; t_0, \dots, t_{n+m}) z_1^{\bar{g}_1 - 1} \cdots z_n^{\bar{g}_n - 1}$$

$$\times \left(1 - \sum_{i=1}^n z_i\right)^{l-3} dz_1 \cdots dz_n.$$

Put $2q - 1 = l - 2 - g$, $g = \max\{\hat{g}_0, \dots, \hat{g}_n, \dots, g_{n+1}, \dots, g_{n+m}\}$. From (3.10), it can be seen that $G_0(z_0) \in C^{2q-1}[-1, 1)$, since $M^{(l-2-g)}(\gamma_0; t_0, \dots, t_{n+m})$ is continuous on $[-1, 1)$. The latter vanishes at the end point $z_0 = 1$ if $x \ne t_0$. However, if $x = t_0$ the $(2q - 1)$th derivative of $G_0(z_0)$ has a singularity when $z_0 = 1$. If we assume $t_0 < t_i$, $i = 1, \dots, n+m$, then $M_G(x)$ vanishes at $x = t_0$ due to (3.1) and no bound is of interest in this case. Hence $G_0(z_0)$ meets the smoothness requirements in (2.10) and $q = \text{integer}\{l - 1 - g)/2\}$.

Taking into account (2.10) and (3.10), after some trivial manipulations we get the bound

(3.11)

$$|R_0(G_0(z_0))| \le C_0' 2(l-1-g)!(x-t_0) \times \sup_{-1 \le z_0 \le 1} \left| \frac{1}{(1-z_0)^{l-g}} \int_{\Delta_{n-1}} M^{(l-1-g)}(\gamma_0; t_0, \dots, t_{n+m}) z_1^{\overline{g}_1 - 1} \cdots z_n^{\overline{g}_n - 1} \times \left(1 - \sum_{i=1}^n z_i\right)^{l-3} dz_1 \cdots dz_n \right|,$$

where C_0' is the same as C in (2.10) with α and β replaced correspondingly by α_0 and β_0.

Let us further consider $G_0(x_{q,i_0}^{(\alpha_0,\beta_0)})$ and make change of variables $z_i = (1-z_1)w_i$, $i = 2, 3, \dots, n$. Assume z_1 in the definition of the simplex Δ_{n-1} is fixed. Then, the latter will transform into $\Delta_{n-2} = \{(w_2, \dots, w_n): w_i \ge 0,\ i = 2, \dots, n,\ w_2 + \cdots + w_n \le 1\}$ and we have

$$G_0(x_{q,i_0}^{(\alpha_0,\beta_0)}) = \int_0^1 z_1^{\overline{g}_1 - 1}(1-z_1)^{l-2+\overline{g}_2+\cdots\overline{g}_n} dz_1 \times \int_{\Delta_{n-2}} M(\tilde{\gamma}_1; t_0, \dots, t_{n+m}) w_2^{\overline{g}_2 - 1} \cdots w_n^{\overline{g}_n - 1} \times \left(1 - \sum_{i=2}^n w_i\right)^{l-2} dw_2 \cdots dw_n, \tag{3.12}$$

$$\text{where } \tilde{\gamma}_1 = \frac{2x - t_0(1 + x_{q,i_0}^{(\alpha_0,\beta_0)}) - (1 - x_{q,i_0}^{(\alpha_0,\beta_0)})\left[z_1 t_1 + (1-z_1)\sum_{i=2}^n t_i w_i\right]}{(1 - x_{q,i_0}^{(\alpha_0,\beta_0)})(1-z_1)\left(1 - \sum_{i=2}^n w_i\right)}$$

Put $z_1 = (1+w_1)/2$. Then rewriting (3.12) we get

(3.13)

$$G_0(x_{q,i_0}^{(\alpha_0,\beta_0)}) = D_1 \int_{-1}^1 (1+w_1)^{\overline{g}_1 - 1}(1-w_1)^{l-2+\overline{g}_2+\cdots+\overline{g}_n} dw_1 \times \int_{\Delta_{n-2}} M(\gamma_1; t_0, \dots, t_{n+m}) w_2^{\overline{g}_2 - 1} \cdots w_n^{\overline{g}_n - 1} \times \left(1 - \sum_{i=2}^n w_i\right)^{l-2} dw_2 \cdots dw_n,$$

where $D_1 = 1/2^{l-2+\overline{g}_1+\cdots+\overline{g}_n}$,

$$\gamma_1 = \frac{4x - 2t_0(1 + x_{q,i_0}^{(\alpha_0,\beta_0)}) - (1 - x_{q,i_0}^{(\alpha_0,\beta_0)})\,[(1 + w_1)t_1 + (1 - w_1)\sum_{i=2}^{n} t_i w_i]}{(1 - x_{q,i_0}^{(\alpha_0,\beta_0)})(1 - w_1)\,(1 - \sum_{i=2}^{n} w_i)}$$

Denote by $G_1(x_{q,i_0}^{(\alpha_0,\beta_0)}, x_{q,i_1}^{(\alpha_1,\beta_1)})$ the last integral in (3.13) evaluated at $w_1 = x_{q,i_1}^{(\alpha_1,\beta_1)}$. Its sth derivative with respect to w_1 is

(3.14)

$$G_1^{(s)}(x_{q,i_0}^{(\alpha_0,\beta_0)}, x_{q,i_1}^{(\alpha_1,\beta_1)}) = L(x, i_0)\frac{s!}{(1 - w_1)^{s+1}}\int_{\Delta_{n-2}} M^{(s)}(\gamma_1)z_2^{\overline{g}_2-1}\cdots z_n^{\overline{g}_n} \times \left(1 - \sum_{i=2}^{n} z_i\right)^{l-3} dz_2\cdots dz_n,$$

where $L(x, i_0) = [4x - 2t_0(1 + x_{q,i_0}^{(\alpha_0,\beta_0)})]/(1 - x_{q,i_0}^{(\alpha_0,\beta_0)}) - 2t_1$.

Substituting $G_0(x_{q,i_0}^{(\alpha_0,\beta_0)})$ from (3.13) in (3.9) and applying again (2.8) we get

(3.15)

$$M_G(x) = D'D_0D_1\sum_{i_0,i_1=1}^{q} H_{q,i_0}^{(\alpha_0,\beta_0)}H_{q,i_1}^{(\alpha_1,\beta_1)}G_1(x_{q,i_0}^{(\alpha_0,\beta_0)}, x_{q,i_1}^{(\alpha_1,\beta_1)}) + D'D_0D_1\sum_{i_0=1}^{q} H_{q,i_0}^{(\alpha_0,\beta_0)}R_1(G_1(x_{q,i_0}^{(\alpha_0,\beta_0)}, w_1)) + D'D_0R_0(G_0),$$

where $x_{q,1}^{(\alpha_1,\beta_1)} < \cdots < x_{q,q}^{(\alpha_1,\beta_1)}$ are the zeros of the Jacobi polynomial $P_q^{(\alpha_1,\beta_1)}$ with $\alpha_1 = l - 2 + \overline{g}_2 + \cdots + \overline{g}_n$, $\beta_1 = \overline{g}_1 - 1$, $H_{q,1}^{(\alpha_1,\beta_1)}, \ldots, H_{q,q}^{(\alpha_1,\beta_1)}$ are the Christoffel numbers.

Using (3.14), and the fact that $M^{(s)}(\gamma_1)$ vanishes when $w_1 = 1$ it can be seen that $G_1(x_{q,i_0}^{(\alpha_0,\beta_0)}, x_{q,i_1}^{(\alpha_1,\beta_1)}) \in C^{2q-1}[-1, 1]$ and again, we can apply (2.10) to the remainder $R_1(\cdot)$ and establish the bound

(3.16)

$$|R_1(G_1(x_{q,i_0}^{(\alpha_0,\beta_0)}, w_1))| \le C_1'|L(x, i_0)| \sup_{-1\le w_1\le 1}\left|\frac{1}{(1 - w_1)^{2q+1}}\int_{\Delta_{n-2}} M^{(2q)}(\gamma_1)z_2^{\overline{g}_2-1}\cdots z_n^{\overline{g}_n} \times \left(1 - \sum_{i=2}^{n} z_i\right)^{l-3} dz_2\cdots dz_n\right|.$$

Further, $G_1(x_{q,i_0}^{(\alpha_0,\beta_0)}, x_{q,i_1}^{(\alpha_1,\beta_1)})$ can be handled analogously, applying again similar transformation of the variables of integration to obtain

(3.17)

$$\begin{aligned}M_G(x) = {} & D'D_0D_1D_2 \sum_{i_0,i_1,i_2=1}^{q} H_{q,i_0}^{(\alpha_0,\beta_0)} H_{q,i_1}^{(\alpha_1,\beta_1)} H_{q,i_2}^{(\alpha_2,\beta_2)} \\ & \times G_2(x_{q,i_0}^{(\alpha_0,\beta_0)}, x_{q,i_1}^{(\alpha_1,\beta_1)}, x_{q,i_2}^{(\alpha_2,\beta_2)}) \\ & + D'D_0D_1D_2 \sum_{i_0,i_1=1}^{q} H_{q,i_0}^{(\alpha_0,\beta_0)} H_{q,i_1}^{(\alpha_1,\beta_1)} R_2(G_2(x_{q,i_0}^{(\alpha_0,\beta_0)}, x_{q,i_1}^{(\alpha_1,\beta_1)}, z_2) \\ & + D'D_0D_1 \sum_{i_0=1}^{q} H_{q,i_0}^{(\alpha_0,\beta_0)} R_1(G_1(x_{q,i_0}^{(\alpha_0,\beta_0)}, z_1)) + D'D_0R_0(G_0),\end{aligned}$$

where $D_2 = 1/2^{l-2+\overline{g}_2+\cdots+\overline{g}_n}$.

Repeated application of the arguments used above yields on the vth stage

(3.18)

$$\begin{aligned}M_G(x) = {} & D' \prod_{k=0}^{v} D_k \sum_{i_0,\ldots,i_v=1}^{q} H_{q,i_0}^{(\alpha_0,\beta_0)} \cdots H_{q,i_v}^{(\alpha_v,\beta_v)} G_v(x_{q,i_0}^{(\alpha_0,\beta_0)}, \ldots, x_{q,i_v}^{(\alpha_v,\beta_v)}) \\ & + D' \sum_{p=0}^{v} \prod_{k=0}^{p} D_k \sum_{i_0,\ldots,i_{p-1}=1}^{q} H_{q,i_0}^{(\alpha_0,\beta_0)} \cdots H_{q,i_{p-1}}^{(\alpha_{p-1},\beta_{p-1})} \\ & \qquad \times R_p(G_p x_{q,i_0}^{(\alpha_0,\beta_0)}, \ldots, x_{q,i_{p-1}}^{(\alpha_{p-1},\beta_{p-1})}, z_p)),\end{aligned}$$

where $D_k = 1/2^{l-2+\overline{g}_k+\cdots+\overline{g}_n}$,

$$\begin{aligned}& G_p(x_{q,i_0}^{(\alpha_0,\beta_0)}, \ldots, x_{q,i_{p-1}}^{(\alpha_{p-1},\beta_{p-1})}, z_p) \\ & \quad = \int_{\Delta_{n-(p+1)}} M(\gamma_p; t_0, \ldots, t_{n+m}) z_{p+1}^{\overline{g}_{p+1}-1} \cdots z_n^{\overline{g}_n-1} \\ & \quad \times \left(1 - \sum_{i=p+1}^{n} z_i\right)^{l-2} dz_{p+1} \cdots dz_n,\end{aligned}$$

γ_p being given in (3.3), $p = 0, \ldots, v$, $\alpha_i = l-2+\overline{g}_{i+1}+\cdots+\overline{g}_n$, $\beta_i = \overline{g}_i - 1$, $i = 0, \ldots, v$, $H_{q,i}^{(\alpha_0,\beta_0)}, \ldots, H_{q,i}^{(\alpha_v,\beta_v)}$, $i = 1, \ldots, q$ the corresponding

Christoffel numbers. Note that the (s)th derivative with respect to z_p

(3.19)

$$G_p^{(s)}(x_{q,i_0}^{(\alpha_0,\beta_0)},\dots,x_{q,i_{p-1}}^{(\alpha_{p-1},\beta_{p-1})}z_p) = L(x,i_0,\dots,i_{p-1})\frac{s!}{(1-z_p)^{s+1}} \times \int_{\Delta_{n-(p+1)}} M^{(s)}(\gamma_p) z_{p+1}^{\overline{g}_{p+1}-1}\cdots z_n^{\overline{g}_n-1}\left(1-\sum_{i=p+1}^{n} z_i\right)^{l-3} dz_{p+1}\cdots dz_n,$$

where $L(x,i_0,\dots,i_{p-1})$ is given in (3.3), $p=0,\dots,v$. From (3.19), we have $G_p(x_{q,i_0}^{(\alpha_0,\beta_0)},\dots,x_{q,i_{p-1}}^{(\alpha_{p-1},\beta_{p-1})},z_p)\in C^{2q-1}[-1,1]$, $p=0,\dots,v$ by the same arguments as in the case $p=1$.

Taking into account (2.10) and (3.19) for the remainder $R_p(\cdot)$ we get

(3.20)

$$\begin{aligned}|R_p(G_p(x_{q,i_0}^{(\alpha_0,\beta_0)},\dots,x_{q,i_{p-1}}^{(\alpha_{p-1},\beta_{p-1})},z_p))| &\le C_p' L(x,i_0,\dots,i_{p-1}) \\ &\times \sup_{-1\le z_p\le 1}\left|\frac{(2q)!}{(1-z_p)^{2q+1}}\int_{\Delta_{n-(p+1)}} M^{(2q)}(\gamma_p) z_{p+1}^{\overline{g}_{p+1}-1}\cdots z_n^{\overline{g}_n-1}\right. \\ &\left.\times\left(1-\sum_{i=p+1}^{n} z_i\right)^{l-3} dz_{p+1}\cdots dz_n\right|,\end{aligned}$$

where C_p' is the same as C in (2.10) with α and β replaced correspondingly by α_p and β_p, $p=0,\dots,v$.

Combining (3.18) with (3.20) for $v=n$ and noting that in this case $G_v(x_{q,i_0}^{(\alpha_0,\beta_0)},\dots,x_{q,i_v}^{(\alpha_v,\beta_v)})$ from (3.18) reads as $M(\gamma;t_0,\dots,t_{n+m})$, γ being given in (3.2) we get the assertion of the Theorem. □

Remark. Let us note that a different bound for the remainder term in (3.1) can be obtained in the course of the proof if we apply (2.11) with $s=2q$ to every $R_p(G_p)$, $p=1,\dots,n$. Thus

$$\begin{aligned}|R_p(G_p)| &\le C_{2q} q^{-2q}(2q)!|L(x,i_0,\dots,i_{p-1})| \\ &\times\int_{-1}^{1}\left|\frac{(1+z_p)^q}{(1-z_p)^{q+1}}\int_{\Delta_{n-(p+1)}} M^{(2q)}(\gamma_p) z_{p+1}^{\overline{g}_{p+1}-1}\cdots z_n^{\overline{g}_n-1}\right. \\ &\left.\times\left(1-\sum_{i=p+1}^{n} z_i\right)^{l-3} dz_{p+1}\cdots dz_n\right| dz_p\end{aligned}$$

and combining the latter with the second term in the sum on the right-hand side of (3.18) we get the bound

(3.21)

$$|R| \le Q \sum_{p=0}^{v} \prod_{k=0}^{p} D_k \sum_{i_0, \dots, i_{p-1}=1}^{q} H_{q, i_0}^{(\alpha_0, \beta_0)} \cdots H_{q, i_{p-1}}^{(\alpha_{p-1}, \beta_{p-1})}$$

$$\times \Bigg| L(x, i_0, \dots, i_{p-1}) \int_{-1}^{1} \frac{(1+z_p)^q}{(1-z_p)^{q+1}}$$

$$\times \int_{\Delta_{n-(p+1)}} M^{(2q)}(\gamma_p) z_{p+1}^{\overline{g}_{p+1}-1} \cdots z_n^{\overline{g}_n - 1} \left(1 - \sum_{i=p+1}^{n} z_i\right)^{l-3} dz_{p+1} \cdots dz_n \, dz_p \Bigg| ,$$

where $Q = D' C_{2q} q^{-2q} (2q)!$.

Note that for this bound there is no need to impose the condition $t_0 < t_i$, $i = 1, \dots, n+m$, which in the case of (3.3) is required to make singularity in the sth derivative of $G_0(z_0)$ (cf 3.10) at $z_0 = 1$ appear at the left most point $x = t_0$.

It can be shown, that a cubature formula, analogous to (3.2) holds for the integral of the generalized B-spline $M_G(x)$. Clearly, we can refer to this integral as the probability distribution function of the random variable S. To give the formula let us introduce the set of knots

(3.22)

$$\begin{aligned}
&\cdots \le \lambda_0 \le \lambda_1 \le \cdots \le \lambda_{l-1} < \lambda_l = \lambda_{l+1} = \cdots = \lambda_{l+\hat{g}_0-1} < \lambda_{l+\hat{g}_0} \\
&= \lambda_{l+\hat{g}_0+1} = \cdots = \lambda_{l+\hat{g}_0+\hat{g}_1-1} < \cdots < \lambda_{l+\hat{g}_0+\cdots+\hat{g}_n} = \lambda_{l+\hat{g}_0+\cdots+\hat{g}_n+1} = \cdots \\
&= \lambda_{l+\hat{g}_0+\cdots+\hat{g}_n+g_{n+1}-1} < \cdots < \lambda_{l+\hat{g}_0+\cdots+\hat{g}_n+g_{n+1}+\cdots+g_{n+m-1}} \\
&= \lambda_{l+\hat{g}_0+\cdots+\hat{g}_n+g_{n+1}+\cdots+g_{n+m-1}+1} = \cdots \\
&= \lambda_{2l-1} < \lambda_{2l} \le \lambda_{2l+1} \le \cdots \le \lambda_{3l-2} \le \lambda_{3l-1} \le \cdots
\end{aligned}$$

in which $\lambda_l = t_0$, $\lambda_{l+\hat{g}_0} = t_1$, $\lambda_{l+\hat{g}_0+\hat{g}_1} = t_2, \dots, \lambda_{l+\hat{g}_0+\cdots+\hat{g}_{n-1}} = t_n, \dots,$ $\lambda_{l+\hat{g}_0+\cdots+\hat{g}_n+g_{n+1}+\cdots+g_{n+m-1}} = t_{n+m}$, the rest of the knots being arbitrary. Let us also introduce the basis of normalized B-splines $N_{i,l}(x)$, $i = 1, \dots, 2l-2$ (cf 2.4) defined on the set (3.22). Then we have

THEOREM 3.2. *The probability distribution function $F_{\mathscr{S}}(x)$ of the random variable S is*

$$F_{\mathscr{S}}(x) = D'' \sum_{i_0, \dots, i_n=1}^{q} H_{q, i_0}^{(\alpha_0', \beta_0)} \cdots H_{q, i_n}^{(\alpha_n', \beta_n)} \sum_{j=l}^{l+\tau} N_{j,l}(\gamma) + R, \tag{3.23}$$

where $D'' = D/2^{(n+1)}$, $\alpha_i' = \alpha_i + 1$, $i = 0, \dots, n$ and α_i, β_i, D, and γ are defined in (3.2), τ is such an integer for which $\gamma \le \lambda_{l+\tau+1}$, and $H_{q, i_j}^{(\alpha_j', \beta_j)}$, $j = 0, \dots, n$ are the Christoffel numbers.

PROOF. We have

$$F_{\mathcal{S}}(x) = \int_{-\infty}^{x} f_{\mathcal{S}}(t)\,dt = \int_{-\infty}^{x} M_G(t)\,dt. \tag{3.24}$$

Substituting $M_G(t)$ from (3.5) in (3.24) and changing the sequence of integration we obtain

$$F_{\mathcal{S}}(x) = D' \int_{\Delta_n} \int_{-\infty}^{x} M(\tilde{\tilde{y}}_0; t_0, \dots, t_{n+m})\,dt\, y_0^{\bar{g}_0 - 1} \cdots y_n^{\bar{g}_n - 1} \times \left(1 - \sum_{i=0}^{n} y_i\right)^{l-2} dy_0 \cdots dy_n, \tag{3.25}$$

where $\tilde{\tilde{y}}_0 = (t - \sum_{i=0}^{n} t_i y_i)/(1 - \sum_{i=0}^{n} y_i)$. Changing variable of integration $t = (1 - \sum_{i=0}^{n} y_i)\tilde{\tilde{y}}_0 + \sum_{i=0}^{n} t_i y_i$ in (3.25) from (2.5) we have
(3.26)

$$F_{\mathcal{S}}(x) = D' \int_{\Delta_n} \sum_{j=l}^{l+\tau} N_{j,l}(\tilde{\tilde{y}}_0) y_0^{\bar{g}_0 - 1} \cdots y_n^{\bar{g}_n - 1} \left(1 - \sum_{i=0}^{n} y_i\right)^{l-1} dy_0 \cdots dy_n$$

where, obviously, the variable t in the expression for $\tilde{\tilde{y}}_0$ should be replaced by x, the normalized B-splines $N_{j,l}(\tilde{\tilde{y}}_0)$ are defined over the set (3.22) and $\tilde{\tilde{y}}_0 \le \lambda_{l+\tau+1}$.

Further, the integral on the left hand side of (3.26) can be handled similarly as (3.5) and using exactly the same arguments as in deriving (3.2) it is a straightforward task to arrive at (3.23). Naturally, bounds for the remainder term in (3.23) of the type (3.3) and (3.21) hold as well. □

Both cubature formulae (3.2) and (3.23) allow for a direct numerical implementation. Zeros of the Jacoby polynomials are simultaneously computed using the efficient Weierstrass-Dochev algorithm. The corresponding Christoffel numbers are easily computed (cf 2.9) using the well known recurrence formula for the Jacoby polynomials. The two routines, combined with a routine for the evaluation of classical B-splines from the package developed by De Boor (1978) are sufficient to allow for the numerical dealing with generalized B-splines. Results of computations, discussion of accuracy and efficiency of the corresponding software will appear elsewhere.

As is known, Monte Carlo methods for the computation of multiple integrals may provide a better alternative to multidimensional quadrature rules, especially for problems of high dimension. We exhibit such a method for the evaluation of generalized B-splines.

Consider the expressions (3.5) and (3.26). Let us multiply and divide (3.5) by $\Gamma(\sum_{i=0}^{n+m} g_i - 1)$. Note that both (3.5) and (3.26) can be viewed as the expectations with respect to certain Dirichlet densities of, correspondingly the classical B-spline $M(\tilde{\tilde{y}}_0; t_0, \dots, t_{n+m})$ and the sum of normalized

B-splines $\sum_{i=l}^{l+\tau} N_{i,l}(\tilde{\bar{\gamma}}_0)$. In other words, we can rewrite (3.5) as

$$M_G(x) = KE_\theta\{M(\gamma(\theta); t_0, \dots, t_{n+m})\},$$

$$\text{where } K = \frac{\Gamma\left(\sum_{i=0}^{n+m} g_i\right)}{\Gamma\left(\sum_{i=0}^{n+m} g_i - 1\right)(l-1)}, \qquad \gamma(\theta) = \frac{x - \sum_{i=0}^{n} t_i \theta_i}{1 - \sum_{i=0}^{n} \theta_i}, \tag{3.28}$$

E_θ denotes the expectation with respect to θ, $\theta = (\theta_0, \dots, \theta_{n+1})$, $(\theta_0, \dots, \theta_{n+1}) \in \mathscr{D}(\overline{g}_0, \dots, \overline{g}_n, l-1)$.

Thus, the expression (3.26) can be rewritten as

$$F_{\mathscr{S}}(x) = E_{\theta'}\left\{\sum_{i=l}^{l+\tau} N_{i,l}(\gamma(\theta'))\right\}, \tag{3.29}$$

where

$$\gamma(\theta') = (x - \sum_{i=0}^{n} t_i\theta_i')/(1 - \sum_{i=0}^{n} \theta_i'), \qquad (\theta_0', \dots, \theta_{n+1}') \in \mathscr{D}(\overline{g}_0, \dots, \overline{g}_n, l).$$

Obviously, expressions (3.28) and (3.29) can be directly implemented as Monte Carlo algorithms. For this purpose a generator of Dirichlet distributed random variables was developed. It is based on the well known fact that if $\xi_0, \dots, \xi_n$ are independent random variables correspondingly Gamma distributed $G(g_0), \dots, G(g_n)$ and $\theta_i = \xi_i/(\xi_0 + \cdots + \xi_n)$, $i = 0, \dots, n$, then $(\theta_0, \dots, \theta_n) \in \mathscr{D}(g_0, \dots, g_n)$. For the Gamma variates the generators RGS and RGKM3 (see Bratley et al. 1983) were used, replacing the uniform generator UNIF by the universal generator UNI due to Marsaglia, et al. (1989). Further details including the convergence with respect to $\mathscr{N}$ of the corresponding, estimates $K\sum_{i=1}^{\mathscr{N}} M(\gamma(\theta(i)); t_0, \dots, t_{n+m})/\mathscr{N}$ and $\sum_{j=1}^{\mathscr{N}}\{\sum_{i=l}^{l+\tau} N_{i,l}(\gamma(\theta'(i))\}/\mathscr{N}$ will be given elsewhere.

4. An application to the computation of the distribution of serial correlation coefficients

As known, SCC represent ratios of two quadratic forms in normal variates. A variety of SCC were introduced to measure the relationship between successive observations in time or space. They have important applications to regression problems (see Watson and Durbin (1951)), and to hypothesis testing in time series analysis (see T. W. Anderson (1971)). Among the large number of contributions to the subject, we will mention the works of R. L. Anderson (1942), and J. von Neumann (1941) in which exact formulae for the distribution of, respectively, the circular and a version of a successive difference SCC were derived. The numerical evaluation of the distribution of SCC is considered in the papers by Pan Jie Jian (1968) and Imhof (1961) whose method is based on the computation of the corresponding Fourier inversion integral. Another approach to study SCC, taken by many authors (see e.g. Ramasubban (1972)) was to seek approximate distributions for them.

However, either of these methods is related with certain inaccuracy, mainly due to errors of 'truncation' and 'approximation'.

Our purpose here will be to illustrate shortly how the results from §3, can be applied to express and compute the unknown distributions of certain SCC, studied by T. W. Anderson (1971).

Let $y_1, \dots, y_N$ be variables which represent a random sample of N successive observations from a population whose distribution is normal $N(0, \sigma^2)$.

A serial correlation coefficient in which successive differences are involved is given as

$$r_j = \sum_{i=1}^{N-j} (\Delta^j y_i)^2 \Big/ \sum_{i=1}^{N} y_i^2, \qquad j < N, \tag{4.1}$$

where Δ^j, the successive difference operator of order j, is given as

$$\Delta^j y_i = \sum_{k=0}^{j} (-1)^{j-k} \binom{j}{k} y_{i+k}.$$

The lag j noncircular serial correlation coefficient in the case of zero population mean is defined as

$$r_j = \sum_{i=j+1}^{N} y_i y_{i-j} \Big/ \sum_{i=1}^{N} y_i^2, \qquad j < N. \tag{4.2}$$

Let us note (see e.g. B. Margolin (1977)) that the distribution of (4.1) and (4.2) coincide with the distribution of the linear combination

$$S = \sum_{i=0}^{p} \lambda_{ji} \theta_i, \qquad j = 1, 2, \dots,$$

where λ_{ji} are the latent roots of the matrix of the quadratic form in the numerator of (4.1) and (4.2) respectively and $(\theta_0, \dots, \theta_p) \in \mathscr{D}(g_0, \dots, g_p)$. The parameters $g_0, \dots, g_p$ are equal to $1/2$ times the corresponding multiplicities of the roots λ_{ji} and hence, some of the $g_i' s$ may be real, some, integer.

In the case of r_j from (4.1) and (4.2) the roots are respectively (see T. W. Anderson (1971), Ch. 6)

$$\cos \pi j i / N, \qquad i = 0, \dots, N-1, \tag{4.3}$$

and

$$\cos \pi j i / (N+1), \qquad i = 1, \dots, N. \tag{4.4}$$

In the Table, we summarize the cases in which the distribution of the SCC r_j from (4.1) and (4.2) coincides with the distribution of a linear combination S, with a number of real and integer $g_i' s$, respectively $n+1 \geq 2$ and $m \geq 0$.

TABLE 1. Multiplicities of the roots of SCC r_j from (4.1) and (4.2).

Roots λ_{ji} defined in	Lag j	Sample size N	Multiplicities of the roots λ_{ji}, $i = 0, \dots, p$	n and m
(4.3)	odd	rj, $(r = 2, 3, \dots)$	$(j-1)/2, \underbrace{j, j, \dots, j}_{r-1}, [[j/2]]$, [[ ]]denotes the smallest integer, greater than $j/2$ $(p = r)$	$r-1$ 1
”	even non-multiple of 4	rj	$j/2, \underbrace{j, j, \dots, j}_{r-1}, j/2$ $(p = r)$	1 $r-1$
”	even non-multiple of 4	even $N \neq rj$	$1, \underbrace{2, 2, \dots, 2}_{(N-2)/2}, 1$ $(p = N/2)$	1 $(N-2)/2$
(4.4)	odd, $(j-1)/2$ odd	$N = rj - 1$	$(j-1)/2, \underbrace{j, j, \dots, j}_{r-1}, (j-1)/2$ $(p = r)$	r 0
”	odd, $(j-1)/2$ even	$N = rj - 1$	$(j-1)/2, \underbrace{j, j, \dots, j}_{r-1}, (j-1)/2$ $(p = r)$	$r-2$ 2

Thus, we can interpret the density of r_j from (4.1) and (4.2) as a generalized B-spline with knots (4.3) and (4.4) of the corresponding multiplicities. Hence, we can apply the cubature formulae (3.2) and (3.23) and the Monte Carlo methods in (3.28) and (3.29) to compute the distribution of r_j. To this end, it suffices to correspondingly replace the constants denoted by t with the roots λ_{ji} defined here, taking as the Dirichlet parameters g_i half times their multiplicities, given in Table 1.

REFERENCES

R. L. Anderson (1942), *Distribution of serial correlation coefficient*, Ann. Math. Statist **13**, 1–13.

T. W. Anderson (1971), *The statistical analysis of time series*, Wiley, New York.

D. A. Bloch and G. S. Watson (1967), *A Bayesian study of the multinomial distribution*, Ann. Math. Statist. **38**, 1423–35.

C. de Boor (1976), *Splines as linear combinations of B-splines*, Approximation theory II, (G. G. Lorentz, C. K. Chui, L. L. Schumaker, and T. D. Ward, eds.) Academic Press, New York, pp. 1–47.

——(1976) *A practical guide to splines*, Springer-Verlag, New York.

P. Bratley, B. L. Fox, and L. E. Schrage (1983), *A guide to simulation*, Springer-Verlag, New York.

A. Chizzetti and A. Ossicini (1970), *Quadrature formulae*, Akademie-verlag Berlin.

H. B. Curry and I. J. Schoenberg (1966), *On Polya frequency functions.* IV *The fundamental spline functions and their limits*, J. d'Analyse Math. **17**, 71–107.

R. A. DeVore and L. R. Scott (1984), *Error bounds for gaussian quadrature and weighted-* L^1 *polynomial approximation*, SIAM J. Numer. Anal. **21**, 400–412.

Z. G. Ignatov and V. K. Kaishev (1985), *B-splines and linear combinations of uniform order statistics*, Math. Research Cent. TSR. #2817, Univ. of Wisconsin, Madison.

——(1987a), *Multivariate B-splines, analysis of contingency tables and serial correlation*, in *Mathematical statistics and probability theory*, vol B. (P. Bauer et al., eds.), 125–137, D. Reidel Publ. Co., Dordrecht, Holland, Proceedings of the 6-th Panonian Symposium on Mathematical Statistics, September 14–20, 1986.

——(1987b), *Some properties of generalized B-splines*, in *Proceedings of the International Conference on Constructive Theory of Functions*, Varna, May 24–31, 1987, Publ. House of the Bulgarian Acad. of Sci., pp. 233–241.

——(1987c), *On the computation of distributions of serial correlation coefficients through B-splines*, Proc. of the Second International Tampere Conference in Statistics (T. Pukkila and S. Puntanen, eds.), Dept. of Math. Sci., Univ. of Tampere, 479–489.

——(1989), *A probabilistic interpretation of multivariate B-splines and some applications*, SERDICA Bulgaricae mathematicae publicationes, **15**, 91–99.

P. J. Imhof (1961), *Computing the distribution of quadratic forms in normal variables*, Biometrika **41** 405–19.

S. Karlin, C. Micchelli, and Y. Rinott (1986), *Multivariate splines: a probabilistic perspective*, J. Multivariate Analysis **56**,

B. H. Margolin (1977), *The distribution of internally studentized statistics via Laplace transform inversion*, Biometrika **64**, **3** 573–582.

G. Marsaglia, A. Zaman, and W. Tsang (1989), *Towards a universal random number generator*, Statistics and Probability Letters, in press.

I. P. Missovskih (1981), *Interpolatory cubature formulae.* Moscow Nauka, (in Russian).

J. von Neumann (1941), *Distribution of the ratio of the mean square successive difference to the variance*, Ann. Math. Statist. **12**, 367–395.

Jie-Jian Pan (1968), *Distribution of the noncircular serial correlation coefficient*, Amer. Math. Soc. and Inst. Math. Statist. Selected Translations in Probability and Statistics, **7**, 281–91.

T. A. Ramasubban (1972), *An approximate distribution of a noncircular serial correlation coefficient*, Biometrika **59**, **1**, 79–84.

A. H. Stroud and D. Secrest (1966), *Gaussian quadrature formulas*, Prentice-Hall, Englewood Cliffs, N. J.

G. S. Watson and J. Durbin (1951), *Exact tests of serial correlation using non-circular statistics*, Ann. Math. Statist. **22**, 446–451.

INSTITUTE OF MATHEMATICS, BULGARIAN ACADEMY OF SCIENCES, P. O. BOX 373, 1090 SOFIA, BULGARIA

Contemporary Mathematics
Volume **115**, 1991

Computational Problems Associated with Minimizing the Risk in a Simple Clinical Trial

J. P. HARDWICK

ABSTRACT. Researchers designing clinical trials often encounter difficulties when trying to determine the best way to allocate patients to treatments so that a specified loss/cost function will be minimized. Ad hoc allocation rules are often used because, in most nontrivial settings, optimal alternatives are essentially impossible to locate. While finding exactly optimal rules is rarely regarded as a necessity, determining the relative and absolute merits of approximately optimal rules does depend on knowing at least something about the optimal ones. In principle, one can obtain *exact* solutions to allocation problems using the method of backwards induction. To date, however, this technique has been underutilized due to the immense computational burden imposed by nearly all realistic model formulations. Here, we examine the potential for taming problems such as these by implementing algorithms designed to take advantage of both vector and multi-processing facilities. As an example of progress one can make in a specific problem, we examine a two treatment clinical trial designed to incorporate both ethical and financial costs.

1. Introduction

Modern computer technology holds great potential for a variety of problems in the statistical sciences. This discussion is offered to illustrate the degree to which progress in statistics can be affected by breakthroughs in computer science. Ideally, the developing dependencies between these fields will act as stimuli to researchers of many persuasions.

For the most part, the statistical sciences underutilize supercomputers. Certainly, we do our share of running *bigger* jobs *longer,* but this leads largely to more accurate solutions to problems that we already knew how to handle

---

1980 *Mathematics Subject Classification* (1985 *Revision*). Primary 62C10, 62L05, 62L10, 62E20, 90C39.

Key words and phrases. Sequential analysis, dynamic programming, backward induction, Bayesian decision theory, clinical trials, ethics.

Research supported in part by National Science Foundation under grant DMS-8914328.

This paper is in final form and no version of it will be submitted for publication elsewhere.

computationally. Still, as a discipline, it seems that we hesitate to try to take advantage of the latest computer technology; for, once we have relegated problems to a home for the computationally intractable, we tend not to renew our attacks despite new arsenals available to us.

In some statistical problems, it is possible to reduce the computational burden by computing selectively or decreasing the number of iterations. However, problems to which these approaches are most suited tend to be those in which approximate solutions have already been deemed acceptable. When computing *exact* solutions, on the other hand, it is usually necessary to examine **all** possible data dependent outcomes. With no magic formulae available to reduce the amount of computation required for a problem in any fixed system, the successful creation of algorithms depends on our ability to make trade-offs between run time and memory requirements. In other words, given space and time as basic resources, it is critical that a programmer have the flexibility to exchange the resources as desired—one for the other. Increasingly, innovative algorithmic techniques inspired by advanced computer architectures offer profoundly improved tools with which even a novice can effect such exchanges. Thus, with persistence on the part of statistical researchers, it may soon be possible for us to experience the benefits of advanced computer technology—to date enjoyed by scientific disciplines other than our own. This discussion, along with the results in §5.3, are presented as an example of potential future directions.

1.1. Sequential experiments. Suppose that we wish to conduct a clinical trial, and that the goal of the trial is to estimate the difference between the efficacy rates, P_1 and P_2, of two therapies, T_1 and T_2. The patient outcomes are dichotomous, independent, and observable without delay. In this simple model, a standard approach might be to sample equally from each population and to use either exact distributions or normal approximations to derive confidence intervals for the means difference, $\theta = P_2 - P_1$. While popular, this approach is not necessarily the best—either for the trial patients or for the researchers seeking information.

A class of alternatives based on sequential designs allows more flexibility in modeling and may, therefore, be preferable. In a sequential experiment, actions taken during the course of a study are allowed to depend on the information already collected. Thus, if for any reason it is suspected that the data themselves can provide information regarding improved sampling methods, it would appear to be in the best interests of all concerned to use a sequential design. However, in spite of the often compelling reasons to prefer sequential experiments to those with predetermined allocation rules, the design and analysis of such experiments are generally more complicated, largely because the sampling distributions of relevant statistics are influenced by the nonstandard allocation procedures and/or stopping rules. Also to be considered are the trial goals themselves. Often these conflict; for, while the main goal may be to collect enough information to infer correctly the superior treatment, it is also important for ethical reasons to try to optimize the treatment of trial patients. These issues notwithstanding, there do exist techniques for deriving solutions, although they will never become familiar or well accepted until practitioners agree to diversify their expression of the problems.

Every trial is different, and no single class of designs can satisfy the great variety of needs facing investigators who are constantly torn between the desire for optimality and the demand for practicality. It seems critical, therefore, that *compromise* be recognized as an ingredient essential to the development of new designs. Consequently, our problem is to unearth classes of designs that are flexible enough to handle the concerns described and to delineate methods of solving them and interpreting the results.

1.2. Decision models. Perhaps the most common method of formalizing complex design goals is to use decision theory. In decision theoretic models, one can specify costs for reaching incorrect conclusions or, more generally, for making 'poor' inferences about parameters. Of particular interest here are Bayesian decision theoretic models that allow the quantification of information that is available *before* the trial as well as that obtained *during* the trial. In Bayesian models, unknown parameters are modeled as random variables. In *sequential* Bayesian models, sampling costs may be added to represent measures of financial loss or even more elusive quantities such as quality of life. Once given these formulations, it is possible to seek optimal procedures that minimize the expected overall cost of a trial.

Since exact solutions are rarely available in realistic formulations, recent research in this area has focussed on improving approximate solutions that may be obtained via a combination of analytic (asymptotic and probabilistic arguments) and computational tools (symbolic manipulation, simulation, numeric integration, and so on). The *exact* methods, however, owing to the great computational burden assumed in them, have thus far been essentially unimplementable. So, while the approximate methods will certainly show manifest improvement through the application of new computing algorithms, it is the exact methods that we hope to see flourish under the influence of new technologies.

What follows is an example of a simple adaptive allocation problem typical of the sort encountered not only in medical research but also in fields such as operations research and economics. Our purpose lies in framing the problem to illustrate that it not only merits, but *requires* the application of advanced computational algorithms such as parallel processing and vectorization. In §2 the problem is specified. In §3, the risk function arising in a standard (non-Bayesian) formulation is examined for both fixed and sequential rules. As a concession to the complexity of the analysis required to approach optimal solutions, the initial model is ultimately embedded in a Bayesian framework.

Once in the Bayesian environment, we use two methods for obtaining the minimum risk for sequential rules.

The first, outlined in §4, describes a method for obtaining large sample approximations in which the expected posterior loss provides the basis for asymptotic expansions. The leading term of the expansion is related to the expression for the minimum risk for fixed (P_1, P_2). Computing this minimum while integrating it with respect to the prior distribution on (P_1, P_2) poses a double problem in numerical analysis since the optimizations and multiple integrations commingle, and must therefore be computed in tandem. Ordinarily, this might not be a problem of interest in a statistics paper, but it is here that we encounter a *statistical multiple integration* problem that ties this work directly to the theme of the conference. In any case, upon accomplishing this numerical task, we have a lower bound for the risk of a broad class of sequential rules. Of equal interest is the slightly smaller class of rules for which the lower bound is achieved. This class, which contains a variety of 'practical' multi-stage rules, goes a long way toward addressing the standing challenge of whether there exists a class of good 'asymptotic rules' that are not totally obscure.

Despite the positive results achieved with large sample approximations, there lingers the question of *how big* a sample must be before the asymptotic results accede. It is in response to this question that we encounter the computational problem that is our major concern—determining *exactly* the minimum integrated risk for sequential rules. Our second approach, therefore, is to use the method of backwards induction to compute exact minima. Preliminary work on this problem indicates that the recursion equations (*dynamic programming*) explicit in the solution of many allocation problems may be structured in such a way that parts or *nodes* in the recursion trees can be computed independently. This basic idea provides the foundation for our belief that there exists an entire class of allocation problems ideally suited for parallel computation. The backward induction technique is discussed in §5.

For the present problem, we have computed exact minima for a variety of sample sizes. An advantage of using two completely different techniques for solving a problem is that they may be used to check each other. In this way, we use the exact results to gauge our approximate answers and use our analytic results to check the dynamic programming algorithms. Results of these comparisons are presented in Table 1 on p. 255.

The preceding discussion should not be interpreted as a suggestion that, in practice, all allocation problems are solvable no matter how complex the model. One inhibiting feature, for example, is the fact that the size of the problem grows exponentially with the total sample size. Still, it is our hope that, with sufficiently large machines, it will be possible to solve problems complex enough that the solutions themselves will provide insight into the nature of solutions to more general problems.

2. A simple model for clinical trials

The problem being considered is quite standard. The sample size for the trial is a fixed number, n, although, the sample sizes for the treatment groups, n_1 for T_1 and n_2 for T_2, may be either fixed or random. It is assumed that X and Y are independent Bernoulli random variables such that

$$(1) \qquad X_1, X_2, \cdots \sim B(1, P_1) \quad \text{and} \quad Y_1, Y_2, \cdots \sim B(1, P_2),$$

where $\mathbf{P} = (P_1, P_2)$ is an element of the parameter space $\Omega = (0, 1) \times (0, 1)$. Let

$$\theta = P_2 - P_1,$$

and define the *ethical* costs of observations from T_1 and T_2 to be $(1 - P_1)$ and $(1 - P_2)$, the respective chances of losing a patient. At the end of the trial, then, the ethical cost of allocating n_1 patients to T_1 and n_2 to T_2 is

$$n_1(1 - P_1) + n_2(1 - P_2).$$

To estimate θ, we use maximum likelihood estimates (noting that, for n_1 and n_2 fixed *or* random, the likelihood function is the same). Suppose that, at any time $m \leq n$, we have observed the patient outcomes

$$X_1, X_2, \ldots X_k, \quad \text{and} \quad Y_1, Y_2, \ldots Y_{m-k}$$

where

$$k = \#\text{ observations on } T_1, \quad m - k = \#\text{ observations on } T_2,$$
$$\text{and, when } m = n \quad (k, m - k) = (n_1, n_2).$$

Then, the maximum likelihood estimator for θ at time m is

$$\hat{\theta}_{k, m-k} = \overline{Y}_{m-k} - \overline{X}_k.$$

The loss due to estimation error is assumed to be squared error, so the *total* loss incurred after n observations is

$$(2) \quad L_n(\mathbf{P}, \hat{\theta}_{n_1, n_2}) = n[\{\sqrt{n}(\theta - \hat{\theta}_{p_n})\}^2 + p_n(1 - P_1) + (1 - p_n)(1 - P_2)],$$

where $$p_n = \frac{n_1}{n} \quad \text{and} \quad \hat{\theta}_{p_n} = \hat{\theta}_{n_1, n_2}.$$

The problem consists of determining how to allocate patients to treatment groups so that, on average, the loss will be as small as possible. In practice, the selection of a loss function should be a careful and thoughtful process. The one presented here is not intended to represent specific concerns other than the struggle between gathering information and saving resources. One aspect worth noting, however, is that a normalizing factor has been built in to distribute the estimation error and ethical costs at least somewhat evenly. Were this not done, the problem would become trivial with one factor dominating the loss as the sample size increases.

2.1. Allocation rules. In allocation problems, the data consist of *two* vectors—the treatment responses and indicators. The latter vector comprises the *allocation rule*—the sequence which specifies treatment selection at each stage of the trail. An allocation rule, $\delta(n)$, is defined to be any sequence $(\delta_1, \delta_2, \cdots, \delta_n)$ for which

$$\delta_i = \begin{cases} 1, & \text{if } T_1; \\ 0, & \text{if } T_2, \end{cases} \qquad i = 1, \cdots, n,$$

where
$$n_1 = \Sigma_{j=1}^n \delta_j \quad \text{and} \quad n_2 = \Sigma_{j=1}^n (1 - \delta_j).$$

In denoting the data, it is assumed that there are two complete sequences of responses

$$X_1, \ldots, X_n \quad \text{and} \quad Y_1, \ldots, Y_n,$$

but that, at each stage, an element from only one of the two sequences is observed. Let

$$W_i = \delta_i X_i + (1 - \delta_i) Y_i, \qquad i = 1, \ldots, n,$$

represent the value observed at Stage i, and let

$$\mathscr{D}_{m(k)} = \{W_1, W_2, \ldots, W_m; \delta_1, \delta_2, \ldots, \delta_m\} \qquad \forall 2 \le m \le n,$$

represent the complete data that have been observed by time m.

We consider two types of allocation rules: *fixed* and *sequential.* For fixed rules, the sample sizes n_1 and n_2 are specified in advance. For sequential rules it is assumed that the decision, δ_{m+1}, at Stage $m+1$, for $m = 1, \ldots, n-1$, depends only on $\mathscr{D}_{m(k)}$, the information accrued so far, for practical reasons, we add one more assumption for both types of rules; i.e., that one observation be taken from each treatment group at the beginning of the trial, i.e. that for each $\delta(n)$, $\delta_1 + \delta_2 = 1$.

3. The risk function

The exchangeability of observations simplifies the problem somewhat, since, despite the availability of 2^n sequences $\delta(n)$, the relationship between the allocation rules and the *expected* loss is not one-to-one. Given an allocation rule $\delta(n)$, the associated risk function is

$$\begin{aligned} \mathrm{R}(\mathbf{P}, \delta(n)) &= \mathbf{E}_{\mathbf{P}}[L_n(\mathbf{P}, \hat{\theta}_{p_n})] \\ &= \mathbf{E}_{\mathbf{P}}\left[n(\{\sqrt{n}(\theta - \hat{\theta}_{p_n})\}^2 + p_n(1 - P_1) + (1 - p_n)(1 - P_2))\right], \end{aligned} \tag{3}$$

where $\mathbf{E}_{\mathbf{P}}$ denotes expectation in the model (1).

Our overall approach to the problem is outlined below (with details available in Hardwick (1991)). For each type of rule we must determine whether

(i) An optimal proportion minimizing the risk function exists,
(ii) Optimal rules *achieving* the minimum exist,
(iii) It is feasible to *construct* optimal or approximately optimal rules,
(iv) Such rules are markedly better than standard procedures.

We begin by examining (3) for fixed rules (§3.1), and then consider sequential rules (§3.2). When $\delta(n)$ is a fixed rule, the risk function is analytically straightforward, involving only simple functions of independent, binomial random variables. When $\delta(n)$ is sequential, however, this is no longer the case because the subscripts on the estimator $\hat{\theta}_{n_1, n_2}$ are random. Thus, while our ultimate goal is to understand the properties of sequential rules, we start with the simpler problem in which, given parameter values $\mathbf{P} \in \Omega$, we seek the *fixed* values n_1 and n_2 that minimize the risk.

3.1. Optimal solutions for fixed allocation rules. For any $\mathbf{P} \in \Omega$ and fixed sample sizes n_1 and n_2, the risk function, (3), can be written as

$$\mathrm{R}(\mathbf{P}, \delta(n)) = n\Psi(\mathbf{P}, p_n) + n(1 - P_2), \tag{4}$$

where for $\mathbf{T} = (t_1, t_2)$ and $t_1, t_2, p \in (0, 1)$,

$$\Psi(\mathbf{T}, p) = \left[\frac{t_1(1-t_1)}{p} + \frac{t_2(1-t_2)}{1-p} + p(t_2 - t_1)\right].$$

The second term on the right side of (4) is independent of $\delta(n)$ so, on its own, it provides a rough lower bound for the risk of any sampling rule. Of far greater interest, however, is the function, Ψ. For all $\mathbf{T}$ and p in the region of integration, Ψ is continuously differentiable and strictly convex in its second argument, p. For fixed $\mathbf{T} \in \Omega$, therefore, we obtain a unique minimum by numerically solving the following quartic

$$\begin{aligned}\frac{\partial\Psi(t_1, t_2, p)}{\partial p} &= (t_2 - t_1)p^4 - 2(t_2 - t_1)p^3 \\ &\quad + (t_2 - t_1)(2 - t_2 - t_1)p^2 + 2t_1(1 - t_1)p - t_1(1 - t_1).\end{aligned} \tag{5}$$

Let $\mathscr{P}$ be the value of $p \in (0, 1)$ that minimizes Ψ and let

$$\Psi^{\mathscr{P}}(\mathbf{P}) = \min_{0<p<1} \Psi(\mathbf{P}, p) = \Psi(\mathbf{P}, \mathscr{P})$$

denote the attained minimum given $\mathbf{P}$. Then, for each $\mathbf{P} \in \Omega$, the function $\mathrm{R}(\mathbf{P}, p)$, possesses a unique minimum, occurring at $p = \mathscr{P} = \mathscr{P}(\mathbf{P})$,

$$\mathrm{R}^{\mathscr{P}}(\mathbf{P}) = n\left[\Psi^{\mathscr{P}}(\mathbf{P}) + (1 - P_2)\right]. \tag{6}$$

Equation (6) defines a lower bound for the risk function

$$\mathrm{R}(\mathbf{P}, \delta(n)) \geq \mathrm{R}^{\mathscr{P}}(\mathbf{P}) \quad \forall \mathbf{P} \in \Omega, \tag{7}$$

with equality if and only if

$$\frac{n_1}{n} = \mathscr{P}(\mathbf{P}).$$

Due to the discrete nature of the proportions available in practice, (7) cannot be met *exactly*, so the actual sampling should be based on a discrete

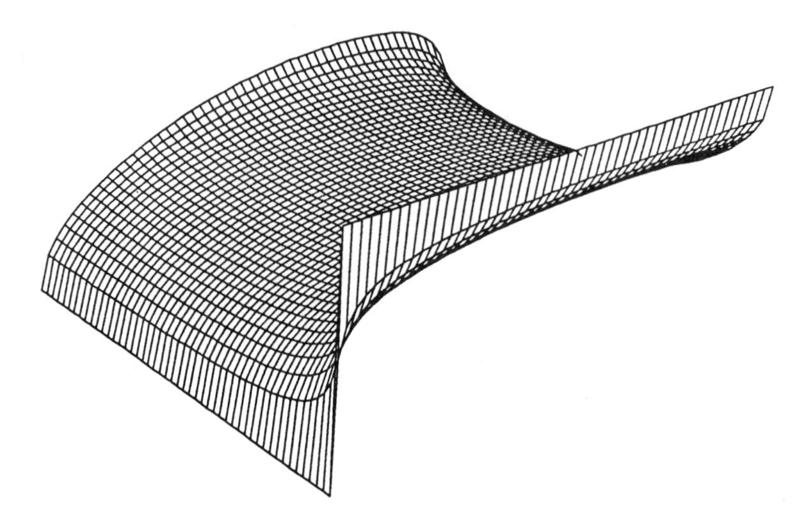

FIGURE 1

approximation to $\mathscr{P}(\mathbf{P})$. For example, if the total sample size is n, allocate according to the 'rule' determined by

$$p_x = \min\left\{\frac{i}{n} : \frac{i}{n} \leq \mathscr{P} \leq \frac{(i+1)}{n}, \quad i = 1, \cdots, n-1\right\}, \tag{8}$$

and assign

- np_x patients to T_1, and
- $n(1-p_x)$ patients to T_2.

Figure 1, illustrates the variation of $\mathscr{P}$ over Ω. That the surface is not symmetric about the line $P_1 = P_2$ emphasizes the two separate components fighting to control the function. Examining plots of this nature can be quite helpful when assigning costs during the design phase of a study. Graphical representations highlight aspects of risk functions that might otherwise go unnoticed.

3.2. The risk for sequential rules. When using fixed allocation rules, we are not in a position to take advantage of the existence of minima since the optimal proportion, p_x, *varies* with the unknown parameter $\mathbf{P}$. This dependency is the crux of the problem, for we must either guess p_x or estimate it on the basis of prior information. This is not to say that the rule determined by (8) is of no use, for it nonetheless provides clues as to how we can proceed in more realistic settings. It is clear, for example, that if we had information equivalent to (8) for *sequential* rules, then we could estimate the optimal proportion during the trial and adjust the sampling proportion accordingly. More precisely, if for a given class of sequential rules, we knew a value, say

$$p_x^s = p_x^s(\mathbf{P}, \delta(n))$$

that minimized the risk function, then we could adapt our sampling proportion in such a way that, at the end of the trial, p_n would be as close as possible to our final estimate of p_x^s. Intuitively, this appears to be a good strategy; however, the same features of the sampling rule that lend flexibility to the process also create some tricky analytic problems. For example, the random subscripts, n_1 and n_2, complicate the sampling distributions of relevant statistics enough to preclude us from expressing the integral, (3), in a form conducive to direct minimization—either numerically or analytically.

Thus, in the sequential case we are faced with a further set of problems. First we must determine how to locate either optimal or near-optimal sampling proportions, p_x^s. Next, we must select good estimators and rules that adapt appropriately. Finally, because we know we will not achieve the minimum risk by using approximations and estimates, we must find the means to assess the rules.

It is not our purpose to answer these questions fully, but rather to provide enough detail so the flavor of the problem and the motivation behind the approach is clear. Also, while the arguments presented are not critical to understanding this application of parallel processing, they do help to explain the origin of the asymptotic lower bound that is used as a benchmark to test the programs.

4. Asymptotic approximations

With no exact methods available, we examine alternatives for approximating the sampling distributions needed. First note that computer-generated approximations appear inadequate since they must be developed for each sampling strategy of interest. Furthermore, for an *entire* grid of parameter values, $\mathbf{P} \in \Omega$, we would need approximate distributions and accurate estimates of the first few moments of each. Withal, such techniques, while possible, seem to involve at least as much approximation and more effort than do large sample methods. We turn, therefore, to asymptotics in hopes of finding analytic approximations from which the required distributional characteristics can be computed more easily.

We begin with standard, large sample theory arguments applied to the pivotal quantity, z_n,

$$z_n = \sqrt{n}(\theta - \hat{\theta}_{p_n}).$$

Previous work suggests that when sequential designs are used, the limiting distribution of z_n is not well approximated by a normal distribution. This indicates that more detailed expansions should be sought—particularly those giving rise to second-order correction terms. In the present problem, however, the expansions one typically expects to provide the desired results (e.g., Edgeworth, Taylor) are extremely messy, and may, in some cases, not even exist.

A good example of how such problems can be circumvented is presented in Woodroofe and Keener (1988), where the authors describe the erratic behavior of Edgeworth expansions for a problem in which the sampling distribution of the pivotal quantity is affected by stopping times. In this work, the authors show that classical expansions would be more functional if they were *averaged*, somehow, over a reasonably defined parameter space. It is in this context that a Bayesian framework is proposed and that expansions of posterior distributions integrated with respect to the data are shown to be equivalent, in an appropriate sense, to the original expansions integrated with respect to the prior distribution utilized in the Bayesian model.

Many statisticians resist Bayesian designs, in the belief that their use implies the existence of an exaggerated level of knowledge regarding design parameters. The example considered in §5.1 is specific in this respect, but may be generalized to incorporate prior distributions with few restrictions. Given this leeway, we believe it is possible to model prior information in almost any fashion desired without violating underlying model assumptions. Furthermore, if the idea of using priors of *any* sort is distressing, one can view the issue from another perspective, vis., how much does it matter? In other words, one can check the extent to which a design affects the analysis in a given problem. The analytic results for data arising from this and similar models are generally insensitive to the choice of prior distribution. Thus, while the Bayesian framework has the advantage of allowing us to determine which designs offer the best results, the design itself has little effect on the eventual interpretation of the data.

The sequential aspect of the present allocation problem makes it amenable to the techniques of Woodroofe and Keener and, with regard to getting asymptotic solutions, our approach is modelled after theirs. The mathematical details for the heuristic arguments that follow may be found for the normal and binomial cases in Woodroofe and Hardwick (1990) and Hardwick (1991), respectively.

To begin, a prior is placed on the parameters $P = (p_1, p_2)$, now presumed random. Next, the loss function is integrated with respect to the conditional distribution of the parameters, and this provides the expected posterior loss or *posterior risk* which serves as a basis for the analysis. As one might expect, deriving expansions for conditional distributions is far easier than working with randomly subscripted variables. Nevertheless, the relief is only temporary. Ultimately, each term of the expansion is required to be integrable with respect to the joint distribution of the data. So, anticipating the difficulties associated with the final integration, we seek expansions with terms that are as free as possible of the data itself. This is accomplished by defining a data dependent transformation of the parameter, the likelihood of which is as close as possible to a normal likelihood. In particular, it is beneficial to locate a transformation that adjusts the picture in such a way that the posterior density of the new variable is expressible as the *product* of a normal density and some other appropriately behaved function. Then, under suitable regularity conditions on both the prior and the allocation rule, the posterior risk is integrated by parts. The consequent expression is a normal integral with manageable correction terms. The regularity conditions mentioned suffice to insure that the terms of the expansion are integrable and that the limiting behavior of the estimates in the expansion are civilized. Depending on the strictness of these conditions, the final result yields unconditional first- or second-order correction terms that provide a lower bound for the integrated risk.

4.1. The integrated risk. Consider the Bayesian model in which $P = (p_1, p_2)$ is a random variable with a joint prior density function ξ, and where, for all $\mathbf{P} \in \Omega$, the model (1) holds conditionally given $P = \mathbf{P}$. In this model, the *integrated* risk for any allocation rule, $\delta(n)$, and any estimate, $\hat{\theta}_n$, of θ is given by

$$(9) \qquad \begin{aligned} \mathscr{R}(\xi, \delta(n)) &= \int_\Omega \mathrm{R}(P, \delta(n))\xi(P)dP \\ &= \mathbf{E}^{\xi}\left[n\left(\{\sqrt{n}(\theta - \hat{\theta}_n)\}^2 + p_n\theta + (1 - P_2)\right)\right], \end{aligned}$$

where $\mathbf{E}^{\xi}$ denotes expectation with respect to the prior density ξ. In this same model, for each $\mathscr{D}_{n(k)}$, the *posterior* risk of $\delta(n)$ and $\hat{\theta}_n$, is defined to be

$$(10) \qquad \begin{aligned} \mathscr{R}(\xi_n, \delta(n)) &= \int_\Omega \mathrm{L}_n(P, \delta(n))\,\xi_n(P)\,dP \\ &= \mathbf{E}^{\xi_n}\left[n\left(\{\sqrt{n}(\theta - \hat{\theta}_n)\}^2 + p_n(\theta) + (1 - p_2)\right)\right], \end{aligned}$$

where
$$\xi_n(P) = \xi(P \mid \mathscr{D}_{n(k)})$$
denotes the posterior density function of P given the data observed by time n. Let
$$\hat{\theta}_{\xi_n} = \mathbf{E}^{\xi_n}[\theta]$$
represent the posterior expected value of θ at time n. Then when the posterior mean of θ is used in place of the maximum likelihood estimate in (1), the integrated risk, (9), becomes the *Bayes* risk. It is this function whose minimum we now seek.

4.2. The limiting solution. Since (7) holds for each $\mathbf{P} \in \Omega$, the minimum integrated risk for fixed allocation rules is simply the integral of the minimum risk
$$\min \int_\Omega \mathrm{R}(P, \delta(n))\xi(P)\,dP = \int_\Omega \mathrm{R}^{\mathscr{P}}(P)\xi(P)\,dP.$$
Generally speaking, this minimum is of little interest unless it is related in a meaningful way to the minimum integrated risk of a class of *sequential* rules. Using arguments from the previous section, one can show that, subject to a

few regularity conditions,

$$\lim_{n\to\infty}\left[\mathscr{R}(\xi\,,\,\delta(n)) - \int_\Omega \mathrm{R}^{\mathscr{P}}(P)\xi(P)\,dP\right] = O(1). \tag{11}$$

Thus, there exists a class of rules Δ, and a set of distributions Ξ, such that, for $\delta(n)\in\Delta$ and $\xi\in\Xi$, the difference between the integrated risk and the integral of the lower bound,

$$\mathbf{E}^{\xi}[\mathrm{R}^{\mathscr{P}}] = \int_\Omega \mathrm{R}^{\mathscr{P}}(P)\xi(P)dP = \int_\Omega n\,[\Psi_{\mathscr{P}}(P) + (1-p_2)]\xi(P)dP\,, \tag{12}$$

is bounded as $n\to\infty$.

In practice, to use (11), we must be able to compute the integrated minimum and satisfy the regularity conditions on the rules and priors. The convoluted numerics associated with calculating (12) have a certain appeal, although, undoubtedly, they present no challenge to numerical analysts. This is, however, the only multiple integration problem we have to offer, so we are obliged to give it an honorable mention. Recall that the optimal proportion $\mathscr{P}(P)$ is the solution to the quartic (5). Of course, repeatedly solving the quartic for various values of $p\in(0\,,\,1)$ poses no difficulty. Nevertheless, the problem becomes more interesting when the double integral must be taken; each time the numerical integration program calls for an evaluation of the integrand at a new value of P, the program must first compute $\mathscr{P}(P)$ by solving the quartic. Here, the difficulty relates to trying to force a library subroutine for multiple integration to go elsewhere for the function evaluation at *every* call. Had we written a dedicated multiple integration routine to handle problems of this nature, a more elegant approach could have been taken.

Of more serious concern, perhaps, are the conditions on $\delta(n)$ and ξ for which the relationship, (11), holds. Roughly speaking, the class Ξ contains smooth distributions with compact support. The exact number of finite moments required varies with the specific class of allocation rules being considered. Two distinguishing features of the rules in Δ are that they give rise to estimates that are consistent and that, in the limit, they never cease allocating to one treatment altogether. Also worth noting is the existence of a sub-class of rules $\Delta_0\subset\Delta$, the elements of which *attain* the integrated lower bound.

All of the conditions mentioned are fairly easy to satisfy. Furthermore, they are natural in that they tend to be built into ad hoc sampling rules proposed by nonstatistical investigators. This subjective statement is supported by both empirical and analytic results (Hardwick (1991) and Woodroofe and Hardwick (1990)). These show that Δ_0 includes a number of intuitively reinforcing and practically implementable allocation rules.

5. Exact solutions by backward induction

Since estimates based on small sample sizes are unlikely to be accurate enough to act as good guides, we are left with the question of what to do when encountering small to moderate samples. Recall that when working with the problem in its original formulation we were unable to analyze the risk function for sequential rules directly. We were also dissatisfied with cursory attempts to generate solutions via resampling or simulation methods (although we do plan to pursue this tact more thoroughly in the future). It happens, however, that the same generalizations adopted to access the tools of Bayesian decision theory, present new options for tackling the problem of computing *exact* solutions. The asymptotic lower bound (12) was derived from expansions of the posterior risk, (10). Equally dependent on the posterior risk are the recursion equations that define a procedure known as *backward induction.*

Backward induction, also referred to as *back tracking,* is a programming technique applicable to a broad variety of resource allocation problems. Bellman's 1957 text on dynamic programming is a classical reference on this topic and more recent discussions may be found in Whittle (1982) and Ferguson (1985). The idea is to infer optimal selections at each stage of an experiment from knowledge of where the process can go at the *next* stage and the probability of its doing so. Typically, this style of programming is underutilized due to its computational intensity. In the present example, the major burden lies in the large multi-dimensional arrays that must be retained in random access memory (RAM) as the program works through the stages of the experiment.

The motivation underlying dynamic programming is the *principle of optimality.* Suppose that $\delta^*(n)$ is an optimal rule. The principle of optimality states that, under regularity conditions (satisfied in the present problem),

$$(13)\quad \mathbf{E}^{\xi_m}\left[\mathscr{R}(\delta^*, \xi_n)\right] = \operatorname{ess}\inf_{\delta'} \mathbf{E}^{\xi_m}\left[\mathscr{R}(\delta', \xi_n)\right] \quad \text{for } m = 3, \ldots, n-1,$$

where the essential infimum extends over all $\delta'(n)$ for which $\delta'_i = \delta^*_i$ for $i = 1, \ldots, m$.

Imagine that the last stage of the trial has just been completed and that we have computed the posterior risk corresponding to every data configuration at time n. Presumably, at the end of the trial, *one* of these risks will represent our best evaluation of the total loss incurred during the experiment. Naturally, we cannot control responses to treatment, so there is no way to guarantee that the trial endpoint will be, say, the smallest posterior risk among the set available at Stage n. Rather, as mentioned, the best we can do is to follow an allocation strategy that will lead *on average* to the smallest stopping risk. This is the job of the backward induction program.

The program begins by providing us with the posterior risk of every possible *state of the system,* $\mathscr{D}_{n(k)}$, at the last stage of the experiment. Next, for each state, $\mathscr{D}_{n-1(k)}$, the program, cycling back a stage, computes the expected loss associated with each treatment alternative and selects the treatment which offers the smallest expected loss. Now, retaining the optimal selection corresponding to each $\mathscr{D}_{n-1(k)}$, the algorithm steps back again and determines the optimal treatment assignments for every state, $\mathscr{D}_{n-2(k)}$ at Stage $n-2$. Further iteration eventually produces an average risk that results from having kept track of the optimal strategy for each data configuration and then weighting it in accordance with the likelihood of having observed that outcome sequence. It is due to the principle of optimality that the final value, obtained recursively, is indeed what we seek—the *minimum Bayes risk*, $\mathscr{R}(\delta^*, \xi_n)$.

5.1. Beta priors. To illustrate the principle, we consider a specific example in which the distributions on P are conjugate priors for independent binomial random variables. Let ξ be the product of two independent beta distributions

$$p_1 \sim Be(a_0, b_0) \quad \text{and} \quad p_2 \sim Be(c_0, d_0),$$

where all four parameters have been set equal to one

$$a_0 = b_0 = c_0 = d_0 = 1.$$

Then at Stage m, the posterior distribution, $\xi_m(P)$, is the product of the two beta posteriors,

$$[p_1 \mid \mathscr{D}_{m(k)}] \sim Be(a, b) \quad \text{and} \quad [p_2 \mid \mathscr{D}_{m(k)}] \sim Be(c, d)$$

where

$$a = i+1, \qquad b = k-i+1, \qquad c = j+1, \quad \text{and} \quad d = m-k-j+1,$$

and where, as before,

$$k = \Sigma_{i=1}^{m} \delta_i, \qquad i = \Sigma_{i=1}^{m} \delta_i W_i, \quad \text{and} \quad j = \Sigma_{i=1}^{m} (1-\delta_i) W_i.$$

The posterior means of p_1, p_2 and θ are

$$\begin{aligned} \mathbf{E}^{\xi_m}[p_1] &= \frac{a}{a+b}, \quad \mathbf{E}^{\xi_m}[p_2] = \frac{c}{c+d}, \\ \text{and } \mathbf{E}^{\xi_m}[\theta] &= \left[\frac{c}{c+d} - \frac{a}{a+b}\right], \end{aligned} \tag{14}$$

and, from (10), the posterior risk is

$$\begin{aligned} &\mathscr{R}(\xi_m, \delta(m)) \\ &= m\left[\frac{1}{p_m}\left(\frac{ab}{(a+b)(a+b+1)}\right) + \frac{1}{1-p_m}\left(\frac{cd}{(c+d)(c+d+1)}\right)\right. \\ &\qquad \left. + p_m\left(\frac{b}{a+b}\right) + (1-p_m)\left(\frac{d}{c+d}\right)\right] \\ &= m\left[\frac{(i+1)(k-i+1)}{p_m(k+2)(k+3)} + \frac{(j+1)(m-k-j+1)}{(1-p_m)(m-k+2)(m-k+3)}\right. \\ &\qquad \left. + p_m\left(\frac{k-i+1}{k+2}\right) + (1-p_m)\left(\frac{m-k-j+1}{m-k+2}\right)\right], \end{aligned} \tag{15}$$

where $p_m = \frac{k}{m}$. For any $m \leq n$, the posterior distribution of P depends on $\mathscr{D}_{m(k)}$ only through a four-dimensional sufficient statistic (m, k, i, j), the vector completely defining the state of the process at m. If at Stage m, we have sampled

$$k \text{ observations from } T_1 \text{ with } \quad \begin{cases} i & \text{Successes (S);} \\ k-i & \text{Failures (F)} \end{cases}$$

and

$$m-k \text{ observations from } T_2 \text{ with } \quad \begin{cases} j & \text{Successes (S);} \\ m-k-j & \text{Failures (F),} \end{cases}$$

we denote the state at m as (k, i, j) and the associated posterior risks and means accordingly as

$$\mathscr{R}_m(k, i, j) = \mathscr{R}(\xi_m, \delta(m)) \quad \text{and} \quad \mathbf{E}_m[p_s \mid (k, i, j)] = \mathbf{E}^{\xi_m}[p_s]$$

for $m = 2, \ldots, n$; $k = 1, \ldots, m-1$; $i = 0, \ldots, k$; $j = 0, \ldots, m-k$, and $s = 1, 2$.

5.2. Optimal strategies. For each $m = 0, \ldots, n$, define the *stopping risk* of state (m, k, i, j)

$$\mathscr{S}\mathscr{R}_m(k, i, j)$$

to be the minimum expected loss incurred if one were to start from this state and then proceed optimally. Now, let

$$\begin{aligned} \mathscr{S}\mathscr{R}^{T_1}_{m+1}&(k, i, j) \\ &= \mathbf{E}_m[p_1 \mid (k, i, j)]\, \mathscr{S}\mathscr{R}_{m+1}(k+1, i+1, j) \\ &\quad + \mathbf{E}_m[1-p_1 \mid (k, i, j)]\, \mathscr{S}\mathscr{R}_{m+1}(k+1, i, j) \end{aligned}$$

be the stopping risk at Stage $m+1$ given that T_1 was selected and

$$\begin{aligned} \mathscr{S}\mathscr{R}^{T_2}_{m+1}&(k, i, j) \\ &= \mathbf{E}_m[p_2 \mid (k, i, j)]\, \mathscr{S}\mathscr{R}_{m+1}(k, i, j+1) \\ &\quad + \mathbf{E}_m[1-p_2 \mid (k, i, j)]\, \mathscr{S}\mathscr{R}_{m+1}(k, i, j) \end{aligned}$$

be the stopping risk at Stage $m+1$ given that T_2 was selected. Then, from (13) and (15), we know that if we are at state (k, i, j) at Stage m, the optimal selection at the next stage is T_i where

$$i = \begin{cases} 1 & \text{if } \mathscr{S}\mathscr{R}^{T_1}_{m+1}(k, i, j) < \mathscr{S}\mathscr{R}^{T_2}_{m+1}(k, i, j); \\ \text{Either} & \text{if } \mathscr{S}\mathscr{R}^{T_1}_{m+1}(k, i, j) = \mathscr{S}\mathscr{R}^{T_2}_{m+1}(k, i, j); \\ 2 & \text{if } \mathscr{S}\mathscr{R}^{T_1}_{m+1}(k, i, j) > \mathscr{S}\mathscr{R}^{T_2}_{m+1}(k, i, j). \end{cases}$$

The heart of the backwards induction program, then, is the recursion expression that allows us to compute each $\mathscr{S}\mathscr{R}_m(k, i, j)$ from stopping risks at stage $m+1$.

(17)

$$\begin{aligned}\mathcal{SR}_m(k, i, j) &= \min\{\mathcal{SR}_{m+1}^{T_1}, \mathcal{SR}_{m+1}^{T_2}\} \\ &= \min\left\{\frac{i+1}{k}\mathcal{SR}_{m+1}(k+1, i+1, j) + \frac{k-i+1}{k}\mathcal{SR}_{m+1}(k+1, i, j);\right. \\ &\qquad \left.\frac{j+1}{m-k}\mathcal{SR}_{m+1}(k, i, j+1) + \frac{m-k-j+1}{m-k}\mathcal{SR}_{m+1}(k, i, j)\right\}.\end{aligned}$$

Note that at Stage n, the *stopping* risk for each (k, i, j) is simply the *posterior* risk for that same state

$$\mathcal{SR}_n(k, i, j) = \mathcal{R}_n(k, i, j).$$

Note also, that—were it not for the constraint that the first two observations in a trial come from different treatments—the program would begin with the posterior risks at Stage n and recursively update the stopping risks until it arrived at Stage 0 with the solution

$$\mathcal{SR}_0 = \mathcal{R}(\delta^*, \xi_n).$$

However, to accommodate the condition that $\delta_1 + \delta_2 = 1$, and without any loss of generality, we set the allocations at Stage 1 and Stage 2 to be T_1 and T_2 respectively.

5.3. Results. The example under consideration is one of the simplest non-trivial allocation problems specifiable. If we were to try to use a sequential algorithm to run a backward induction program based on (17) on a typical PC, we would be doing well to handle sample sizes as big as say 10 or 15. Moving to a typical workstation or campus mainframe, we might manage samples as big as 200, which is pretty good, but not good enough when you consider the simplicity of the set-up. The main burden is encountered at the beginning of the program because the posterior risks for virtually all data configurations must be stored in an array required for the first backwards step. In our example, an initial array would be the cube of the sample size. Thus, for a problem of size $n = 100$, we have an initial requirement that two arrays of 100^3 be retained in RAM. At each successive step, the arrays become smaller, but the roadblock of handling the first arrays—without using virtual memory—remains.

Note that there is considerable redundancy in the parameter settings required for the evaluation of certain states in the backward induction pathway. If we spell out the amount of information needed at the lowest (i.e. the $m = n^{th}$) stage incorporating the redundancy of the required parameter settings then we need only q pieces of information (i.e., memory or storage locations), where

$$q = \sum_{i=1}^{n-1} i \cdot (n - (i-1))$$

is roughly $\frac{n^3}{6}$. A minor simplification for this problem, then, would be to utilize arrays containing only the q pieces of information needed. This strategy is still memory intensive, but, to some extent, the code can be vectorized and also adjusted to run on a parallel-loop compiler. A version of this scheme was implemented and compiled on the CNSF's IBM 3090 with parallel loop facilities. We found that, once n had reached, say, 50, the elapsed run time approach the $\frac{i}{c}$ *single processor* time limit one might hope for using c processors to compute the loops. The size of this test problem did not warrant the utilization of more than $c = 6$ processors, and there remain questions as to how best to program a problem such as this for larger parallel machines. We hope to examine this and other more advanced algorithmic techniques in the future. For now, however, we are satisfied to see that this *style* of programming is amenable to simple applications of parallel processing.

The following table shows exact and asymptotic results for sample sizes up to 360, with the values at 360 coming from a parallel algorithm. The values up to 160 were carried out on a variety of computers, and then computed again in parallel. An extrapolation technique was used to estimate the exact solution at 640. Of greatest interest is the second column of the table, which contains the minimum integrated risk (MIR) divided by the sample size. A review of the loss function will show that it was originally scaled up by a factor of n in order to carry out the asymptotics. Thus, what we aimed to see, and do in fact, see, is that the values in column two converge rather well to $\frac{1}{n} \cdot$ AMIR: $\frac{1}{n}$ times the asymptotic minimum integrated risk which was computed using the integration routine referred to in §4.1.

TABLE 1

n	$\frac{MIR}{n}$	MIR	AMIR	% Difference
10	1.1531	11.5310	11.214	0.0275
20	1.1437	22.8732	22.428	0.0195
30	1.1391	34.1731	33.642	0.0155
40	1.1364	45.4543	44.856	0.0132
50	1.1344	56.7222	56.070	0.0115
60	1.1330	67.9817	67.284	0.0103
70	1.1319	79.2345	78.498	0.0093
80	1.1310	90.4813	89.712	0.0085
160	1.1273	180.372	179.424	0.0053
320	1.1248	359.921	358.848	0.0030
640	1.1230*	718.688	717.696	0.0014
↓	↓			
∞	1.1214†			

* = Estimated values based on extrapolation.
† = $\frac{1}{n}$ AMIR.

Note that column 4 of the table merely contains $n \cdot 1.1214$. Finally, an important matter to consider at this time is the degree to which any of this matters. For example, how much of the total loss is attributable to the variation in p_x? In a certain sense, this ought to be examined, since, if p_x creates only slight variation we need not worry about sequential rules because we have little to gain and a lot of added complexity to lose. Nevertheless, our major concern here has been to emphasize a new computational technique which may be applied to similar problems in which the effect of using different allocation strategies may be greater. There is a great deal of experimentation still to be done—even on this simple problem. It is nice, however, *for once,* to have been able to take a backward induction routine far enough that the exact results can be used seriously to assess the asymptotics.

Acknowledgments

The author is grateful to the NSF along with the AMS, IMS, and SIAM for sponsoring and organizing this fascinating conference. Most of all, thanks are due to AMS publications and the editors of this volume for being so exceedingly patient as the delays in this submission accumulated beyond all reason.

References

R. Bellman (1957), *Dynamic programming,* Princeton University Press, New Jersey.

T. Ferguson (1985), *Lecture Notes: Sequential analysis and optimal stopping* (manuscript in development).

J. Hardwick (1990), *Practical adaptive allocation rules with loss functions that incorporate ethical costs,* University of Michigan, technical report.

P. Whittle (1982), *Optimization over time,* Wiley, New York.

M. Woodroofe and R. Keener (1987), *Asymptotic expansions in boundary crossing problems,* Ann. Prob. **15**, 102–114.

M. Woodroofe and J. Hardwick (1990), *Sequential allocation for an estimation problem with ethical costs,* Ann. Statist. **18**, 1358–1377.

M. Woodroofe (1989), *Very weak expansions for sequentially designed experiments: Linear models,* Ann. Statist. **17**, 1087–1102.

Department of Statistics, University of Michigan, Ann Arbor, Michigan 48109
E-mail address: Janis_Hardwick@ub.cc.umich.edu

Contemporary Mathematics
Volume **115**, 1991

Discussion on Papers by Geweke, Wolpert, Evans, Oh, and Kass, Tierney, and Kadane

JAMES H. ALBERT

In the last ten years, a substantive amount of research on Bayesian computing algorithms has been published. It seems that three main approaches to the integration problem have been presented. One general approach, discussed in Naylor and Smith (1982), (1988) and Smith, Skene, Shaw, Naylor, and Dransfield (1985), is based on the application of a Gauss–Hermite quadrature rule on a product grid. A second approach, outlined by the paper of Kass, et al., is based on intelligent asymptotic expansions of the integrands needed in the computation of posterior expectations and marginal posterior densities. The papers of Geweke, Wolpert, Evans, and Oh discuss the third general approach—approximating multidimensional integrals using Monte Carlo methods.

Geweke introduces the Monte Carlo algorithm and summarizes the success that he has had using this methodology. The basic problem is to compute the expectation

$$\overline{g} = E[g(\theta)] = \int g(\theta)f(\theta)\,d\theta \Big/ \int f(\theta)\,d\theta,$$

where f is the unnormalized posterior density and g is a function of interest. In the case where it is difficult to sample from the posterior distribution, importance sampling is typically used. To use this method, one first finds a suitable importance sampling density I which is easy to simulate and mimics the posterior distribution. A sequence of random variables $\{\theta_i\}$ is simulated from I, and the approximation to the expectation is given by

$$\overline{g}_n = \sum_{i=1}^{n} g(\theta_i)w(\theta_i) \Big/ \sum_{i=1}^{n} w(\theta_i),$$

where $w(\theta) = f(\theta)/I(\theta)$.

The main features of this approximation, pointed out by Geweke, are

1. Applying the Central Limit Theorem, $\overline{g_n}$ is approximately normally distributed with mean $\overline{g}$ and variance σ^2/n. The estimated standard error of $\overline{g_n}$ provides a useful measure of accuracy of the approximation.
2. One can improve the accuracy of the estimate by increasing the number of simulations n or decreasing the variance σ^2. Since the standard error of $\overline{g_n}$ decreases at the slow rate of $n^{-1/2}$, the effective Monte Carlo techniques are designed to reduce the variance σ^2.
3. One can reduce the variance by the choice of a good importance sampling density. For a large class of posterior distributions, Geweke has found the multivariate split-Student family to be an efficient class of importance densities.
4. Other variance-reduction techniques can significantly accelerate the rate of convergence. For example, some posterior distributions exhibit certain types of symmetry, and antithetic acceleration in these cases can improve convergence.

Wolpert, in his review of Monte Carlo integration, makes some relevant comments about why there is renewed interest in this method for Bayesian calculations. First, after a suitable transformation has been performed, posterior distributions are often well approximated by the multivariate Student or multivariate Gaussian forms. Second, Monte Carlo methods appear to be significantly more effective than product-rule or asymptotic methods for high-dimensional problems. The computation time for product-rule quadrature algorithms (such as that of Naylor and Smith, 1982) increase exponentially in the number of parameters. In contrast, the dimension of the problem for Monte Carlo does not appear to be a crucial aspect of its convergence behavior. Finally, the method is easy to implement when there are many functions g of interest, such as in the evaluation of a kernel density estimate. All the integrals can be approximated using the same sequence of random deviates $\{\theta_i\}$. Wolpert demonstrates these desirable features of Monte Carlo in his bioassay example.

Oh and Evans' papers are excellent illustrations of the current research in Monte Carlo Bayesian integration. It is not clear in the basic description of Monte Carlo how the accuracy depends on the number of parameters p. Oh shows that, if a good importance function is selected, the accuracy of the Monte Carlo estimate, measured by the variance of the weight function $w(\theta)$, decreases only linearly as a function of p. However, if a poor importance function is chosen, the variance of $w(\theta)$ can increase exponentially in p. This result reaffirms the value of a good importance function $I(\theta)$. A natural question to ask is: how does one tell if one has selected a good importance function? Geweke calculates a statistic RNE which compares the estimated standard error using the importance sampling density with the estimated standard error if the posterior could be sampled. Low values of RNE indicate that a more efficient importance sampler than the present one may be found.

As Geweke indicates in his summary, there is a need for more robust and intelligent algorithms for constructing an importance sampling density. Geweke's split-t importance sampling density is selected on the basis of an initial calculation of the posterior mode and the Hessian matrix evaluated at the mode. Once this sampling density is selected, it is not adapted further in the simulation process. In contrast, Oh describes an alternative algorithm which continually adapts the importance sampling density on the basis of the simulated values. Evans describes a similar method of adapting the importance function to the posterior distribution. The special feature of his approach is the concept of chaining. Suppose a class of integration problems $\int f_v(\theta)\,d\theta$ is considered, indexed by ν and let $\{l_\lambda(\theta),\ \lambda \in \Lambda\}$ denote a family of importance functions. Let l_{λ_i} denote the best fitting importance function for the integration problem where $\nu = \nu_i$. Evans considers a sequence of problems $\{\nu_1, \dots, \nu_n\}$ and the corresponding adaptive functions $\{l_{\lambda_1}, \dots, l_{\lambda_n}\}$. If we have some confidence in the solution of the initial problem (ν_1) and the changes in the $\{\nu_i\}$ are small, we can be somewhat confident in the solution of the ν_n problem.

Kass, et al. summarize a number of recent results on the approximate solutions to Bayesian integral problems using asymptotic expansions of integrals. This methodology has some attractive features. The most attractive feature is the ease and quickness of computation. Existing algorithms to find maximum likelihood estimators can be applied and symbolic algebra programs like *Maple* and *Mathematica* can be used in calculating the derivatives needed. A second attractive feature is that one can derive analytical approximations to posterior moments and marginal densities. For example, Kass and Steffey (1989) use these expansions to derive attractive approximations to moments for hierarchical models. Probably the biggest problem in the application of these approximations is the lack of any error estimate. Laplace methods, like asymptotic maximum likelihood theory, can perform poorly for various problems and the user needs some diagnostic statistic which indicates that the method may be inaccurate.

These papers indicate that there has been rapid progress in the development of good general-purpose algorithms for Bayesian computation. There are, however, a number of issues that still need to be addressed—two of these are described below.

1. One implicit issue in all of the computing algorithms is parameterization. The accuracy and efficiency of a given algorithm can often be improved by the "right" reparameterization. Hills (1987) and Achcar and Smith (1990) illustrate this point with many examples. Since a common assumption is an underlying multivariate Gaussian or Student t distribution, a simple reparameterization is to change the support of a p-dimensional parameter to

$\mathbf{R}^p$. Thus, for example, a positive parameter is transformed by a logarithm and a proportion is transformed by a logit. It would be useful to consider a family of possible transformations and, for a particular problem, choose a member of this family for which the computing algorithm on the transformed parameters is most efficient. Graphical methods and various diagnostic statistics need to be developed which indicate poor parameterizations.

2. It seems that typically a particular algorithm is used in isolation and few comparisons are made with competing algorithms. One would like to visualize a computing environment where product-rules, Laplace methods, and Monte Carlo methods (with a wide choice of importance functions) are all available. One would be able to easily compare the use of various algorithms for a given problem. Also, one could use mixed strategies, where, for example, Laplace methods are used to integrate out some well-behaved nuisance parameters and Gaussian quadrature is used to construct a contour plot for two parameters of interest. The Bayesian computing system that is closest to this view is Bayes 4, developed at the University of Nottingham (Naylor and Shaw, (1985)). A more recent version of this program incorporates adaptive product rules, spherical rules, and Monte Carlo methods. Unfortunately, this system is currently not widely available in the United States. Also, the user needs to be familiar with programming and the behavior of the computing algorithms. For example, the adaptive quadrature algorithm will overflow if a poor starting value is given and the algorithm on a small grid may take a long time to stabilize. A good Bayesian computer program should provide some guidance on the correct algorithm and parameterization to use for a given problem.

References

J. A. Achcar and A. F. M. Smith (1990), *Aspects of reparametrization in approximate Bayesian inference*, Essays in Honor of George Bernard (S. Geisser, J. S. Hodges, S. J. Press, and A. Zellner, eds.), North-Holland, Amsterdam.

S. Hills (1987), *Parameter transformations*, Technical report, Department of Mathematics, University of Nottingham.

R. E. Kass and D. Steffey (1989), *Approximate Bayesian inference in conditionally independent hierarchical models (parametric empirical Bayes models)*, J. Amer. Statist. Assoc. **84**, 717–726.

J. C. Naylor and J. E. H. Shaw (1985), *Bayes Four—User Guide*, Technical report, Department of Mathematics, University of Nottingham.

J. C. Naylor and A. F. M. Smith (1982), *Applications of a method for the efficient computation of posterior distributions*, Appl. Stat. **31**, 214–225.

——(1988), *Econometric illustrations of novel numerical integration strategies for Bayesian inference*, J. Econometrics **38**, 103–125.

A. F. M. Smith, A. M. Skene, J. E. H. Shaw, J. C. Naylor, and M. Dransfield (1985), *The implementation of the Bayesian paradigm*, Communications in Stat. **14**, 1079–1102.

Department of Mathematics and Statistics, Bowling Green State University, Bowling Green, Ohio 43403

E-mail address: albert@andy.bgsv.edu

Contemporary Mathematics
Volume **115**, 1991

Comments on Computational Conveniences Discussed in the Articles by Evans, Geweke, Müller, and Kass–Tierney–Kadane

RAMALINGAM SHANMUGAM

ABSTRACT. The intent of this article is to comment on computational conveniences and disadvantages of the approaches discussed by Evans, Geweke, Müller, and Kass–Tierney–Kadane in this volume. My discussion concentrates on computing the entropy of a gamma population.

1. Introduction

First, let us briefly recap various techniques that are suggested by Evans, Geweke, Müller, and Kass–Tierney–Kadane in this volume. In a Bayesian analysis with interest on the expectation of a function $g(\theta)$ of the parameter θ in a chance mechanism, one encounters integrals of the following type:

$$Eg(\theta) = \left(\int_{-\infty}^{\infty} g(\theta)f(\theta)\,d\theta\right)\left(\int_{-\infty}^{\infty} f(\theta)\,d\theta\right)^{-1}. \tag{1}$$

When the function $f(\theta)$ is $p(\theta)l(\theta, x_{(n)})$, with $p(\theta)$ and $l(\theta, x_{(n)})$ denoting, respectively, prior and likelihood density for θ based on observed sample data $x_{(n)}$ of size n, the expectation in (1) is data oriented and hence is denoted by $E[g(\theta)|x_{(n)}]$. To assess the significance of information in data $x_{(n)}$, Bayesian analysts compare $E[g(\theta)|x_{(n)}]$ with its counterpart $E[g(\theta)]$, the amount before the instrument of data $x_{(n)}$, and in that case, for $E[g(\theta)]$, the function $f(\theta)$ is simply a prior $p(\theta)$. Sources of difficulty in finding an analytically assimilable closed form for the above mentioned expectations trace to either $f(\theta)$ or $g(\theta)$. When a prior $p(\theta)$ is improper, $f(\theta)$ could inherit the difficulty causing the second integral in (1) not to be amenable to a closed form. Another occasion is when $g(\theta)$ is structurally incompatible

Key words and phrases. Monte Carlo, numerical integration, Laplace's method, importance sampling, entropy.

with that of $f(\theta)$ causing the first integral in (1) also to be nonsolvable into a closed form, as an example in §2 illustrates. Different techniques have been suggested in this volume to numerically approximate analytically unsolvable integrals of the type in (1), and they are briefly summarized below.

Monte Carlo technique is more accurate and generic (cf. Geweke (1989)) but could be misleading in irregular cases and also inefficient sometime. A remedy is to go for an adaptive or a chaining algorithm (cf. Evans (1989)). To approximate $E[g(\theta)|x_{(n)}]$, the Monte Carlo principle is practiced based on either random drawings $\theta_1, \theta_2, \ldots, \theta_n$ from the likelihood function $\ell(\theta, x_{(n)})/\int \ell(\theta, x_{(n)})\,d\theta$ or drawings $\theta'_1, \theta'_2, \ldots, \theta'_n$ from the posterior function $\ell(\theta, x_{(n)})p(\theta)/\int \ell(\theta, x_{(n)})p(\theta)\,d\theta$. In the former case, the approximation is

$$E_\ell[g(\theta)|x_{(n)}] = \sum_{i=1}^{n} g(\theta_i)p(\theta_i) \Big/ \sum p(\theta_i), \tag{2}$$

where the subscript ℓ in $E(\cdot)$ denotes the function from which drawings are made, and in the latter case, it is

$$E_f[g(\theta)|x_{(n)}] = \sum g(\theta'_i)/n. \tag{3}$$

Expression (2) is preferable over (3) if the likelihood function based variance $\mathrm{Var}_\ell[g(\theta)p(\theta)]$ is smaller than the posterior density based variance $\mathrm{Var}_{p(\theta|x)}[g(\theta)]$. Note that there is no choice but to use drawings $\theta'_1, \ldots, \theta'_n$ from just the prior $p(\theta)$ to approximate $E_p[g(\theta)]$.

When it is possible to come up with a density called importance density $I(\theta)$ for each and efficient drawings $\theta_1, \ldots, \theta_n$, where $I(\theta)$ is such that $f(\theta) = p(\theta) = w(\theta)I(\theta)$, the expectation of $g(\theta)$ prior to the availability of data is

$$E_I[g(\theta)] = \sum g(\theta_i)w(\theta_i) \Big/ \sum w(\theta_i). \tag{4}$$

In the posterior case, it is

$$E_{I'}[g(\theta)|x_{(n)}] = \sum g(\theta'_i)w'(\theta'_i) \Big/ \sum w'(\theta'_i) \tag{5}$$

as $f(\theta) = p(\theta)\ell(\theta, x_{(n)}) = w'(\theta)I'(\theta)$ and the drawings $\theta'_1, \ldots, \theta'_n$ are from $I'(\theta)$ not $I(\theta)$. An advantage of the importance sampling is that it is fairly accurate although it is more time-consuming, and success depends on a good choice of I. The support for I should contain the support of f. The efficiency of importance sampling is further improved by adding a principle called "acceleration" (cf. Geweke (1989); Müller (1989)).

Another method due to Lindley (1980) provides an approximation

$$E_{L_i}[g(\theta)] = g(\hat{\theta}) + \frac{1}{2}\sum_{i,j}(g_{ij}(\hat{\theta}) + 2g_i(\hat{\theta})\rho_i(\hat{\theta}))\sigma_{ij} + \frac{1}{2}\sum_{i,j,k,m} L_{ijm}g_k\sigma_{ij}\sigma_{mk}o(n^{-1}) \tag{6}$$

to an order of n^{-1}, where $\rho(\theta) = \ln p_k(\hat{\theta})$, $\hat{\theta}$ is the mode of $L = \ln \ell(\theta, x_{(n)})$ and the subscripts in g, L, ρ, and σ indicate the order of differentiation with respect to specific components of θ. A disadvantage in Lindley's method is that it requires evaluations up to third-order derivatives.

A method called "Laplace method" discussed by Kass–Tierney–Kadane exponentializes Lindley's method and it requires evaluations up to only two derivatives. The method is more time-consuming when $g(\theta)$ is nonpositive. The Laplace method is interactive and asymptotic. The fully exponential Laplace approximation for positive $g(\theta)$ is

$$(7) \qquad E_{L_a}[g(\theta)|x_{(n)}] = \{\Sigma_N|/|\Sigma_D|\}^{1/2} \exp[n\{\sup h_D(\hat{\theta}_D) - \sup h_N(\hat{\theta}_N)\}],$$

where

$$(8) \qquad h_N(\theta) = h_D(\theta) - \tfrac{1}{n} \ln g(\theta),$$

$$(9) \qquad h_D(\theta) = -\tfrac{1}{n} \ln f(\theta),$$

and Σ_K is the inverse of the minus of the second derivative $\partial_\theta^2 h_K(\theta)$ for $K = D, N$.

An interesting dividend of the Laplace method is its ability to assess the value of one prior $q(\theta)$ over another $p(\theta)$. Calling

$$(10) \qquad \rho(\theta) = \ln[q(\theta)/p(\theta)]$$

a perturbation function, Kass, Tierney, and Kadane have identified that

(11)

$$E_q[g(\theta)|x_{(n)}] = \left(\int_{-\infty}^{\infty} g(\theta) q(\theta) \ell(\theta, x_{(n)})\, d\theta\right) \left(\int_{-\infty}^{\infty} q(\theta) \ell(\theta, x_{(n)})\, d\theta\right)^{-1}$$
$$= E_p[g(\theta)|x_{(n)}] + d(\rho, g, \tilde{\theta}) + o(n^{-2}),$$

where $d(\cdot)$ denotes a drift in the posterior expectation of $g(\theta)$ due to changing prior $p(\theta)$ to $q(\theta)$ and it is to be estimated using (12) below.

$$(12) \qquad d_g(p, q, \tilde{\theta}_p) = [\partial_\theta \rho]' \{-\partial_\theta^2 \ln f_p(\theta)\}^{-1} [\partial_\theta g]|_{\theta = \tilde{\theta}_p},$$

where $\tilde{\theta}_p$ denotes the mode of $f_p(\theta) = \ell(\theta, x_{(n)}) p(\theta)$. With this summary, we examine in §2 the performance of these techniques in computing the entropy of a gamma population.

2. Gamma entropy

Recently, Shanmugam (1989) characterized a linear mean exponential family. In the discussions of this article, we consider only one member of the family, namely gamma population. Results pertaining to other members of the family are reported separably elsewhere.

A positive valued natural observation X is called gamma type if its likelihood function $\ell(\theta, \gamma, x)$ is of the format

$$(13) \qquad \ell(\theta, \gamma, x) = \frac{x^{\gamma - 1}}{\Gamma(\gamma)} \exp[x\theta - \gamma \psi_\theta], \qquad x \in (0, \infty)$$

with respect to a Lebesgue measure $m(x)$ where $\theta \in (-\infty, 0)$ is the natural parameter of the chance mechanism, $\psi_\theta = -\ln(-\theta)$ and $\gamma \in (0, \infty)$ is the reproductive parameter as it catalogs an additivity of observations. That is, for observed data $x_{(n)} = (x_1, \ldots, x_n)$ from n independent, not necessarily identical gamma populations with likelihood functions $\{\ell(\theta, \gamma_i, x_i), i = 1, 2, \ldots, n\}$, the data total $t_n = \Sigma x_i = n\overline{x}$ is also a gamma with likelihood function $\ell(\theta, \Sigma'\gamma_i, n\overline{x})$. In real life applications of the gamma model, γ needs to be estimated. We, here, assume γ to be known for the sake of simplicity needed to understand discussions about θ. In a health application, the natural parameter θ is inversely related to the life expectancy of AIDS patients (see in (14) below) assuming that x denotes survived days of a patient since the time illness was diagnosed.

$$E(X|\theta, \gamma) = \int_0^\infty x\ell(\theta, \gamma, x)\,dx = -\gamma/\theta = \gamma e^{-\psi_\theta}. \tag{14}$$

Realizing that at least in an AIDS application, the average in (14) is not large, we note that either γ has to be smaller or $|\theta|$ is larger. Corresponding to the second possibility is that ψ_θ is too large, causing its computation to be difficult. Now we may ask: What is the attraction to ψ_θ? The answer is that it is a part of the gamma entropy

$$-\int_0^\infty \ell(\theta, \gamma, x)\ln\ell(\theta, \gamma, x)\,dx = \psi_\theta + a(\gamma), \tag{15}$$

where

$$a(\gamma) = \gamma + \ln\Gamma(\gamma) + (1-\gamma)\partial_\gamma \ln\Gamma(\gamma). \tag{16}$$

It is well known that an entropy measure is unique and hence plays a valuable role in displaying statistical information. See Golomb (1966) for details. To find the average entropy, a stochastic prior specification of one's experience on the random parameter θ is needed. The specification could follow any one of the principles: conjugation, noninformation, or locally invariance.

Conjugate prior is a convenient "building block" due to its structural similarity with the likelihood function, and it, for (13) is

$$p_c(\theta) = \frac{(n_0\overline{x}_0)^{n_0\gamma+1}}{\Gamma(n_0\gamma+1)} \exp[n_0(\overline{x}_0\theta - \gamma\psi_\theta)], \qquad \theta \in (-\infty, 0) \tag{17}$$

with its parameters n_0 and $\overline{x}_0$ denoting the number of observations considered thus far and their average, respectively.

The noninformational prior is locally uniform (a synonimity for a total ignorance). A prescription (see Jeffreys (1961)) for finding it is to have it propositional to the square root of Fisher's information, which is the likelihood based expectation $E[\partial_\theta \ln\ell(\theta, \gamma, x)]^2$. For gamma population, the noninformational prior is

$$p_{ni}(\theta) \propto \exp(-2\psi_\theta), \qquad \theta \in (-\infty, 0) \tag{18}$$

which is quite improper probability density but not flat.

Locally invariant prior possesses desirable properties like S-labelling invariance, Ω-labelling invariance, Ω restriction invariance, sufficiency, direct product and repeated product (cf. Hartigan (1964)). Hartigan's prescription to find it is, in our notation.

$$p_H(\theta) \propto \exp\left[-\int\left\{\frac{E[\partial_\theta \ln \ell(\theta, \gamma, x)\partial_\theta^2 \ln \ell(\theta, \gamma, x)]}{E[\partial_\theta^2 \ln(\theta, \gamma, x)]}\right\} d\theta\right]$$

provided the likelihood based average influence $E[\partial_\theta \ln(\theta, \gamma, x)]$ is zero. For the gamma likelihood in (13), the above requirement is satisfied and hence the locally invariant prior is

$$p_H(\theta) \propto 1, \qquad \theta \in (-\infty, 0) \tag{19}$$

which is flat and improper.

It is easy to see that the posterior density

$$f(\theta) = p(\theta)\ell(\theta, n\gamma, n\overline{x}) / \int_{-\infty}^{0} p(\theta)\ell(\theta, n\gamma, n\overline{x})\, d\theta$$

is structurally the same as in (17) in the case of using conjugate prior in (17) for $p(\theta)$, but with $n_0\overline{x}_0$ updated by $\overline{x}_c = n_0\overline{x}_0 + n\overline{x}$ and $n_0\gamma$ by $n_c = (n_0 + n)\gamma$. That is,

$$f_c(\theta) = [(\overline{x}_c)^{n_c+1}/\Gamma(n_c + 1)] \exp[\overline{x}_c\theta - n_c\psi_\theta]. \tag{20}$$

With noninformational prior in (18) for $p(\theta)$, the posterior density becomes

$$f_{ni}(\theta) = [(\overline{x}_{ni})^{n_{ni}+1}/\Gamma(n_{ni} + 1)] \exp[\overline{x}_{ni}\theta - n_{ni}\psi_\theta], \tag{21}$$

where $\overline{x}_{ni} = n\overline{x}$ and $n_{ni} = n\gamma + 2$. With locally invariant prior in (19), the posterior density is

$$f_H(\theta) = [(\overline{x}_H)^{n_H}/\Gamma(n_H + 1)] \exp[(\overline{x}_H\theta - n_H\psi_\theta)], \tag{22}$$

where $\overline{x}_H = n\overline{x}$ and $n_H = n\gamma$, a special case of (20) with $n_0 = 1$. Except for the conjugate prior in (17), the other two priors (noninformational and locally invariant) are improper, contributing to computational difficulties on two counts in finding the average entropy for gamma population

$$E_K[\psi_\theta + a(\gamma)] = \left(\int_{-\infty}^{0} p_K(\theta) \ln(-\theta)\, d\theta\right)\left(\int_{-\infty}^{\infty} p_K(\theta)\, d\theta\right)^{-1} + a(\gamma), \tag{23}$$

prior to the availability of data $x_{(n)}$, where $K = ni$, H, respectively, according to (1); first, the second integral in (23) is unyielding to a closed form due to improperness in $p_K(\theta)$ when $K = ni$ or H. In the absence of a likelihood function, the deficiency of $p_K(\theta)$ is absorbed by $f(\theta)$ as $f(\theta)$ identifies entirely with $p_K(\theta)$; secondly, the structural incompatibility between the nonlinear function ψ_θ and $f(\theta) = p_K(\theta)$ for $K = ni$ and H

causes the first integral in (23) also not to be amenable to a closed form. But, the conjugate prior based average entropy

$$(24)\qquad E_c[\psi_\theta + a(\gamma)] = \left[\frac{\Gamma(n_0\gamma+1)}{(n_0\overline{x}_0)^{n_0\gamma+1}}\right]\int_{-\infty}^{0} p_c(\theta)\ln(-\theta)\,d\theta + a(\gamma)$$

prior to the availability of data $x_{(n)}$ is not amenable to a closed form or only one count, namely, the noncompatibility between the nonlinear ψ_θ and $p_c(\theta)$, where $p_c(\theta)$ is as given in (17).

However, for posterior average entropy $E[\psi_\theta + a(\gamma)|x_{(n)}]$ which incorporates the data likelihood via $f(\theta) = \ell(\theta, n\gamma, n\overline{x})p(\theta)$ as shown in (20), (21), and (22), the computational complexity is reduced to only one count: the structural noncompatibility between nonlinear ψ_θ and $f(\theta)$, no matter which type of prior only among the three mentioned has been provoked. The exponential structure of the gamma likelihood shown in (13) does help as it filters out the difficulty passed on by the improperness of the prior $p_{ni}(\theta)$ or $p_H(\theta)$ as the case may be, making the posterior $f(\theta)$ immune to it in both cases. With conjugate prior $p_c(\theta)$, this filtering is unnecessary as the prior itself is proper and compatible with the data likelihood $\ell(\theta, n\gamma, n\overline{x})$. Hence, making use of the format in (1), we obtain the posterior entropy as

$$(25a)\quad E_K[\psi_\theta + a(\gamma)|x_{(n)}] = \beta_K(\overline{x}_K, n_K)\int_{-\infty}^{0} \psi_\theta \exp[-n_K\psi_\theta]\{\exp(\overline{x}_K\theta)\}\,d\theta + a(\gamma),$$

where the constant

$$(25b)\qquad \beta_K(\overline{x}_K, n_K) = (\overline{x}_K)^{n_K+1}/\Gamma(n_K+1)$$

for $K = c$, ni, or H, respectively, for conjugate, noninformational, or locally invariant cases. The structuralism helps so far as the second integral in (1) is concerned only in the posterior case for all three priors. Solving the integral in (25a) needs a method among those discussed in §1. Which technique is best for our need in approximating (25a) is the theme for §3 below.

3. A comparison of entropy computation

Computing the average entropy in (23) of gamma population prior to data $x_{(n)}$ is surrounded with problems due to improperness in the function $f(\theta) = p(\theta)$ no matter whether the Monte Carlo, Importance Sampling, Lindley's or Laplace method is used unless the function $p(\theta)$ is the conjugate prior in (17). The result for conjugate prior $p_c(\theta)$ based average entropy $E[\psi_\theta + a(\gamma)]$ before the availability of data $x_{(n)}$ is a special case, with $n = 0 = \overline{x}$, of the result for conjugate posterior $p_c(\theta)\ell(\theta, n\gamma, n\overline{x})$ based average $E[\psi_\theta + a(\gamma)|x_{(n)}]$ which follows subsequently.

Applying the Monte Carlo formula in (2), we obtain an approximation for entropy in (25a) as

$$(26)\quad E_{\ell}[\psi_\theta + a(\gamma)|x_{(n)}] = \beta_K(\overline{x}_K, n_K)\left[\sum_{i=1}^{m} \psi_{\theta_i} p_K(\hat{\theta}_i) \Big/ \sum_{i=1}^{m} p_K(\hat{\theta}_i)\right] + a(\gamma)$$

with drawings $\theta_1, \ldots, \theta_m$ from the likelihood $\ell(\theta, n\gamma, n\overline{x})$ as specified in (13), and the approximation is different from the result

$$(27)\qquad E_f[\psi_\theta + a(\gamma)|x_{(n)}] = \beta_K(\overline{x}_K, n_K)\left[\sum_{j=1}^{m} \psi_{\theta'_j}/n\right] + a(\gamma),$$

due to the formula in (3), except for the case involving locally invariant prior $p_H(\theta)$, where the drawings $\theta'_1, \ldots, \theta'_m$ are from the posterior density $f_K(\theta)$ as specified in (20), (21), or (22) for $K = c$, ni, or H, respectively. In the case involving locally invariant prior $p_H(\theta)$, results of (26) and (27) are the same.

Noting that the part $I(\theta) = \exp(\overline{x}_K\theta)$ is a proper probability density in the range $\theta \in (-\infty, 0)$, we could use it as an importance density for drawings $\theta''_1, \theta''_2, \ldots, \theta''_m$ so that an approximation in (28) for the entropy is obtainable. According to (5), the approximation is

$$(28)\qquad E_I[\psi_\theta + a(\gamma)|x_{(n)}] = \beta_K(\overline{x}_K, n_K)\sum_{i=1}^{m} \psi_{\theta''_i} w_K(\theta''_i) + a(\gamma),$$

with the weight function

$$(29)\qquad w_K(\theta) = \exp(-n_K\psi_\theta)$$

as $K = c$, ni, or H. After comparing $w_K(\theta)$ in (29) with an appropriate $p_K(\theta)$ in (17), (18), or (19), it is evident that the approximations in (26) are different from their counter results due to (29).

Now, we try Lindley's approximation in (6) for the gamma entropy in (25a). It is easily seen that $\hat{\theta} = -\gamma/\overline{x}$ is the mode of the likelihood in (13). Hence, the approximations are

$$E_{Li}[\psi_\theta + a(\gamma)|x_{(n)}]$$

$$(30a)\quad = \beta_c(\overline{x}_c, n_c\left[\ln\left(\frac{\overline{x}}{\gamma}\right) + \frac{1}{2n\gamma} + \left(\frac{n_0}{n}\right)\left(\frac{\overline{x}_0}{\overline{x}} - 1\right)\right] + a(\gamma) + o(n^{-1})$$

in the case of using conjugate prior in (17),

$$(30b)\quad E_{Li}[\psi_\theta + a(\gamma)|x_{(n)}] = \beta_H(\overline{x}_H, n_H)\left[\ln\left(\frac{\overline{x}}{\gamma}\right) + \frac{1}{2n\gamma}\right] + a(\gamma) + o(n^{-1})$$

in the case of using locally invariant prior in (19), and

$$(30c)\quad E_{Li}[\psi_\theta + a(\gamma)|x_{(n)}] = \beta_{ni}(\overline{x}_{ni}, n_{ni})\left[\ln\left(\frac{\overline{x}}{\gamma}\right) - \frac{3}{2n\gamma}\right] + a(\gamma) + o(n^{-1})$$

in the case of using noninformational prior.

Lastly, we approximate using the Laplace method. The Laplace method is in a sense an inconvenient one for finding an average entropy as $\psi_\theta \in (-\infty, \infty)$ is not always positive. For the negative possibilities of ψ_θ, use of its moment generating function needs to be found first (see Kase–Tierney–Kadane for details). Applying (17), we note that even for the positive part of ψ_θ the approximation $E_{La}[\psi_\theta + a(\gamma)|x_{(n)}]$ is difficult due to the nonlinearity in the estimating equation

$$\partial_\theta \ln f_K(\theta) = 0 = \overline{x}_K + \frac{n_K}{\theta} - \frac{1}{\theta \ln(-\theta)} \tag{31}$$

for finding $\hat{\theta}_N$, although $\hat{\theta}_D$ is easily found to be $\hat{\theta}_D = n_K/\overline{x}_K$, for $K = c$, ni, and H. It is not clear how much an approximation of the third term in (31) would distort the result of $E_{La}[\psi_\theta|x_{(n)}]$, and remember it is only for the positive case of ψ_θ.

However difficult the finding of $E_{La}[\psi_\theta + a(\gamma)|x_{(n)}]$ may be, the dividend of the Laplace method discussed in (12) is much more useful for assessing a drift in the average entropy $E[\psi_\theta + a(\gamma)|x_{(n)}]$ due to changing a prior $p(\theta)$ to another $q(\theta)$. In our discussions of this article, we have three pairwise comparisons among conjugate, noninformational, and locally invariant priors stated in (17), (18), and (19), respectively. Designating $p_c(\theta)$ in (17) for $p(\theta)$ and $p_H(\theta)$ in (19) for $q(\theta)$, we note that the drift in average entropy is

$$d_{\psi_\theta + a(\gamma)}[p_c, p_{ni}, \tilde{\theta}_p] = -\gamma n n_0(\overline{x}_0 - \overline{x})/n_c \overline{x}_c \tag{32}$$

as $\tilde{\theta}_{pc} = -n_c/\overline{x}_c$ is the mode of the posterior $f_c(\theta)$ in (20). The drift would have been inclining only if a current sample mean $\overline{x}$ is smaller than the prior sample mean $\overline{x}_0$. Instead, if we have designated prior $p_{ni}(\theta)$ in (18) as an alternate choice for $q(\theta)$, the average entropy would have drifted by an amount

$$d_{\psi_\theta + a(\gamma)}[p_c, p_H, \tilde{\theta}_p] = -\gamma n n_0(\overline{x}_0 - \overline{x})/n_c \overline{x}_c \tag{33}$$

which is the same drift in (32) between p_c and p_{ni}, implying a relationship only to the order of $o(n^{-2})$, that

$$\begin{aligned} E_{p_H}[\psi_\theta + a(\gamma)|x_{(n)}] &= E_{p_c}[\psi_\theta + a(\gamma)|x_{(n)}] + a(\gamma) + \gamma n n_0(\overline{x} - \overline{x}_0)/n_c \overline{x}_c \\ &= E_{p_{ni}}[\psi_\theta + a(\gamma)|x_{(n)}], \end{aligned} \tag{34}$$

which makes sense from another view. That is, locally invariant prior based posterior $f_H(\theta)$ in (22) is being a special case, with $n_0 = 0$ of conjugate prior based posterior $f_c(\theta)$ in (20), the drift $d_{\psi_\theta + a\gamma}[p_H, p_{ni}, \tilde{\theta}_H]$ is zero, validating (34). But, by choosing prior $p(\theta)$ in (19) for $q(\theta)$ and prior $p_{ni}(\theta)$ in (18) for $p(\theta)$ in the formula (11), we obtain a relationship only to the order of $o(n^{-2})$, that

$$E_{p_H}[\psi_\theta + a(\gamma)|x_{(n)}] = E_{p_{ni}}[\psi_\theta + a(\gamma)|x_{(n)}] + (2/(n\gamma + 2)). \tag{35}$$

Noticing that the relationships in (34) and (35) between $E_{pH}(\cdot)$ and $E_{p_{ni}}(\cdot)$ are not exact, where both relations are to the same order $o(n^{-2})$, the author ponders over the mystery behind the term $(2/(n\gamma+2))$. Is it an anomoly in the Laplace method, promoted by Kass, Tierney, and Kadane?

Acknowledgment

I dedicate this work, with respect, to my teacher and friend Professor Jagbir Singh.

References

M. Evans (1989), *Adaptive importance sampling and chaining*, this volume, pp. 137–143.

J. Geweke (1989), *Generic, algorithmic approaches to Monte Carlo integration in Bayesian inference*, this volume, pp. 117–135.

S. W. Golomb (1966), *The information generating function of a probability distribution*, I.E.E.E. Trans. Inform. Theory **12**, 75–77.

J. A. Hartigan (1964), *Invariant prior distributions*, Ann. of Math. Stat. **35**, 836–845.

H. Jeffreys (1961), *Theory of Probability*, Oxford University Press, London.

R. E. Kass, L. Tierney, and J. B. Kadane (1989), *Laplace's method in Bayesian analysis*, this volume, pp. 89–99.

D. V. Lindley (1980), *Approximate Bayesian methods*, Bayesian Statistics (J. M. Bernardo et al., eds.), University Press, Valencia.

P. Müller (1989), *Monte Carlo integration in general dynamic models*, this volume, pp. 145–163.

R. Shanmugam (1989), *Asymptotic homogeneity tests for mean exponential family distributions*, J. Stat. Planning and Inference **23**, 227–241.

Department of Mathematics, University of Colorado at Denver, Denver, Colorado 80465

E-mail address: RShanmugam@cudnor.denver.colorado.edu

Contemporary Mathematics
Volume **115**, 1991

A Discussion of Papers by Genz, Tsutakawa, and Tong

INGRAM OLKIN

The task of computing the probability of falling in an interval for many univariate distributions has been well documented. There are a variety of procedures that yield very accurate results even for relatively bizarre univariate distributions. However, in the multivariate case this problem still remains somewhat unresolved, especially in high dimensions.

In the statistical and probability literature there have been a number of directions that the research has taken. Suppose that we know an underlying p-variate distribution—how do we determine the probability of falling in a region $\mathscr{A}$? When the underlying distribution is normal, tables for $p = 2$ and $p = 3$ are available for common regions $\mathscr{A}$ such as rectangles or cubes. For higher dimensions results are available only for very special covariance matrices. Exact values for the probability of falling into orthants remains a difficult problem, even in the case of normality. The focus of the three papers lies in what to do when the dimensionality is moderate to large.

The paper by Genz provides a general discussion of algorithms for the evaluation of $P\{a_i \le X_i \le b_i,\ i = 1, \dots, n\}$. The choice of method is intimately connected with the size of n. The main discussion focuses on subregion adaptive algorithms. An important aspect is that error estimates are provided. The present results are encouraging, but the fact that the algorithms are currently most useful for values of rather small values of n $(2 \le n \le 10)$ suggests that we still await the development of reliable methods for larger n.

The paper by Tsutakawa addresses a Bayesian approach to item response theory. The essence of the problem lies in several maximizations and an integration. The maximizations are

$$\max_{\xi} p(x, y \mid \xi)\, p(\xi) \equiv \max_{\xi} h(x, y, \xi),$$

$$\max_{\xi} E(\theta \mid x, \xi)\, h(x, y, \xi),$$

0271-4132/91 $1.00 + $.25 per page

and the integrations are

$$\int h(x, y, \xi)\, d\xi, \qquad \int E(\theta \mid x, \xi)\, h(x, y, \xi)\, d\xi.$$

The vectors ξ, x are p-dimensional; θ is n-dimensional. The order of magnitude in practical situations may be $n = 400$ or 2000 and $p = 30$ or 40. Although this model relates to a particular item response model, the difficulties posed relate to many Bayesian analyses. What is interesting here is that we have a concrete example in which the effects of dimensionality become clear.

An alternative direction, and one to which considerable effort has been expended, is to obtain bounds for falling into rectangles and ellipsoids. With only moment assumptions, but without assumptions of the underlying distributions, these bounds are called Chebychev inequalities. The theory of how to develop these inequalities is now well established. (See e.g., Karlin and Studden (1966).) Because Chebychev inequalities serve for any distribution they generally provide weak bounds.

A second general approach for obtaining bounds is to assume a known underlying density, say $f(x; \theta)$, and to consider $P\{X \in \mathscr{A}\} = \int_{\mathscr{A}} f(x, \theta)\, dx$ as a function, $g(\theta)$, of θ. Alternatively, the density need not depend on a parameter θ, but instead the region is a function of the parameter, in which case $g(\theta) = P\{X \in \mathscr{A}(\theta)\} = \int_{\mathscr{A}(\theta)} f(x)\, dx$. If $g(\theta)$ is an order preserving function for some order $\prec$, then we obtain bounds $g(\alpha) \le g(\theta) \le g(\beta)$ whenever $\alpha \prec \theta \prec \beta$ in this order. The paper by Tong makes use of a particular ordering; namely, the majorization ordering. The breadth of results arising from this ordering is surprising, although other orderings have been found to be fruitful in different contexts.

References

S. J. Karlin and W. J. Studden (1966), *Tchebycheff systems: with applications in analysis and statistics*, Wiley, New York.

Department of Statistics, Stanford University, Stanford, California 94305
E-mail address: iolkin@playfair.stanford.edu

Contemporary Mathematics
Volume 115, 1991

A Discussion of Papers by Luzar and Olkin, Kaishev, and Monahan and Liddle

NANCY FLOURNOY

Problems in multiple integration have captured the attention of many in the Bayesian community of statisticians; and, therefore, they are having a strong influence on the classes of problems and solutions that are currently being studied. Considerable overlap exists between the integration requirements that arise from Bayesian and non-Bayesian problems, but the later merit highlighting so as to provide a robust view of multiple integration problems from the statisicians' perspective.

This discussion focuses on the multiple integration problems of Luzar and Olkin, Kaishev, and Monahan and Liddle that were motivated by non-Bayesian problems in statistics. Originally, we sought to identify common elements and characteristics, but an elegant characterization of multiple integration problems in non-Bayesian statistics proved elusive. Each motivation is different; and, therefore, the class of techniques considered and the strategies used in their implementation have distinct elements. Furthermore, our sense is that the problems involving multiple integration included in this volume represent a sample from a larger population of problems of unknown size and character. It seems that statistical requirements for multiple integration are still emerging. Therefore, this discussion does not strive for cohesiveness, but attempts to elucidate the diversity of statistical multiple integration problems. We will enjoy seeing additional problems exposed and characterized.

1. Eigenvalue estimation

One major thrust in statistics involves relaxing the independence assumptions in probability models so as to extend multivariate analysis techniques to accommodate more natural complexities. The work of Luzar and Olkin falls

0271-4132/91 $1.00 + $.25 per page

into this category. Their paper is distinctive in this collection in that a direct approach to estimating the characteristic roots of a covariance matrix Σ does not require any integration at all. They demonstrate, however, that procedures involving multivariate integration can yield stable estimates with fewer replications than does direct simulation. They compare estimates obtained by direct simulation and computation with estimates involving multivariate integration, including estimates based on importance sampling with a variety of sampling densities and estimates obtained using ratio and regression methods.

Multivariate analysis classically has dealt with random variables that are distributed according to a multivariate normal distribution. Dependencies between p multivariate normal random variables are described by the $p \times p$ population covariance matrix Σ, and significant information regarding these dependencies is contained in the characteristic roots of Σ. One goal is to make inferences concerning Σ by estimating the characteristic roots of a sample covariance matrix. Luzar and Olkin point out that estimates based on the sample covariance matrix are biased, and they focus on methods for eliminating bias and comparing the relative performance of resulting estimates.

Luzar and Olkin begin this study for the simplest case where $\Sigma = I$ and so the characteristic roots are all identically one. That they obtained estimates on the order of 1.56 instead of 1.00 is not the focus of the paper. Rather, as they come to appreciate the variation in the behavior of estimation procedures, they recognize that their problem depends on the development of a strategy for comparing and contrasting the numerical procedures as well as analyzing bias in the resulting estimates. Thus this paper analyzes the methods by which estimates of the characteristic roots can be obtained. The methods all involve simulation and tend to agree in the second decimal point, but they vary greatly in terms of the number of replications required to obtain a stable estimate.

It is interesting that the most direct approach, without any integration at all, requires a large number of replications before stabilizing relative to other approaches. Importance sampling was employed using the exponential and Chi-square densities and mixtures of Chi-square densities. The Chi-square density with the largest degrees of freedom (i.e. 15) stabilized within 20,000 replications whereas 100,000 replications were insufficient for 2 degrees of freedom. The Chi-square with 15 degrees of freedom outperformed the mixtures studied as well. Clearly the performance of importance sampling depends on the density sampled, and yet we do not have easy, robust strategies for identifying good densities for a wide variety of problems.

Whereas several authors' reference to Monte Carlo methods has implied importance sampling, Luzar and Olkin also use Monte Carlo methods as part of the ratio, gap and regression methods that they compare. They show that the early high variation of estimates based on importance sampling can be curtailed considerably by regressing the estimates obtained by importance sampling on other information that is known about the problem.

Unfortunately, the analysis of integration strategies and convergence criteria for the simple case where $\Sigma = I$ does not yield a simple path for extensions to more general Σ. Luzar and Olkin's comparisons focus on the number random variables that must be generated to stabilize the variability of the estimate. Monte Carlo simulation provides a way to tackle many difficult computational problems in statistics. However, Luzar and Olkin's work clearly indicates that results from every simulation should include an analysis of the convergent behavior of the calculations performed. Papers reporting a fixed number of replications without justification or analysis should not be accepted for publication, and proposals to perform such simulations should not be funded.

2. The distribution of serial correlation coefficients

Kaishev develops Gaussian cubature formulae for representing generalized B-splines and integrals of them. Importance sampling methods are also developed that take advantage of the generalized B-splines representation as the probability density function of a linear combination of Dirichlet distributed random variables. Kaishev then shows how the density of serial correlation coefficients r_j can be interpreted as a generalized B-spline with knots and, hence, how to use his results to compute the distribution of r_j.

3. Optimizing integrals of weighted densities

Monahan and Liddle propose the use of a stationary version of stochastic approximation to optimize nonlinear equations that are defined by an integral over r variables x. Stochastic approximation uses Monte Carlo methods to solve the combined integration and optimization problems. Specifically, Monahan and Liddle consider integrands that factor into two components: $g(x; \theta)$ a function of both x and θ, and $w(x)$, a density function of x alone. Thus the problem is to find the minimum (or maximum) of

$$\int g(x;\theta)\, w(x)\, dx \tag{1}$$

by setting the derivative in (1) equal to zero, assuming that the derivative may be passed under the integral sign. There are many problems in statistics of this form, including minimizing an expected loss functions and (motivating Monahan and Liddle) finding the minimum Hellinger distance between a density estimate $\hat{f}(x)$ and a parametrized density $f_\theta(x)$ as proposed by Tamura and Boos (1986), that is, finding the minimum (or maximum) of

$$\int [\hat{f}(x)^{1/2} - f_\theta(x)^{1/2}]^2\, dx.$$

Stochastic approximation was developed by Robbins and Monro (1951) to find the root t^* of a regression function $E\{Y|t\} = m(t)$, where $Y = h(X; t)$

and the random variables X are independently identically distributed with density $w(x)$. The random variables are generated by Monte Carlo as for importance sampling, and the root is obtained iteratively by

$$t_{k+1} = t_k - a_k A Y_k,$$

where t_k and Y_k are p dimensional vectors, and a_k is scaler and A is $p \times p$. The root is determined by estimating the slope of the regression equation and estimating where the regression line crosses zero. Monahan and Liddle point out problems with stochastic approximation, namely that convergence is slow with least-squares methods overestimating the slope initially, and this bias dissipates slowly.

Monahan and Liddle claim to improve the $p \times p$ matrix estimate of the slope by imposing stationarity ($a_k = a$ and A fixed). However, t_n does not converge, so Monahan and Liddle discuss alternative estimates of the root t^*, dismissing the root of the estimated regression function as an option, and they recommend using the process average $\bar{t}_n$ instead. A confidence region is also obtained for this estimate.

Monahan and Liddle conclude with an algorithm for stationary stochastic approximation and a consideration of performance in several applications. It is not surprising that the smoothness of the surface to be optimized is a major factor in the performance of the algorithm, although it should be straightforward to modify to procedure to accommodate advances in searching techniques for bumpy surfaces (another current thrust for statistical computation).

References

R. N. Tamura and D. D. Boos (1986), *Minimum Hellinger distance estimation for multivariate location and covariance*, J. Amer. Statist. Assoc. **67**, 223–229.

H. Robbins and S. Monro (1951), *A stochastic approximation method*, Ann. of Math. Statist. **22**, 400–407.

Mathematics and Statistics Department, The American University, Washington, DC 20016-8050

E-mail address: flournoy@auvm.bitnet